23

3^00^

FIELD GUIDE TO
THE SNAKES AND OTHER REPTILES
OF SOUTHERN AFRICA

To my sons, Robbie, Matthew and Tom, with love.
May we continue to have as much fun in the veld.

BILL BRANCH'S

FIELD GUIDE TO THE SNAKES AND OTHER REPTILES OF SOUTHERN AFRICA

STRUIK

Cover photographs: (top left) Geometric tortoise; (top right) Boomslang; (bottom left) Egyptian cobra; (bottom right) Smith's dwarf chamaeleon. (All photographs by Bill Branch).

Struik Publishers
(an operating division of Struik Holdings (Pty) Ltd)
Struik House, Oswald Pirow Street
Foreshore,
8001 Cape Town

Reg. No.: 80/02842/07

First published 1988

Edited by Tracey Hawthorne, Cape Town
Colour section designed by Joan Sutton, Cape Town
Photoset by McManus Bros (Pty) Ltd, Cape Town
Reproduction by Hirt & Carter (Pty) Ltd, Cape Town
Printed and bound by National Book Printers, Goodwood

ISBN 0 86977 639 8 (hard-cover)
ISBN 0 86977 641 X (soft-cover)

ɔntents

Acknowledgements

It is difficult to acknowledge fully the many friends and colleagues who have made this book possible. First and foremost I extend my tremendous gratitude to Don Broadley (Natural History Museum of Zimbabwe, Bulawayo) and Wulf Haacke (Transvaal Museum, Pretoria), who have patiently nurtured my herpetological education over the last 15 years.

To Rod, Richard and Harold, who have shared fatigue and crushed fingers for the ludicrous pleasure of exposing some small, insignificant 'gem', my thanks for their comradeship. May we share a campfire and a beer, and go frogging again.

More specifically for this book, I thank John Akester, Graham Alexander, Tony Bannister, Hartwig Berger-Dell'mour, Richard Boycott, Don Broadley, John Coates-Palgrave, Atherton de Villiers, Mike Griffin, Wulf Haacke, Leonard Hoffman, George Hughes, Niels Jacobsen, Johan Marais, Chris Mattison, Chris McCartney, Geoff McLachlan, Le Fras Mouton, National Parks Board, C. Schlettwein, Steve Spawls, Chris Stuart and Colin Tilbury, for letting me use their photographs of rare and interesting species (specifically credited on page 318).

Unpublished observations, which helped in many ways to improve the scope and accuracy of the guide, were freely given by Ron Auerbach, Ernst Baard, Hartwig Berger-Dell'mour, Don Broadley, Richard Boycott, Atherton de Villiers, Mike Griffin, Wulf Haacke, Niels Jacobsen, Geoff McLachlan, Le Fras Mouton, Steve Spawls and Colin Tilbury. I am greatly in their debt.

Most of the lizard distribution maps in this guide are a result of the diligence and labour of Geoff McLachlan (South African Museum), whose atlas of lizard distributions in southern Africa is as yet unpublished. Similarly, most of the species boundaries for the thick-toed geckos (*Pachydactylus*), a notoriously complicated group, follow those determined by Geoff in his revision (as yet unpublished) of the genus; oversights or unusual interpretations remain my own. Ron Auerbach, Hartwig Berger-Dell'mour, Mike Griffin, Don Broadley and Niels Jacobsen allowed me to incorporate new information from unpublished manuscripts.

Graham Alexander, Orty Bourquin, Harold Braack, Gerald Haagner, Johan Hurter, and all of my colleagues noted above, plus many others I have overlooked, have helped me in obtaining live specimens of rare and wondrous 'beasties' for my chromosomal and hemipenial studies. Forgive me if I also took the opportunity to capture their beauty on film.

I thank the Director and Board of the Port Elizabeth Museum, and my fellows at that enjoyable institution, for making my work so pleasant and worthwhile. Shantal Koch ably prepared most of the illustrations.

To Tracey Hawthorne at Struik, my warm thanks for editing the text.

My family has tolerated, and perhaps at times even enjoyed, the endless stream of herpetological oddities (human and otherwise) that have crossed our doorstep. They have also accepted the financial sacrifices of a museum scientist.

Finally, to all my herpetological colleagues, both in South Africa and overseas, I offer this first guide. Darts need dartboards, and I am aware of its limitations. I know their studies will necessitate many changes in future editions.

Bill Branch
Port Elizabeth, 1988

Preface

While this field guide is long overdue, it is also, in many ways, premature.

Until now, no book has existed that deals with all the reptiles of southern Africa; a number of texts have dealt with the region's snakes, the most nota being Vivian FitzSimons' monograph (1962), and its scholarly revision by D Broadley (1983). In addition, a few smaller, popular books have dealt with t more common snakes. The only treatment of our lizards, also by Vivian FitzSimons, appeared in the Memoirs of the Transvaal Museum (1943); it ha long been out of print, and was always difficult to obtain. Although it remain the scientific base for studies in the region, such have been the advances ir our understanding of the subcontinent's lizards that it is now extremely out (date. By 1981, no less than 59% of the lizards of the Cape Province alone h been affected by some form of taxonomic change since FitzSimons' monograph; since then, a further two lizards have been described from the Cape, and others have been discovered or have new names. Changes in th taxonomy of lizards from other parts of the subcontinent are just as extensiv As a result, it became obvious that some form of updated checklist was needed.

Now more than ever before, people are becoming interested in all forms (wildlife, not just large mammals, colourful birds and pretty flowers. If ungair vultures and thick-skinned rhinos can become conservation 'stars', then su there is hope that people will begin to take reptiles as seriously. To this end, the public needs an introductory text that will display reptiles in all their diversity and beauty. In southern Africa we have nearly 400 species of repti including the world's richest diversity of land tortoises. More than half of the species are endemic to the subcontinent, while many others occupy most o their range here; if guides for other parts of Africa existed (which they don't) they would still deal with only a fraction of the fauna of southern Africa.

The reason I describe this guide as premature is that we are at present undergoing a renaissance in herpetological studies in southern Africa, and more herpetologists are active in the field and laboratory than ever before. New species, new distributions and new insights have become almost commonplace. I am aware of two new species of leaf-toed gecko (*Phyllodactylus*), four new dwarf chamaeleons (*Bradypodion*), a new Weste legless skink (*Typhlacontias*), several new flat lizards (*Platysaurus*), and thr new frogs. (A few of these 'novelties' are illustrated on Plate 3.) All are being described at present, and the publication of these descriptions will date this field guide. In addition, Niels Jacobsen's work in the Transvaal, previously thought to be herpetologically well covered, has led to the discovery of numerous new lizard populations that do not fit easily into the current taxonomy. Eventually, this will lead to a greater understanding, but for the moment there is 'chaos' in numerous groups, including the dwarf geckos (*Lygcdactylus*), the flat geckos (*Afroedura*), the flat lizards (*Platysaurus*), the crag lizards (*Pseudocordylus*), and Warren's girdled lizard and its relatives (*Cordylus warreni* complex). Anyone collecting specimens in these groups from the northern and eastern Transvaal should regard the species account in this book as temporary arrangements. However, despite these numerous proposed name changes, the taxonomy of the region's herpetofauna is in a healthy state, particularly when compared with that of the rest of Africa.

Unfortunately, the state of ecological research on reptiles in southern Afric is pitiful; many South African scientists, and Nature Conservation Departme in particular, have had an antiquated preoccupation with large mammal research, to the neglect of the other groups. Although there are signs that th is changing, there remains an almost total lack of natural history data for our reptiles. Only a handful of detailed lizard ecological studies, plus a couple o tortoises, are available. Nearly all of these have been done by visiting

American scientists, who appreciate better the value of small reptiles as models for ecological research. No local snake has ever been studied, although many are venomous and responsible for large numbers of bites (eg. the Mozambique spitting cobra, in Zululand), while others are endangered and yet acknowledged to be useful in the control of agricultural pests (eg. the rock python, in Natal). The crocodile is the only reptile for which we have more than a superficial knowledge of reproduction, growth, behaviour and habitat.

The following changes, adopted in this field guide, have yet to be described in the formal literature.

1. The following, all previously treated as races of more widespread species, are now treated as full species: *Pachydactylus gaiasensis, Pachydactylus labialis, Pachydactylus oculatus, Pachydactylus sansteyni, Scelotes sexlineatus, Cordylus tasmani, Cordylus minor, Cordylus jordani* and *Bitis inornata*.
2. *Homopus bergeri* and *Atractaspis duerdeni* were previously considered to be unusual specimens of more common species, but are now recognized as full species.
3. The following species have recently been discovered in northern Namibia, and are additions to the region's herpetofauna: *Pachydactylus caraculicus, Mabuya chimbana, Cordylus vittifer machadoi* and *Prosymna visseri*. (A few other species are also known to enter the region.)
4. Four species, known from single type specimens and which have never been rediscovered, have been excluded. All were probably based on specimens from other regions, and their possible true identity is noted in brackets: *Rhoptropus braconnieri* (*R. afer*), *Scelotes bicolor* (*S. arenicola*), *Melanoseps schebeni* (*M. occidentalis*), and *Latastia kidwelli* (*L. johnstoni*).

A similar case could be argued for excluding *Phyllodactylus peringueyi*; this small gecko is generally not considered to be African, but as yet has no other suitable identity. It has eluded all attempts to rediscover its habitat. Until it can be assigned to another species, it must remain provisionally placed in our herpetofauna.

It is my hope that this guide will stimulate naturalists to look more closely at the common and small reptiles that live, largely unnoticed, all around us.

How to use this book

This guide provides an easy method of identifying the reptiles found in southern Africa. It is primarily a photographic guide, designed for quick identification by comparison of the reptile with its illustration in the colour plates. The plates do not follow the systematic arrangement of the text, but have instead been arranged according to the reptile's appearance.

All the legless squamates (snakes, amphisbaenians and some lizards) ha been placed together, and have been arranged by colour pattern so that all the species that appear similar (eg. striped, blotched, green, etc.) are grouped together. The lizard plates (excluding the legless forms) have bee grouped by general body shape. For economic and practical reasons it has not been possible to illustrate every reptile species occurring in southern Africa. For some, no photograph exists. A representative of every genus occurring on the subcontinent has been included, and where a species is n illustrated, the reader is referred to a similar species where possible. Althou many species have very little colour variation, others occur in such a bewildering array of colour patterns that it is not possible to illustrate all phases. Colour polymorphism may be linked to sex (adult males are usually more brightly coloured than females), season (some males develop bright colours in the breeding season), age (juveniles may have a different colour and/or pattern from adults), or locality.

When identifying a wild reptile for the first time, take careful note of its general shape (Does it have legs or a long, smooth tail? Is the body fat and covered with spines? Is there an obvious neck?). Note whether the head, bo and tail differ in colour and pattern. Record the habitat in which it is living (Is on a granite rock face high in mountain grassland, or crawling on a sand du in the desert?), and the behaviour of the animal (Did it run up a tree or into a hole in the ground?). Now scan through the photographic plates in the guid When you find an illustration that is similar to your unknown species, check whether it occurs in the region. If there is a big difference between where yo have found it and where it should be, try another picture. If there seems to b no better alternative, match the body shape of the specimen as closely as possible to an illustration in the guide, and read the descriptions of species the same genus that do occur in the area. You may simply have a colour variant of a common species, such as a striped puff adder (Plate 3), a xanth brown house snake (Plate 3), or a yellow-throated plated lizard with a blue throat; other unusual specimens may be hybrids, such as a puff adder/Gaboon adder hybrid (Plate 3). Check also that your habitat and biology notes match those of the species. For example, if you think you have found an Austen's gecko, but it was hiding under a rock flake high on a gran outcrop, the chances are that you have misidentified it; Austen's geckos live on the ground in coastal sand dunes. If you cannot match the specimen wit anything in the guide, you may have found something new, such as a new species of dwarf chamaeleon or leaf-toed gecko (Plate 3), or at least a new distribution record. Either way, it should be brought to the attention of a herpetologist (see page 317) or a game ranger or teacher, etc. If it is permitted the specimen may be collected; if not, try to get a good photograph.

The terminology used in the guide has been kept simple. Common words are used rather than their more specific, but often obscure, scientific equivalents, eg. 'burrowing' instead of 'fossorial', and 'overlapping' rather th 'imbricate'. However, it is not always possible, or even desirable, to avoid scientific terms, particularly for descriptions of the head scales. The illustrations provided include the scientific terms. The glossary gives simple definitions of technical and scientific terms.

Keys are provided to families and genera. They are designed for local use and may not work elsewhere in Africa. A scientific key is a branching sequence of decisions, presented as alternatives. Those who are unfamiliar with their use may find the following simple example helpful:

1. With legs.. 2
   Without legs.. 3

Does the reptile have legs? If so, then go to number 2; if not, then go to number 3. The key is read in this way until the name of a family or genus is reached. Page references are provided for each family or genus.

To avoid repetition, general information common to a number of related species is covered in the introductions to the families and genera. These discuss wider aspects of the distribution and biology of the group, to give a better understanding of the significance and relationships of our local species.

Each species account emphasizes those features that distinguish species within the same genus. However, identifying closely related species may require a good microscope and a working knowledge of scale counting.

There exists no standardized list of common names for the subcontinent's reptiles. Some names have been used regularly in the popular literature and most of these have been maintained. One major change is the use of the more descriptive and traditionally derived name 'burrowing asps' for mole vipers.

The following classification of the southern rock python, *Python sebae natalensis*, illustrates the hierarchy of scientific names:

Class: Reptilia (Reptiles)
Order: Squamata (Scaled reptiles)
Suborder: Serpentes (Snakes)
Family: Boidae (Boas and Pythons)
Subfamily: Pythoninae (Pythons)
Genus: *Python* (Typical pythons)
Species: *sebae* (African rock python)
Subspecies: *natalensis* (Southern rock python)

For each species, an average and maximum size are given. The latter is the length of the largest specimen recorded locally. Snakes, lizards, amphisbaenians and crocodiles are measured in a straight line along the backbone, from the tip of the snout to the tip of the tail. Where the length of the tail is discussed in relation to the body, the former is the original tail (untruncated and not regenerated), and the latter is the snout-vent length, measured from the tip of the snout (thus including the head) to the rear margin of the cloacal opening. Only the shells of chelonians are measured. In tortoises and terrapins, the length is taken straight along the midline of the carapace, while sea turtles are measured along the curve of the midline.

The distribution maps give the known range of a species. However, these serve only as rough guides; all animals prefer specific habitats to which they are usually restricted within a given geographical area. In addition, our knowledge of species' distributions is very poor, and new range extensions are commonplace. Some maps include isolated records, away from the species' main distribution. Occasionally these have been inadvertently introduced by man. Some are relict populations, survivors of a formerly widespread species, whose range has contracted due to climatic changes or for other reasons. Most, however, probably reflect our ignorance of the intervening gaps, as many regions have been poorly collected.

As much information as possible is given about the biology of a species. Too often field guides simply identify a species without giving any insight into its natural history. Unfortunately, few species have been properly studied and for many only scanty information is available. Where no details are given, the general information in the generic introduction may help.

'SA RDB' refers to the South African Red Data Book, a national publication that lists those reptiles and amphibians in South Africa that are threatened with extinction. Most seriously threatened are 'endangered' species; these are likely to become extinct within the near future unless urgent steps are taker protect them and their habitats. 'Vulnerable' species are those that are declining in numbers due to various threats (pollution, habitat destruction, collecting for food or the pet trade, etc.), and will soon become endangere unless action is taken. Species that are classified 'Rare', 'Restricted' or 'Peripheral' may become threatened with extinction because of low popula numbers or restricted ranges. To protect threatened species, many countr are signatories of CITES (the Convention for International Trade in Endangered Species). This legislation prohibits the movement of endange species unless special export and import permits are obtained.

The need for and acceptance of subspecies is a controversial concept. Subspecies (races) are populations of a species that are different and geographically restricted. They usually differ in appearance, but other crite such as the sounds they make or their chromosomes, may be used in identification. Where subspecies have been described, they are listed for each species. However, the criteria used to distinguish some of them are b minor and variable, and these poorly defined races may not have general acceptance. They are included only for completeness.

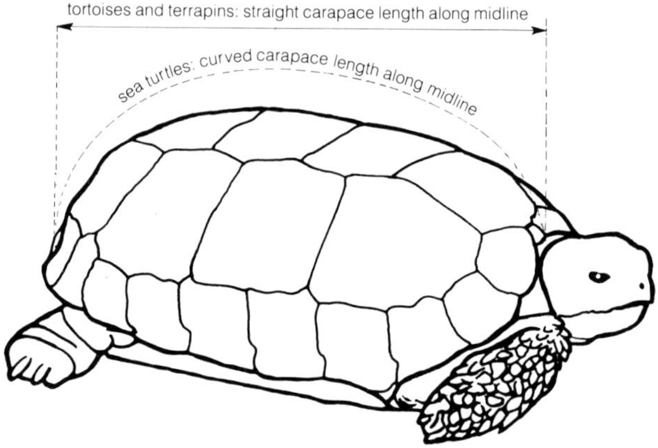

Measuring chelonians

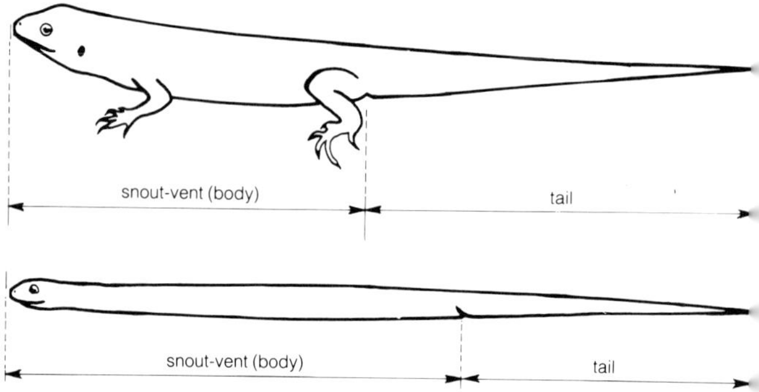

Measuring lizards and snakes

Observing and collecting reptiles

It must be stressed at the outset that most reptiles are protected by legislation throughout the subcontinent, and permission must be obtained to collect, transport and possess them. In addition, there is no reason why reptiles cannot be enjoyed in the field as much as birds or mammals; there is no need to collect every specimen. However, it is sometimes necessary to collect specimens, such as for regional surveys or education purposes. The following notes give a few hints.

HOW TO FIND REPTILES

Since most reptiles are shy, and scurry to safety at the first sign of danger, it takes patience to find them. Knowledge of a species' preferred habitat will save fruitless searching – but search you must, even in prime habitat. The traditional methods are to look under boulders, rotting logs, grass piles, etc. Conservationists are occasionally embarrassed to discover that suburban wasteland, with its associated rubbish, is often 'prime habitat' for snakes and lizards; old sheets of corrugated iron make ideal basking sites as they conduct heat, allowing reptiles to warm themselves without being exposed to danger.

Burrowing reptiles can be found by digging in loose soil under boulders and rotting logs, or wind-blown sand at the base of grass tussocks. Many are exposed by farm ploughs or during the construction of new roads. They are often washed downstream when rivers flood. Holes in the ground may act as pit-fall traps and should be investigated closely. Man-made holes are bigger and better than anything found in nature and many interesting specimens, including the first specimens of the dwarf wolf snake, *Cryptolycus nanus* (page 61), were found trapped in the trench dug for the Beira-Mutare pipeline. Thousands of reptiles and other animals have been trapped in the new Eastern National Water Carrier in Namibia. Stormwater culverts in most towns also act as traps and should be investigated.

In the cool regions of southern Africa, snakes and lizards often hibernate during winter in old termite nests. Rock-living snakes may also gather in deep rock cracks to avoid winter snows. Many specimens have been collected by smashing open termite nests or dismantling rock outcrops with crowbars. Unfortunately, once flattened, these retreats cannot be used again, and no true conservationist willingly resorts to such destruction.

Do not limit your search to daylight. Noctural reptiles often crawl on to tarred roads at night, to absorb the residual heat from the sun-warmed surface. They can be collected by slowly driving on quiet roads on warm evenings (however, the results in southern Africa never seem to approach the success achieved in America). Geckos can be collected around night lights, where they feed on moths and other insects. Chamaeleons are easier to find at night as they sit in a prominent position and stand out against the dark foliage in the torchlight.

Early in the morning, before the wind blows and obliterates the evidence, the trails of snakes and burrowing legless lizards can be followed in loose sand. This is best done in spring when males are very active, searching for mates. The homes of some lizards can be located by looking for their droppings around rock cracks.

COLLECTING HINTS

When disturbed, snakes and lizards usually retreat to safety, climbing into trees, deep cracks or holes. A mechanical noose or 'grab-stick' may be used to catch venomous species such as mambas and cobras, which are too alert and dangerous to be grabbed with the unprotected hand. Most lizards are too small and fragile to be collected with grab-sticks, and a noose is more useful. This can be made from thin nylon or cotton, attached to a hollow-fibreglass fishing rod or an extending radio aerial. Many agamas and skinks will let you stalk to within a few metres of them, and the noose can be lowered over thei head and pulled tight. (Agamas are easy to catch as they have thin necks, b skinks do not, and the noose may slip.) Rock-living lizards, particularly cordylids, often retreat into a crack, but can be urged out with wire or, bette still, by working a strong nylon noose over their head. A one-metre section o 8-mm copper tubing (eg. fuel pipe or old refrigerator cooling tube), threade with 20-kg fishing line, is ideal. After noosing the lizard, the crack can be gently expanded with a crowbar and the lizard pulled out. There is no need demolish the whole rock outcrop.

Reptiles in inaccessible spots can sometimes be 'encouraged' to move within range by squirting them with unpleasant fluids, for example chlorofor or alcohol (do not use petrol or any other liquid that can cause permanent damage to the lungs). A water-pistol, loaded with water, can be used to kno geckos from walls and rock faces. Lizards can be stunned with a large elast band. A range of different sizes can easily be carried and matched to the target. They are hooked on the tip of the thumb, stretched and shot off, just a all schoolboys have done since elastic bands were invented. A variation is t cut long strips from a car inner tube and to 'twang' a strip at a small lizard, while holding one end of the strip. The effective range is reduced to about o metre, but it is more accurate and can be used with more control to stun the victim. Practise on a matchbox first, or you may squash and kill your specimen.

The use of guns and traps is illegal unless special permits are obtained; they should always be used with care and restraint. If specimens are not required alive, they can be killed with 0.22 dust shot. A six- or eight-chambe revolver is best as it is lighter than a rifle. Dust shot is not suitable for a pistol it does not have sufficient back pressure to eject the spent case. The shot pattern groups at about 30 cm at seven metres (depending on the barrel length) and is sufficient to kill even small lizards without obliterating them in process.

Reptiles may be collected in pit-fall and funnel traps. An empty five-litre paint tin sunk level with the soil surface acts as an efficient pit-fall trap, and small snakes and lizards (but not most geckos) that fall in cannot crawl out. Pit-fall traps are particularly effective when used with drift fences. Long plan or lengths of thick polythene sheeting or shade cloth fixed flush with the ground will shepherd small reptiles into the mouth of a pit-fall or funnel trap. These traps should always be protected from the sun as the captives may rapidly overheat and die. Pit-fall traps should be attended daily, and should never be discarded in the field.

Reptiles are often found dead on the road, ploughed up during farming or road construction, or drowned in the swimming pool. They can be preserve in formalin (10%) or methylated spirits. Specimens that are thicker than 2 cm should be injected with a preservative or the belly slit open to prevent them rotting in the centre. A label, written in soft pencil, should list the date, time, locality and circumstances in which the specimen was killed or found, and c be tied to or placed with the specimen; if desired, the specimen may be sen to a suitable institution for identification (for a list of addresses, see page 31

Finally, I would like to emphasize that collecting is rarely necessary. It is usually far preferable to sit and enjoy the reptile, and then both go in peace.

The southern African environment

The geographical scope of this guide covers the area south of a line connecting the Cunene and Zambezi rivers. This region (over 3,5 million square kilometres in extent) covers the habitats of almost 400 reptile species, nearly a third of all those occurring in Africa. This diversity is a reflection of the varied climates and habitats of the region, and exceeds that of both the USA (273 species in 9,3 million square kilometres) and Europe (excluding the USSR) and SW. Asia combined (380 species in 11,25 million square kilometres).

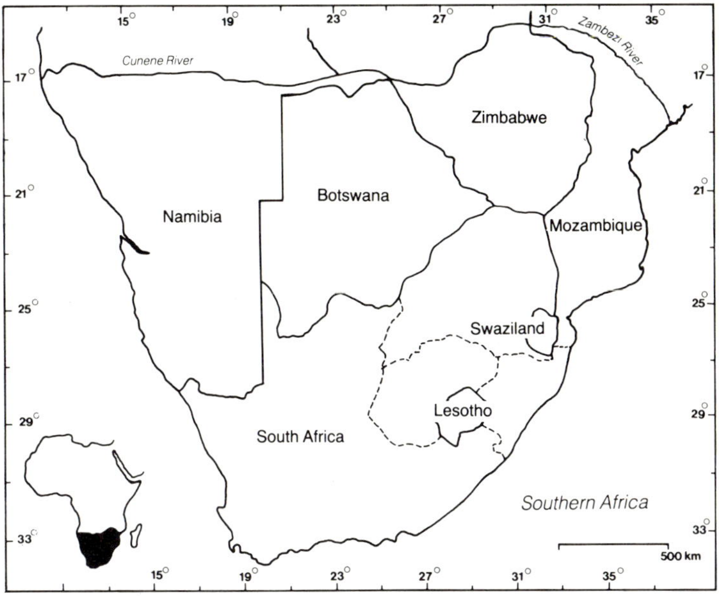

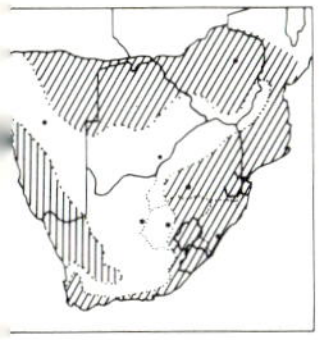

Annual rainfall
more than 500 mm p.a.
125-500 mm p.a.
less than 125 mm p.a.

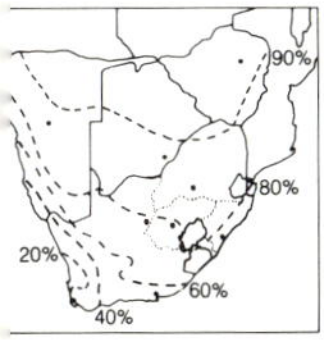

sonal rainfall
lling in October-March

CLIMATE

Southern Africa has a mainly temperate climate, with wide seasonal contrasts in rainfall and temperature.

The rainfall is strongly influenced by the cold Atlantic Ocean (Benguela) and warm Indian Ocean (Agulhas) currents that sweep up and down the west and east coasts respectively. The annual rainfall increases considerably in the north and east, whereas the west coast and adjacent regions are the driest parts. Rainfall is largely dependent on the prevailing winds, which in summer sweep anticlockwise across the continent, carrying moist air in the form of thunderstorms from the Indian Ocean across the eastern regions. Because of the high altitude of the interior, little rain remains to fall in the west. In winter, the winds bring rain to the west coast, whereas the rest of the country is dry.

The temperature is affected by both the position (between 17 °S and nearly 35 °S) and the high altitude (1 000-2 000 m) of much of the interior. The cold Benguela sea current sweeping up the west coast keeps the air temperatures cool, and causes local fogs that sweep across the Namib Desert, and on which many of the region's reptiles depend for their water. Summer temperatures are highest in the Kalahari and surrounding regions, becoming less intense with altitude on the highveld of the Transvaal and Orange Free State and the Zimbabwean and Namibian plateaus. Coolest areas at this time are the high mountains of the eastern escarpment and the south-western Cape. Winters are mild along the east coast (the Mozambique Plain and adjacent lowveld of the eastern Transvaal, and the Limpopo and Zambezi r valleys). The coldest winter regions are the highveld and the mountains of t Great Escarpment, which regularly have night frosts and where (with the exception of the Namibian escarpment) heavy snowfalls are frequent. The of the region has warm, sometimes pleasantly sunny, winter days with cool occasionally cold, nights.

HABITATS

Ecologically, southern Africa is made up of a complex mosaic of different vegetation types. Seven major categories, called biomes, are recognized, these have numerous subdivisions.

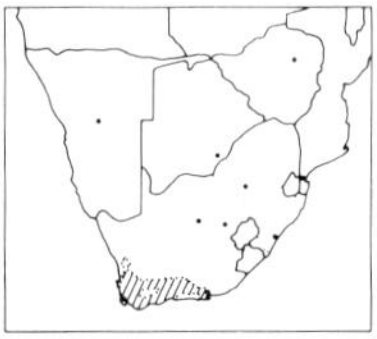

Fynbos is the local name for the Mediterranean-like heathland that covers the south-western and southern Cape, from the Cedarberg to Port Elizabet It is composed of an amazing diversity of plants, dominated by heathers, proteas and restios, that grow on poor, rocky soils. Receiving mainly winter rain (300-2 500 mm per annum), they form a low (1-3 m), woody scrubland that has adapted to frequent fire. There are various subdivisions of fynbos, including rhenosterbosveld, which has more grass and fewer heathers anc proteas. Less than one per cent of rhenosterbosveld remains, the rest havi been ploughed under for wheat and grape farming.

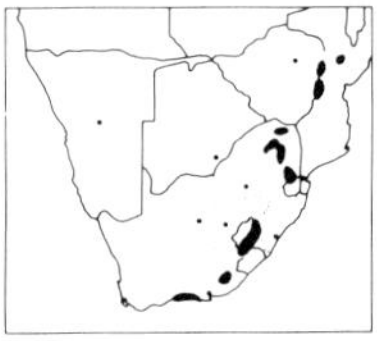

Afromontane vegetation is restricted to the high rainfall regions. At high altitudes, it occurs on cool mountain slopes and consists of heathers and sometimes proteas, mixed with grassland. At lower altitudes, it takes the fo of tall, thick, cool forest (yellowwoods, stinkwoods, etc.) that grows down tc sea level in the southern Cape (the Tsitsikamma Coastal Forest). To the nor this vegetation type exists as a scattered 'archipelago' of isolated forests associated with the Great Escarpment. These isolated pockets were abutti and more extensive during cooler glacial periods, but have contracted ove the last 16 000 years as the climate has warmed, and have suffered further the last 200 years as the remaining vestiges have been felled for timber anc replaced with pine plantation.

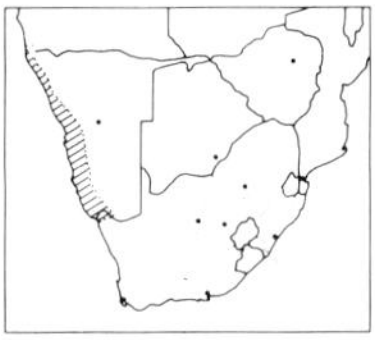

Desert occurs as a narrow strip along the west coast. The Namib is an anci desert, composed of shifting sand dunes along the coastal strip, and hard, gravel plains inland. Plant cover is sparse, with scattered grass and specialized succulent plants on the sand and gravel plains, and stunted acacia trees in the river courses. It is not unusual for less than 10 mm of rair fall in a year, and droughts lasting 4-5 years may occur. The offshore Benguela Current is responsible for the cold, moisture-laden fogs that may extend up to 50 km inland.

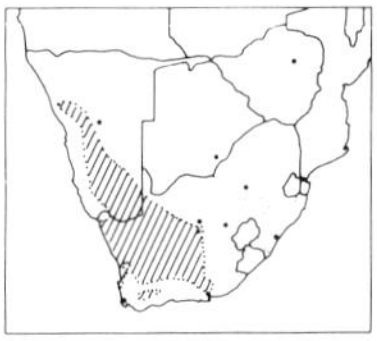

Karoo and semi-desert covers the central and western Cape, extending through Namibia, inland of the Namib Desert. This biome ranges from sea level to 2 700 m, with numerous mountainous regions and a complex geolc In the north, it merges into the arid savannah of the Kalahari. Rainfall (occurring mainly in winter in the west, and in summer in the east) rarely exceeds 250 mm. The soils are poor, shallow and rocky, and support a sparse, dwarf, woody scrub. A succulent Karoo shrubland occurs on the sandy coastal plain of Namaqualand, with outlying pockets in the rain shac valleys behind the Cedarberg (the Tanquwa Karoo) and within the Cape fo mountains (the Little Karoo).

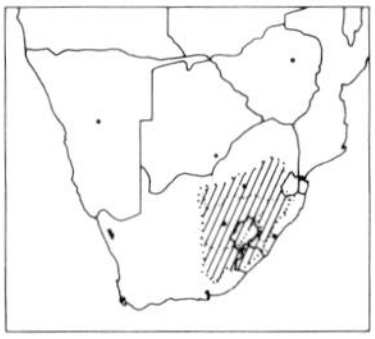

Grassland occurs on the interior plateau from 1 200 to 2 100 m, covering tr highveld of the southern Transvaal, Orange Free State and north-eastern Cape. The growth of trees is inhibited by dry, extremely frosty winters, and possibly by the regular winter fires. Rainfall ranges from 250 to 500 mm per year and occurs mainly in summer. Grassland merges into the next biome the north-east, via the hardveld of eastern Botswana (the Kalahari-Highvelc transitional zone).

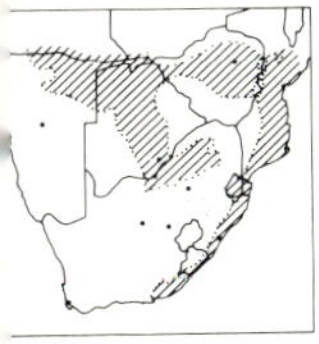

Arid savannah is open, grassy savannah, adapted to low rainfall and cold, dry winters. As flat, open, acacia woodland it covers the Kalahari sands of southern and central Botswana, extending into northern Namibia (sandveld). As dry mopane woodland it extends along the Limpopo Valley, the eastern Transvaal lowveld and adjacent Mozambique.

Mesic savannah covers the eastern regions of higher rainfall and warmer winters. As the Indian Ocean belt, it extends as a narrow strip along the east coast as far south as Port Elizabeth. Here, the warm Mozambique Current causes high moisture and warm weather; mean annual temperatures range from 26 °C in the Zambezi Valley to 17 °C at Port Elizabeth, and frost is absent. In the north, the belt extends 240 km inland across the sandy soil of the Mozambique Plain, and comprises rich *Brachystegia* woodland (now largely replaced with secondary wooded grassland and cultivation), with patches of mangrove and swamp forest along the coast. From Natal southwards, the terrain is more rugged and the belt may narrow to as little as eight kilometres. Here, complex soil types occur, and rainfall is moderately high; although it falls mainly in summer, there is no obvious dry season. The vegetation is a mosaic of succulent and moist thicket, which may be dense and impenetrable. A secondary belt of more open *Brachystegia* woodland covers much of the Zimbabwe plateau, the Caprivi Strip and adjacent northern Botswana.

Some areas of special interest

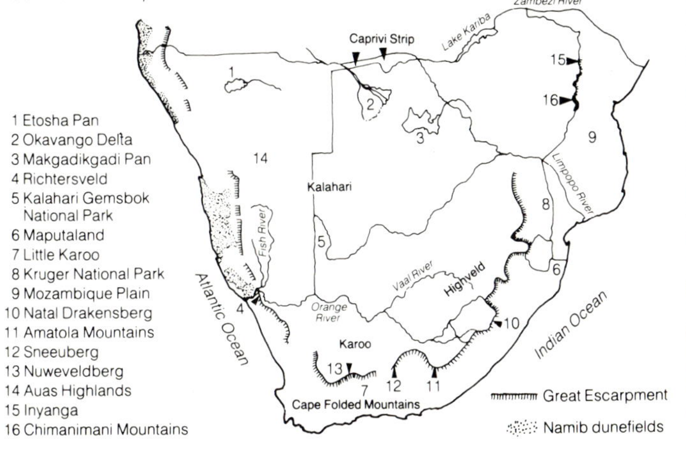

AREAS OF SPECIAL INTEREST

The Great Escarpment forms the boundary of the interior plateau, and rises from a minimum of 900 m in the Kalahari depression to over 3 482 m in the Lesotho mountains. It is composed of a number of distinct and imposing mountain ranges. As the Drakensberg of Natal and the eastern Transvaal, it forms a sheer and continuous edge, but elsewhere it is broken into a chain of isolated highlands. To the north, it extends as the Zimbabwe Eastern Highlands, including Inyanga and the Chimanimani range. In the Cape, it forms the isolated mountains of the Amatolas, Sneeuberg, Nuweveldberg and Roggeveldberg, and in Namibia it exists as the Aus highlands. In the east, it receives high rainfall and is covered with afromontane vegetation. Numerous endemic species are found on the isolated ranges, including the berg adder, the plain mountain adder, and some crag and mountain lizards.

The Great Karoo is an area of low rainfall, but flash floods can occur, and with devastating effect. The soils are poor, shallow and stony, with a stunted, woody, scrub vegetation. Most of the Karoo lies above the Great Escarpme but the southern plain falls below it, in the rain shadow of the Cape fold mountains, within which the Little Karoo lies as an isolated pocket. Among t few characteristic species found here are the golden spotted gecko and th tent tortoise.

The Kalahari is a huge, inland drainage basin with a deep cover of old sanc which has extended much further east in past dry periods. Although it is sometimes called a desert, no part of the Kalahari receives less than 100 m of rain a year. The southern region is dry and cold in winter, and covered wi arid acacia scrub and grass. The south-western part, including the Kalahar Gemsbok National Park, is very dry, and consists of ancient, long sand dun The central Kalahari is flat, and has a number of shallow basins that may periodically flood, including the extensive pans of the Etosha and Makarika depressions. Both areas are the home of endemic lizard species. Between these pans run the Okavango and Chobe rivers, which bring waters from th Angolan highlands. These never reach the sea, but drain into the deep sanc forming the lush Okavango and Chobe swamps. Among the numerous endemic species found here are the Okavango hinged terrapin, the Barotse water snake, and the striped swamp snake.

The Namib Desert runs along the whole coast of Namibia, just entering southern Angola. Over most of its length it is approximately 100 km wide; shifting sands are restricted mainly to the coastal region, with two main dunefields lying north and south of Swakopmund. Much of the inner Namib consists of hard gravel plain, broken by rugged, barren mountains. Annual rainfall in the coastal region is only 10 mm, but for 100-200 days each year, fog on which many reptiles depend for their water rolls in from the sea. Here many unique species can be found. The web-footed gecko, Péringuey's adder, the shovel-snouted lizard, and the desert plated lizard, have all adapted to the shifting sands. The Namib day geckos and many thick-toed geckos, desert lizards and sand lizards, are restricted to the Namib gravel plains and rock outcrops.

Namaqualand straddles the lower Orange River. It includes the Richtersvelc small area of dry, rugged mountains that nestles between Great and Little Namaqualand (which sit north and south of the Orange River, respectively). It is dry (sometimes devastatingly so), but occasionally experiences good winter rains (up to 250 mm). This area is justly famous for the beauty of its spring flowers, and its many unusual reptiles make it a herpetological paradise. Among these are Austen's gecko, Schneider's adder, the speckle padloper, and the Namaqua day gecko.

The Highveld of the Orange Free State and the southern Transvaal is an are of high altitude grassland on deep soils. It experiences irregular and often violent summer rains, and cold, dry winters. There has been extensive agricultural and urban development in this region. One of the few endemic reptiles found here is the giant girdled lizard or sungazer.

The Cape fold mountains run parallel to the south-western and southern Ca coast, and are responsible for the region's high winter rainfall, as well as its scenic grandeur. Most of the mountains in the range have altitudes of betwe 1 000 and 1 500 m, with the highest rising to more than 2 000 m. The sandstone has weathered to form rugged gorges and spectacular rock forn The soils are poor and rocky, and sparsely covered with fynbos vegetation. Unusual endemics found here include the southern rock lizard, the small-scale leaf-toed gecko, the Cape mountain lizard, and the graceful crag lizard.

Maputaland and the **Mozambique Plain** of the east coast are composed of Cretaceous and Tertiary sandstones and limestones, extensively covered w sand. The vegetation is varied, with dry mopane woodland occurring in area of seasonal rainfall and good drainage. In the low-lying coastal strip, swamp and mangrove forest occur. The climate is mild, and the region forms an important corridor for tropical east African species such as the green mamb the Gaboon adder, and Moreau's tropical house gecko.

REPTILES

Class Reptilia

Public knowledge of reptiles is full of misconceptions. They are considered to be slimy, cold, slow or cumbersome, and most are thought to be highly venomous. Even among many scientists and naturalists, who perhaps should know better, reptiles are perceived as being inherently inefficient. This is quite untrue. Reptiles are as successful, in terms of their diversity and the varied habitats in which they live, as their furred and feathered descendants. In fact, in southern Africa there are more reptile species than there are mammals, and in the western deserts they match birds in number if not diversity.

In addition to the living forms, there is a bewildering array of extinct reptiles, including creatures as diverse as the giant saurischian and ornithischian dinosaurs, the aquatic plesiosaurs and dolphin-like ichthyosaurs, and the flying pterodactyls. Given this diversity, it is not easy to list simple features that distinguish all reptiles. In many ways they are easier to describe by what they lack rather than by what they possess, and they could be defined as tetrapods (four-legged – at least in their ancestors – vertebrates) that lack fur, feathers, and a tadpole stage in their development.

Perhaps the most obvious characteristic of reptiles is their dry, horny skin that is usually modified into scales or plates. This prevents rapid water loss and has allowed them to move on to dry land, whereas amphibians are still restricted to moist habitats. They are also amniotes (like birds and mammals), having foetal membranes that surround the developing embryo. These contain the embryo in its own 'pond'; amniotes do not have a free tadpole stage like amphibians. In addition, many reptiles lay eggs that have thick shells and yolk stores, and which undergo development independent of water or parents. Such self-contained, sealed eggs are called cleiodic eggs.

A primitive feature of the skull of all reptiles is the single occipital condyle – the knob on the back of the skull that articulates with the backbone. Mammals and birds have paired condyles. Like primitive amphibians and birds, reptiles have only a single bone in the ear, and each half of the lower jaw is composed of several bones.

The relationships of living and extinct reptiles is a complex problem and involves much controversy. The details are beyond the scope of this book, and it is sufficient to note that much of the major grouping is based on skull anatomy, particularly the nature of the openings in the side of the skull. Primitive reptiles and chelonians have a solid skull with no large openings in the side (an anapsid skull). All other living reptiles have two openings in the sides of the skull (a diapsid condition), as did the dinosaurs.

All reptiles are cold-blooded, although this is a confusing term, as most are active at body temperatures that exceed those of mammals. The subtle difference between the warm blood of reptiles and that of mammals and birds is that reptiles obtain their heat externally, usually from the sun. All reptiles bask, absorbing warmth from the environment. Many simply sit in the sun until their body reaches the correct temperature, and subsequently shuttle between sun and shade, maintaining a constant optimal temperature. Burrowing species crawl beneath a sun-warmed rock or into the surface layers of the soil to gain heat. Reptiles, therefore, do not generate heat internally b metabolizing food as do mammals and birds; reptiles can more correctly be called ectotherms (meaning 'outer warmth') rather than endotherms ('inner warmth'). The advantage of ectothermy is fuel efficiency. Mammals and bir convert 90% of the food they eat into heat in order to maintain muscle and biochemical efficiency at all times. This allows them to operate at times (late night and winter) and in climates (Arctic tundra and cold seas) that reptiles cannot. However, this requires a constant supply of food. Reptiles, on the other hand, become temporarily dormant in cold weather, and provided the are protected from danger, they need not waste energy. As an extreme example, many snakes survive, and indeed grow, on perhaps 10 meals a year

Reptiles have internal fertilization. Sperm is tranferred to the female via a single penis in crocodilians and chelonians, and paired hemipenes in squamates (although only one hemipenis is used at a time). The tuatara lac an intromittent organ and suffices with close contact, rather as birds do. The majority of reptiles lay eggs, and these may be either soft or hard-shelled. They are laid in a warm, moist, hidden spot and left to incubate on their own Only crocodiles and a few lizards and snakes stay with their eggs and prote them during development. In all crocodiles and most chelonians the sex of t embryo is determined by the incubation temperature. In crocodiles, males develop in eggs at high temperatures, while the same temperatures in chelonians produce females.

Various degrees of viviparity have been achieved by squamates, but by r other living reptile. The eggs may be retained in the body for most or all of th development, and hatch within weeks, days or even minutes of being laid. Some lizards and snakes have developed a placenta, similar to that of mammals, that allows the mother to transfer food to the developing foetus. Live-bearing reptiles are usually found in cool climates.

Reptiles arose from amphibians during the Carboniferous period, and the earliest reptile fossils (small, lizard-like creatures, found inside fossil tree stumps) are about 315 million years old. During the subsequent aeons, reptiles evolved a truly bewildering array of forms, including many giant species, and for over 150 million years the dinosaurs and their relatives dominated the earth.

Living reptiles are either remnants of this period, or a recent flowering tha has taken place, for the most part, since the dinosaurs became extinct. The are divided into four orders, including the crocodilians (order Crocodylia), tortoises, terrapins and turtles (order Chelonii), snakes, lizards and amphisbaenians (order Squamata), and the tuatara (order Rhynchocephali These orders arose as separate evolutionary lineages 200-300 million years ago, very early in the radiation of reptiles; lizards, crocodiles and chelonians can all be distinguished as distinct groups from before the great radiation o dinosaurs. Their fossils can be found along with those of the giant dinosaurs

With the exception of the lizard-like tuatara, which is restricted to a few islands on the north coast of New Zealand, the orders are well represented i southern Africa. In total, the world contains more than 6 550 species of repti that are arranged in 900 genera in 48 families. Southern Africa is blessed wi nearly 400 species; these are described in the following species accounts.

CHELONIANS

Order Chelonii

The shield reptiles – tortoises, terrapins and turtles – are all instantly recognizable by virtue of their characteristic shell. This may be soft, leathery, hard, flat, knobbled or hinged, but is unlike anything else, and chelonians can be mistaken for no other reptiles. Chelonians (the correct scientific term) are characterized by a number of features, notably an anapsid skull and a bony protective shell which is divided into an upper carapace and lower plastron. There is confusion about the common names; the terms 'tortoise', 'turtle' and 'terrapin' have no strict scientific meaning, and are simply used to differentiate those species that live on land, in the sea and in fresh water, respectively.

The first fossil chelonians date from early Triassic rocks in Germany – 210 million years ago. Similar fossils have recently been discovered in South Africa. It is difficult to appreciate the immensity of such a long period; the chelonians have seen the rise and fall of the dinosaurs, the explosive success of mammals and birds, and man's brief but devastating reign.

The chelonian shell is a complex structure, composed of an outer horny layer covering a bone case which is fused to the rib cage. This entails some radical rearrangements in tortoise anatomy, not the least of which is the placing of the shoulder blades and hips inside the rib cage. Early tortoises had teeth, but these are absent in all living forms. Instead they have a horny beak, similar in appearance and function to that of a parrot. In aquatic habitats the feet may be webbed or modified into flippers, and the bony shell is reduced. The weight of the protective shell limits locomotion and chelonians are not famed for their speed. Food is therefore usually sedentary, and most chelonians are herbivorous. Some aquatic species with lighter shells capture prey such as fish, but this is usually accompanied by concealment and ambush; others feed on invertebrates such as sponges, molluscs and jellyfish.

All chelonians lay eggs, which are usually soft-shelled in aquatic forms (sea turtles and side-necked terrapins), but hard-shelled in tortoises and soft-shelled terrapins. The female takes great care in finding a suitable, moist yet sunny spot in which to lay her eggs. Sea turtles come ashore to lay their eggs. A small vertical pit is dug with the hind legs, and after the eggs are laid they are covered and left to incubate. This is the extent of maternal care, and after hatching the young must fend for themselves. The time from laying to hatching can vary from four to 15 months, and is in part dependent on the season, as eggs laid in autumn undergo very little development during winter, and do not hatch much earlier than those laid in the following spring. Clutch size varies with the species, the greatest numbers being laid by sea turtles. In many species the sex is determined by the egg incubation temperature, females being produced at higher temperatures (31-34 °C).

Chelonians are found in almost every environment – aquatic, oceanic and terrestrial – throughout the tropical and temperate zones. Four suborders containing 20 families are recognized, and half of this number is extinct. Living forms are placed in 66 genera, with some 230 species. Southern Africa has a very rich chelonian fauna, with five sea turtles in coastal waters, nine terrapins in freshwater rivers and vleis, and at least 12 land tortoises.

Key to the Chelonian families in southern Africa

1.	Head withdrawn into shell sideways	Pelomedusidae (Side-necked terrapins, page 38)
	Head withdrawn straight back or cannot be withdrawn	2
2.	Limbs not modified as flippers, with 3-5 claws	3
	Limbs modified as flippers, with 0-2 claws; marine	4
3.	Carapace with horny shields; feet with 4-5 claws	5
	Carapace soft, without horny shields; feet with 3 claws	Trionychidae (Soft-shelled terrapins, page 36)
4.	Carapace with horny shields; flippers with 1-2 claws	Cheloniidae (Modern sea turtles, page 32)
	Carapace leathery, without horny shields; flippers clawless	Dermochelyidae (Leatherback turtles, page 32)
5.	Skin of head divided into large and small shields; toes not webbed; terrestrial	Testudinidae (Land tortoises, page 22)
	Skin of head smooth and undivided; toes webbed; aquatic	Emydidae (Pond terrapins (introduced), page 37)

Modern Chelonians

Suborder Cryptodira

This suborder comprises the majority of living forms. 'Cryptodira' means 'hidden neck' and refers to the way the head is withdrawn, by a vertical, S-shaped flexure of the neck. In terrestrial forms, this is usually accompanie[d] by retraction of the front legs so that the head is completely hidden and thus protected. All sea turtles and land tortoises, and many terrapins, are cryptodirans.

LAND TORTOISES
Family Testudinidae

These chelonians are highly modified for terrestrial life. Most have domed, thick shells (except for the pancake tortoise, *Malocochersus tornieri*, from E. Africa, which has a reduced and very thin shell). The top of the head is covered with several distinct shields. The hind feet are webless and elephant-like; they walk on the tips of their heavily armoured and clawed forefeet. Buttock tubercles are often present. They are found in the temperat[e] and tropical regions of all continents except Australia, as well as on the islan[ds] off E. Africa and the Galapagos Islands. There are some 40 (or more) specie[s] in 10-15 genera (depending on the extent to which the large genus *Geochelone* is subdivided). The world's greatest diversity of land tortoises occurs on the subcontinent. Of the five genera and 12 species occurring in t[he] region, three genera and 10 species are endemic.

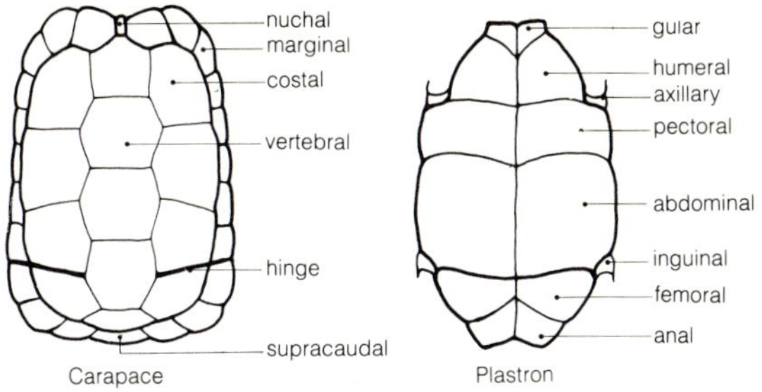

The shields of a tortoise shell

y to the southern
rican genera in the
studinidae

1. Rear of carapace hinged (in adults); outer edge of 4th costal wider than that of 3rd costal (all ages) .... *Kinixys* (Hinged tortoises, page 30)

   Rear of carapace not hinged; outer edge of 4th costal equal to or smaller than that of 3rd costal 2

2. Gular single and protruding .... *Chersina* (Angulate tortoise, page 26)

   Gulars paired 3

3. Nuchal shield absent .... *Geochelone* (Giant land tortoises, page 25)

   Nuchal shield present 4

4. Shell usually flat and vertebral shields never conical; each gular usually broader than long .... *Homopus* (Padlopers, page 23)

   Shell usually domed, often with conical vertebral shields (knobs); gulars usually as long as or longer than broad .... *Psammobates* (Geometric and Tent tortoises, page 27)

DLOPERS
mopus

This is a group of five small to very small tortoises that includes the world's smallest species. The carapace is not hinged and is relatively flat, with the scutes never raised into knobs, and often with depressed centres. The plastron has paired, thickened gulars that are wider than they are long. A nuchal is present. Shell abnormalities (extra and misshapen scutes, etc.) are relatively common. The generic name means 'same foot' as both front and hind feet of two species have four claws (all other tortoises, including three padlopers, have five claws on the front feet).

Padlopers are endemic to southern Africa. They are restricted mainly to the Cape, with two species occurring in adjacent regions.

Greater Padloper *Homopus femoralis* (Pl. 6)

(Endemic) 10-13 cm; max. 16 cm
A small tortoise that weighs only 200-300 g, but is, nonetheless, the largest padloper. It lacks a hinge, but has a nuchal and paired gulars, and usually 11 marginals. The beak is not hooked, is tricuspid, and has a serrated edge. The nostrils are below the level of the eye. The forelimbs are covered with

large, overlapping scales and have four claws. Buttock tubercles are prese but may be small or absent in juveniles. The tail does not have a terminal spine. The carapace is olive or reddish-brown, the individual scutes of juveniles often having black margins. The plastron is uniform dirty yellow-green in adults, often with darker areas restricted to the anterior edge of the scutes in juveniles. The skin of the body is yellow-brown, and sometimes orange-pink. Males do not have concave plastrons, but have long tails. Females grow larger than males. **Biology and breeding:** These chelonians are common in suitable habitat, where they may be found sheltering under rock slabs or in old, hollowed-out termitaria. They hibernate deep in rock cracks when winter snows blanket the peaks. Enemies include crows and birds of prey, jackals and the rock leguaan. The female lays 1-3 oval, hard-shelled eggs (29-35 x 25-27 mm) in summer. Hatchlings measure 25-30 mr and weigh 6-8 g. **Habitat:** Grasslands of mountain plateaus, particularly of old escarpment. **Range:** Inland mountains of E. Cape, extending into S. anc central OFS, N. Cape around Kimberley, and mountains of SE. Lesotho. Re populations occur along the old escarpment edge in the Karoo, from Murraysburg to Sutherland.

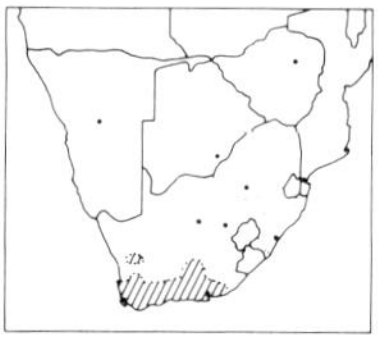

Parrot-beaked Tortoise *Homopus areolatus* (Pl. 5)

(Endemic) 7-9 cm; max. 11,4 cm

A small tortoise that lacks a hinge, but does have a nuchal and paired gular The carapace is often attractively sculptured, with the scute margins deeply etched and the centres indented. There are usually 11 marginals. The beak strongly hooked (hence the common name), tricuspid, and has a weakly serrated edge. The nostrils are situated high on the snout. The forelimbs, which are covered with very large, overlapping scales, each have four claw Buttock tubercles are absent or very small. The tail does not have a termina spine. Mature males have flat plastrons, long tails and, in breeding season, bright orange nasal scales. Females grow slightly larger than males. The carapace is yellowish-olive to green, usually with red-brown areolae and da brown to black margins in juveniles and adult females; males remain uniforr orange-brown, except for the orange nasal scales in breeding season. The plastron is yellowish with a brownish centre. **Biology and breeding:** Becau of their small size, these tortoises rarely forage in the open, favouring sunny spots around the edge of thick cover. They often shelter under rocks or in disused animal burrows. They can climb well up steep slopes. They struggl when restrained and will readily eject the contents of their cloacal water reservoir. They are commonly eaten by crows and even secretary birds. The have lived for longer than 28 years in captivity. The female usually lays 2-3 eggs (very rarely up to four) in a small nest hole dug in sandy soil. Elongate and oval (27-33 x 20-23 mm), the eggs take 150-300 days to hatch. The yo weigh 7-8 g and measure about 30 mm. **Habitat:** Varied; coastal fynbos, karroid broken veld and valley bushveld. **Range:** Cape coastal region from East London to Klawer, extending inland to Pearston in the east and Middelpos in the Roggeveldberge. Usually occurs below 600 m, but extenc inland to higher altitudes via the moist corridor of the Cradock gap.

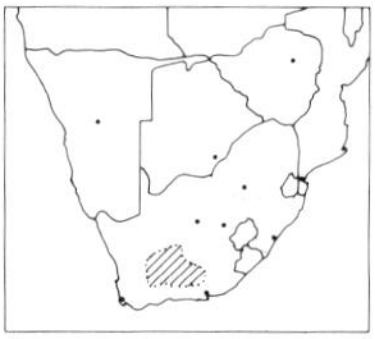

Karoo or Boulenger's Padloper *Homopus boulengeri* (Pl. 6)

(Endemic) 10-13 cm; max. 16 cm

A small tortoise that lacks a hinge, but has a nuchal and paired gulars. The flattened shell has a rounded bridge, and usually 12 marginals. The beak is weakly hooked or not hooked at all, is tricuspid, and has a weakly serrated edge. The forelimbs are covered with very large, overlapping scales; each forelimb has five claws. Buttock tubercles are absent. The tail does not have terminal spine. Males have long tails and a deep concavity in the plastron. T carapace varies in colour from dark red to yellow-brown, and sometimes oli

The plastron is similar but lighter in colour. The skin of the neck and limbs is dull yellow, and sometimes bright yellow with orange scales. **Biology and breeding:** This very secretive species shelters under rock slabs on rocky outcrops, plateaus and ironstone ridges. It is active on cool summer days, particularly when thunderstorms threaten (hence the common name 'donderweerskilpad'; it is also known as the 'klipskilpad' and the 'rooiskilpadjie'). They are killed and eaten by crows, which break their shells by dropping them on to rocks. Breeding is poorly known; the female lays one egg (32-39 x 22-23 mm, 10 g) in January. **Habitat:** Karroid regions. **Range:** Great Karoo, from Pearston and Wolverfontein in the east to Sutherland and Carnarvon in the west.

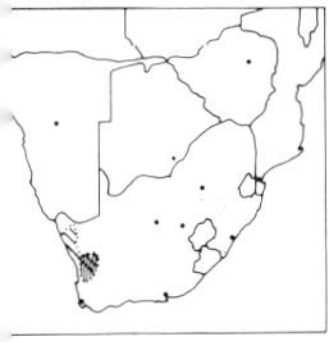

H.s. signatus
H.s. cafer

Speckled Padloper *Homopus signatus* (Pl. 6)

(Endemic) 6-8 cm; max. 9,6 cm
The world's smallest tortoise. It lacks a hinge, but has a nuchal and paired gulars. The flattened shell has a rounded bridge, and usually 12 marginals that may be serrated (see Subspecies). The beak is weakly hooked or not hooked at all. The forelimbs are covered with very large, overlapping scales; each forelimb has five claws. Buttock tubercles are present. Males have a well-developed concavity in the plastron, and a long tail. Coloration is varied (see Subspecies). **Biology and breeding:** These tortoises are active in the early morning in rocky areas, foraging for small succulent plants among granite slabs. They are very common in suitable habitat, and several may be found sheltering together under a rock slab. Breeding is poorly known; the female lays a single egg in summer. **Habitat:** Varied; mainly western succulent karoo, but extending into fynbos in the south. **Range:** W. Cape to S. Namibia. **Subspecies:** Two races occur. *H.s. signatus* has serrated marginals, the nuchal wider than it is long, and the carapace shields raised with sunken centres; the carapace is light brown with extensive black splashes. It occurs in Little Namaqualand and S. Namibia. *H.s. cafer* has marginals that are not serrated, the nuchal narrower than it is long, and smooth carapace shields; the carapace is orange-red to salmon-pink, with fine black spots and stippling. It occurs in W. Cape from Piketberg to Klawer and Calvinia.

Nama or Berger's Padloper *Homopus bergeri* (Pl. 6)

(Endemic) 10-15 cm
A very small tortoise that lacks a hinge, but has a nuchal and paired gulars. The flattened shell has a rounded bridge, and usually 12 smooth marginals. The beak is weakly hooked or not hooked at all. The forelimbs are covered with very large, overlapping scales; each forelimb has five claws. Buttock tubercles are absent. The carapace scutes are chestnut to red-brown, with pale centres and black edges. The plastron is uniform light yellow, sometimes with a greenish tinge. **Biology and breeding:** Not much is known about this species, as it has only recently been rediscovered. It appears to be similar to the Karoo padloper, *H. boulengeri* (page 24). **Habitat:** Rocky semi-desert. **Range:** Small area around Aus in S. Namibia.

ANT LAND
RTOISES
ochelone

This is a large and ancient genus, with numerous living and fossil species. The taxonomy is very confused, as some researchers divide these tortoises into six living genera that are distinguished by subtle internal and skeletal characteristics. The genus includes the largest living tortoises, *Geochelone gigantea* (measuring 1,4 m and weighing 250 kg), from Aldabra Island in the Indian Ocean, and *G. elephantopus* from the Galapagos Islands in the eastern Pacific. Even these giants are surpassed by the fossil species *G. gigas* that lived in the East Indies two million years ago; it measured 2,4 m and weighed more than 850 kg. The genus *Geochelone* contains approximately 15 living species, two of which occur in Africa.

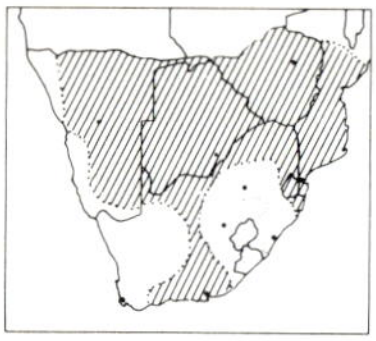

Leopard Tortoise *Geochelone pardalis* (Pl. 4)

30-45 cm; max. 72 cm
This tortoise weighs between eight and 12 kg, but grows much larger in E. Cape, where adults average 15-20 kg and may exceptionally exceed 70 in length and 40 kg in weight. The carapace is domed and not hinged, with scutes not, or only faintly, raised. The gulars are paired and are as long as t are wide. The nuchal is absent. There are 10-12 marginals, with those on th rear edge usually serrated and often upturned. The beak is sometimes hooked, is unicuspid and often serrated. Each of the front feet has five claw There are 2-3 buttock tubercles on each side. The tail does not have a termi spine. The carapace of hatchlings is yellow, with central paired or single bla spots, the ground colour becoming darker and heavily blotched and streak in black with age; old adults are often uniform dark grey-brown. The plastro yellowish, often with black radiating streaks and spots. Males have a longer tail than females, and a well-developed plastral concavity.

Biology and breeding: Growth in the first years of life is relatively slow, b increases rapidly as the young tortoise gets too large for small carnivores tc kill, and can feed further from cover. It weighs about 1 kg by 7-8 years, and thereafter body mass may double every 2-3 years. Sexual maturity is proba reached in 15 years, by which time growth has slowed considerably. These tortoises may live in captivity for at least 30 and possibly up to 75 years. The eat a wide variety of plants, including grasses, annuals and succulents. The also gnaw bones, and even hyaena faeces, to obtain calcium for shell grow and egg shell development. Many predators, including ants, rock leguaans storks, crows and small carnivores, feed on hatchlings and juveniles. Adults are relatively immune to predation, except by man; the latter is probably responsible for the absence of large tortoises from many areas, particularly Transkei. Many adults have cracked shells from falls in rocky areas, while others are killed in veld fires. They are usually well infested with ticks in the s groin skin. During cold rain and in winter they shelter deep in thick bush. Th occupy a large home range (1-2 km^2), and have been known to undertake long return journeys (5-10 km) when translocated from their territories. Two males meeting, particularly in breeding season, will engage in combat (pushing, butting and sometimes overturning one another). Mating is simila robust, the male pursuing and butting the female into submission. Copulatic is often accompanied by much straining and 'asthmatic' wheezing by the m

The gravid female selects a sunny, well-drained site and excavates a flas shaped pit (up to 25 cm across and deep) with her hind feet; she urinates copiously to soften hard soil. She usually lays 6-15 (exceptionally up to 30) large, almost spherical, hard-shelled eggs (32-41 x 35-44 mm, similar in siz and shape to pingpong balls). The hole is refilled and the female may tamp down the soil by lifting and dropping her shell regularly on the spot.
A large female may lay 3-6 similar-sized clutches at monthly intervals durinc the summer. Incubation takes 10-15 months, depending on the temperature (in captivity, eggs incubated at 28 °C hatch in eight months); eggs laid early the season take longer to develop as they lie dormant during the winter. During the long development, the ground may become very hard; hatchling may then have to wait for days, even weeks, for rain to soften the soil, before they can burrow to freedom. Hatchlings weigh 23-50 g and measure 40-50 mm. **Habitat:** Varied; not restricted to montane grassland, also occurrinc fynbos, valley bushveld, and arid and mesic savannah.
Range: Found throughout the savannahs of Africa, from Sudan to S. Cape. Historically absent from SW. Cape, but now introduced there.

ANGULATE TORTOISE *Chersina*

This tortoise is endemic to the tip of southern Africa, extending just into Namibia. It is unique among African tortoises in having an undivided gular a the front of the plastron. Various aspects of its biology are also unusual. It is possibly distantly related to the hinged tortoises, *Kinixys* (page 30).

Angulate Tortoise *Chersina angulata* (Pl. 4)

(Endemic) 15-25 cm; max. 30 cm
A medium-sized tortoise that grows larger in the western regions. In the east males rarely exceed 22 cm (1 kg), while on Dassen Island and the adjacent west coast mainland even females may exceed 24 cm (1,5 kg). The carapace is never hinged; it is elongate and flared at the front and behind in mature males. The scutes are slightly raised. There are 10-12 (usually 11) marginals. A nuchal is present. The gular is single and protrudes in mature males. The beak is weakly hooked, bi- or tricuspid, and is rarely serrated. There are five claws on each of the forefeet. Buttock tubercles are absent. The tail does not have a terminal spine. The carapace is light straw-yellow in colour. The top scutes have dark brown areolae and black edges. The marginals have a black triangle on the posterior edges. The plastron has a dark, irregular centre, often with white sutures in old animals, and the abdominals are light orange to bright red (particularly in W. Cape 'rooipens' form). Old adults become smooth-shelled and a uniform dirty straw colour. Unlike most other tortoises, males grow larger than females; the males have a 'peanut' shape, an elongate gular and concavity in the plastron.

Biology and breeding: The diet includes grasses, annuals and succulents. Angulate tortoises drink through the nose from rock pools; on sandy soils they raise the rear legs and extend the neck, pushing the snout into the soil, and filtering water from that which runs off the shell and collects around the head. They are active during the early morning, and retreat to cover when the sun becomes too hot. They can withstand high body temperatures (higher than 40 °C), but prefer to be cooler (around 30 °C). This tortoise has lived for 32 years in captivity. It readily ejects the liquid contents of its bowels when handled, often spraying them up to one metre, and this with surprising accuracy. Enemies include small carnivores, rock leguaans, secretary birds, sea gulls and crows. Hatchlings have even been found skewered on tree thorns by fiscal shrikes. Juvenile angulate tortoises wander slowly from place to place, and select a suitable territory only when they reach maturity. Males have much larger home ranges than females. The males defend territories, driving off all other mature males and preventing them from mating with the females. Combat involves butting and using the enlarged gular to overturn one another. A single 'fall' usually resolves the dispute, but the opponent may right himself and re-enter the fray.

Copulation is a noisy affair. After rain, when the soil is soft and moist, the female digs a shallow depression about 10 cm wide, with a small chamber 4 cm wide and deep at its bottom. The claws of the hind foot are used to break up the soil, which is then pushed away with the side of the lower leg. Only a single (very rarely two) hard-shelled, spherical egg (30-35 x 37-42 mm, 20-25 g) is laid, and this may occur 4-6 times a year. After laying, the soil is tamped down by the female with her shell. The whole procedure may take 2-3 hours. Although the area is carefully camouflaged, a large number of eggs are excavated and eaten by mongooses. Incubation takes 90-200 days, depending on the season. Eggs may crack 6-10 days before young emerge. Hatchlings weigh 8-12 g and are 32-35 mm long. Growth is rapid in the first 8-10 years, slowing thereafter; there is little growth after 20 years. Sexual maturity is reached in 9-12 years.

Habitat: Sandy coastal regions, including valley bushveld and coastal fynbos; scarcer in the arid hinterland. **Range:** Cape coastal regions, from East London to Oranjemund, extending inland as far as Cradock in association with karoo broken veld. There are isolated records in S. Namibia, and a relict population at Beaufort West.

OMETRIC AND
NT TORTOISES
ammobates

These small, attractive tortoises usually have their carapace scutes raised into knobs, with beautiful, radially arranged light and dark bands giving a striking geometric pattern. The carapace is domed and is never hinged. A nuchal is present. The paired gulars are longer than they are broad. Buttock tubercle are present in some species. Males are much smaller than females.

'Psammobates' means 'sand-loving', and most of these tortoises inhabit arid interior of the subcontinent.

They are endemic to southern Africa. At present three species are recognized, of which one is very varied with local races, and another endangered due to habitat destruction.

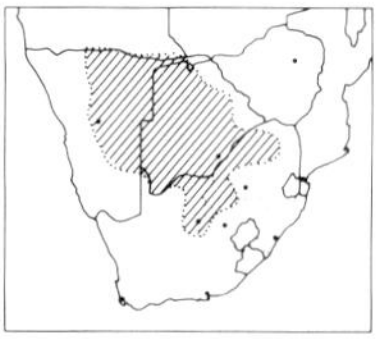

Serrated or Kalahari Tent Tortoise
Psammobates oculifer (Pl. 7)

(Endemic) 8-12 cm; max. 14 cm
A small tortoise with a low, domed carapace with a strongly serrated edge. lacks a hinge. The nuchal is broad and often divided, and the paired gulars are longer than they are broad. There are usually 11 (sometimes 10 or 12) marginals that are strongly serrated at the front and back. The scutes are o slightly raised. There is a single axillary scale. The beak is hooked and tricuspid. The forelimbs each have five claws and a few large, and one extremely large, scales. Buttock tubercles are present. The tail does not ha a terminal spine. The carapace is light brown-yellow, with each scute beautifully marked with a radial pattern of 6-10 dark brown to black rays. Th plastron is yellowish, with radiating dark rays. Males have a longer tail and flatter shell than females, and an obvious plastral concavity. **Biology and breeding:** Despite its wide distribution, this tortoise is poorly known. It may into loose soil at the base of scrub, or retreat into small mammal burrows. It common in some areas, eg. Khuis. Historically, the shell was decorated an used by Bushmen to make buchu pouches and, despite protection, large numbers of this species are still killed to supply the tourist trade. It feeds or small succulents and grasses, and also the droppings of sheep and game The female lays 1-2 eggs (28-31 x 40-42 mm) in December. **Habitat:** Arid savannah and scrub desert. **Range:** Kalahari and adjacent regions; not fou south of Orange River.

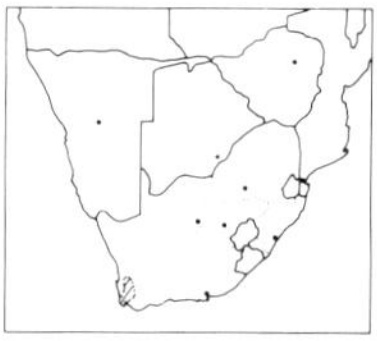

Geometric Tortoise *Psammobates geometricus* (Pl. 7)

(Endemic) 8-12 cm; max. 14,5 cm
A small tortoise with a high, domed carapace that lacks a hinge. A nuchal is present. The paired gulars are longer than they are broad. There are 11-12 marginals (numbers 4-7 being higher than they are broad) that are slightly upturned at the back. The scutes are only slightly raised. There is a single axillary scale. The beak is hooked. The forelimbs each have five claws and covered with scattered, large scales that are separated by smaller ones. Buttock tubercles are absent. The tail does not have a terminal spine. The carapace is beautifully marked with geometric patterns; the scutes have yellow centres from which yellow rays (8-15 on the vertebrals, 9-12 on the costals and 2-4 on the marginals) radiate, separated by black. The plastron yellow, with radiating faint black rays and bands. Many juveniles have a sin yellow 'X' on each carapace scute (but this may be absent) and a black plastron; the radial pattern develops with age. Females are more common than males, and are slightly larger (males average 10,6 cm and 207 g; females average 12,5 cm and 436 g). **Biology and breeding:** This is an endangered species (CITES Appendix 1; SA RDB, Endangered), due to destruction of habitat for urban development and the cultivation of wheat ar vineyards. Possibly as few as 2 000-3 000 specimens remain. Veld fires kill some tortoises, but hatchlings emerge soon after these fires to feed on you plants. Geometric tortoises feed on a variety of succulent and perennial plants; however, snails and even the remains of young parrot-beaked tortoi have been found in the faeces. They are active in the early morning and afternoon, often in relatively cold conditions. Enemies include crows and

secretary birds, small carnivores and man. This tortoise probably lives for longer than 30 years. The female lays 2-4 eggs (32 x 24 mm) in September-November, and these hatch in March-May, after 150-210 days' incubation; hatchlings (30-40 mm) emerge with the first winter rains. Growth is relatively rapid, with two growth rings being laid down each year. Sexual maturity is attained in 7-8 years. **Habitat:** Restricted to low-lying coastal renosterbosveld on the west coast. **Range:** Previously from Cape Town to Eendekuil, 160 km north along W. Cape forelands, with isolated populations around Ceres and Worcester. Now restricted to a few patches of natural veld (96% of its habitat has been destroyed). A number of reserves have been proclaimed (eg. at Strand).

P.t. tentorius
P.t. verroxii
P.t. trimeni

Tent Tortoise ***Psammobates tentorius*** (Pl. 7)

(Endemic) 8-10 cm; max. 14,5 cm
A small tortoise that comes in a bewildering range of shapes and colours. This single species has at times been divided into no less than six species with 22 races. The carapace is domed or flat, with or without raised scutes (see Subspecies), and is not hinged. A nuchal is present, typically broader than it is long, and is often minute but rarely absent. The paired gulars are longer than they are broad. There are usually 11 (sometimes 10 or 12) marginals (numbers 4-7 being broader than they are high). Two to three (rarely one) axillaries are present. The beak is usually hooked and bi- or tricuspid. The forelimbs each have five claws and are covered with large, abutting scales. Buttock tubercles are typically present, but may be reduced or absent in the western race. The tail does not have a terminal spine. Coloration is varied; the carapace has geometric patterning (see Subspecies). The males are much smaller than the females, rarely exceeding 10 cm, and they have longer tails and a shallow plastral concavity. **Biology and breeding:** Despite its wide range, this species is not easily found and often occurs in low densities. During droughts they burrow into sandy soil at the base of low shrub, emerging after the onset of rains. They are active in the cool of early morning and evening, when they feed on small succulents. Their geometric patterning provides very effective camouflage in broken shade at the base of bush. They drink by raising the rear legs and sipping the water that drains along the shell grooves to the forelimbs. They rarely do well out of their natural range, and have lived for only 7-8 years in captivity. Enemies include small carnivores, rock leguaans, eagles, crows, and even ostriches. Very few details are known about their breeding. The female lays a few eggs: only 1-2 (24 x 35 mm) in the western races, and 2-3 in the typical race (21-24 x 27-31 mm). These are laid in summer (September-January) and hatch after about 220 days. Hatchlings (25-30 mm) are more circular in shape than adults. **Habitat:** Varied; usually arid karroid areas or rocky sandveld. **Range:** Throughout Great and Little Karoo, to succulent Karoo of Cape west coast and into S. Namibia. **Subspecies:** Only three races are recognized at present; the species is very variable and intermediate conditions occur. *P.t. tentorius* is the largest race. Its carapace scutes are strongly raised into knobs, with a well-marked pattern of yellow to orange radiating stripes on a black background. The plastron has solid mahogany markings in the centre and a yellow to orange edge. It occurs in SE. Cape, from Grahamstown to Matjiesfontein, intergrading with the next race in central Karoo. *P.t. verroxii* has a carapace with smooth or faintly raised scutes, and is a drab colour, with dull orange stripes on a dark brown background. The plastron has diffuse or indistinct dark brown patterning. It occurs in N. and central Karoo, intergrading with the next race in Bushmanland and S. Namibia. *P.t. trimeni* is the most attractive race, its carapace having small knobs and being well-marked with rich orange-yellow stripes (shading to red around the marginals) on a black background. The dark brown centre of the plastron is broken by lighter patches. It occurs in Namaqualand and S. Namibia, intergrading with the previous race in the east.

HINGED TORTOISES
Kinixys

These unusual tortoises have a unique hinge in the carapace (in adults) tha allows the rear of the shell to close, protecting the hind feet and tail region. Juveniles in which the hinge has not developed can be distinguished from padlopers, *Homopus* (page 23), as the outer edge of the third costal is long than that of the fourth (these two genera are not sympatric, so this feature is academic in the field). The hinge develops first at the edge between the seventh and eighth marginals and then spreads inwards, separating the second and third costals, the sutures being replaced with fibrous cartilage In some adults the hinge fails to develop. Nuchal and submarginal scutes a present. The paired gulars are very thickened at the front; they are sometin longer than they are broad. There are usually five claws on each of the forefeet.

Four species are recognized. Two of these (*K. homeana* and *K. erosa*) ha spiny and upturned marginals, and are restricted to the rain forests of W. a central Africa. The other two species inhabit savannah; both occur on the subcontinent, and one is endemic.

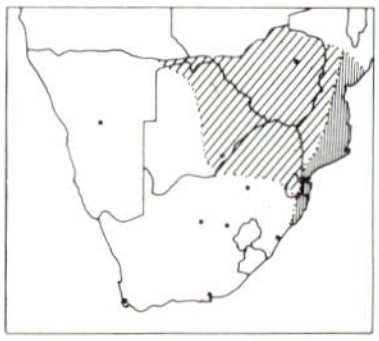

K.b. *belliana*
K.b. *spekii*

Bell's Hinged Tortoise *Kinixys belliana* (Pl. 5)

12-17 cm; max. 21 cm
A medium-sized tortoise having either a smooth, convex or depressed carapace (see Subspecies) and a prominent hinge. A nuchal is present. The gulars (together) are less than twice as wide as they are long. There ar 11 marginals that are not serrated or upturned. The supracaudal is undivid There are 2-3 axillaries. The beak is unicuspid. The forefeet each have five claws (but four in *K.b. nogueyi* from W. Africa) and are covered with large, overlapping, sometimes pointed scales. Buttock tubercles are absent. The has a terminal spine which is large in males. Coloration is varied (see Subspecies); it may fade in adults, particularly in males. Males are easily distinguished by this fading, the plastral concavity and the very large tail. **Biology and breeding:** These tortoises are active in the wet summer mont aestivating underground during the cold dry season (April-September) in o termitaria or small burrows which they scrape into earth embankments. The diet is very varied, and includes *Syzygium* and other fruit, mushrooms, giar land snails (*Achatina*) and pill millipedes. This species lives for up to 22 yea in captivity. It prefers humid conditions. Some specimens struggle when captured, and may empty the bowels before retreating into their shells and closing the hinge. Breeding is poorly known locally. The female lays 2-7 (sometimes 10) elongate eggs (39-48 x 32-36 mm, 23-32 g) throughout summer (November-April); she may lay clutches at 40-day intervals. Incubation takes 90-110 days; hatchlings (38-40 mm) have been found in September-October and March-April. **Habitat:** Savannah, coastal plain an dune forest, entering thornveld. **Range:** Savannahs of central and E. Africa through Zimbabwe to Transvaal and Lobatsi in SE. Botswana, and along th Mozambique coastal plain as far south as Zululand. Also on Madagascar, b possibly introduced there.

Subspecies: Four races recognized, two in southern Africa. *K.b. noguey* has four claws on each of its forefeet, and occurs in W. Africa. *K.b. mertens* poorly defined, being distinguished only by a short median pectoral sulcus occurs in N. Zaire and Uganda. *K.b. belliana* has a convex carapace, with a yellow and black radial pattern (4-6 rays) on the vertebrals and costals. The plastron is varied in colour, but is largely black in juveniles, except for a bro yellow band. The patterning becomes fainter and vaguely radial in adults. T race occurs on the E. African coastal plain, as far south as Zululand. *K.b. spekii* has a depressed carapace with zonary patterning, the areolae being dark brown with concentric light and dark zones (these may break up into radiations in adult females). The plastron is uniform dirty yellow in male with vague zonary patterning in females and juveniles. This race occurs on central plateau of central and southern Africa.

Natal Hinged Tortoise *Kinixys natalensis* (Pl. 5)

(Endemic) 8-12 cm; max. 15,5 cm
A small tortoise with a slightly convex carapace and a prominent hinge. A nuchal is present. The gulars (together) are more than twice as wide as they are long. There are 12-13 marginals, and these are not serrated or upturned. The supracaudal is usually divided. There are 2-3 axillaries. The beak is tricuspid. The forefeet each have five claws and are covered with large, overlapping, sometimes pointed scales. Buttock tubercles are absent. The tail has a terminal spine which is large in males. The carapace shield displays concentric patterning, with the areolae light to dark brown with concentric rings of orange-yellow and black-brown; there is a similar ringed pattern on the plastron shield that may fade in adults. Males have a shallowly concave plastron and are smaller than females. **Biology and breeding:** This has only recently been considered a valid species, and is still poorly known (SA RDB, Rare). It inhabits dry, rocky areas in an altitude range of 300-1 000 m. Copulation in captivity has been observed in February. **Habitat:** Tropical lowveld, entering valley bushveld in the south. **Range:** Natal lowlands, through Lebombo range to E. Transvaal.

A TURTLES
perfamily
elonioidea

The first chelonians lived in swamps, but by the Cretaceous period at least four families had moved into the oceans. Many of these early sea turtles were large, the largest being the giant *Archelon* which was over 3 m long. Sea turtles are now less diverse than in those times, and are represented by only eight species in two families, one of which (the leatherback turtle) is highly aberrant. They retain certain primitive features, such as being unable to retract the head or limbs, and they have a robust skull and a row of inframarginals across the bridge. Other unusual features which are adaptations to marine life include the excretion of excess salt via the tear ducts and the modification of the limbs into flippers, which retain only one or two claws and are 'rowed' in unison rather than used alternately as do the terrapins.

All sea turtles are tied to land for reproduction and must brave dangers as they haul themselves ashore to lay their eggs. Clumsy and almost helpless on land, the breeding females are easy to kill. Their numbers have been decimated and all are now endangered. All breed in a similar fashion, crawling ashore on to sheltered, sandy beaches, usually on moonless nights. They dig a deep pit and lay large numbers of spherical, soft-shelled eggs. They emerge from the water one to five times at 10- to 15-day intervals, and may lay up to 1 000 eggs during a breeding season. Most females breed at three- to four-year intervals, but this varies with the species and the individual. Hatchling sex is determined by incubation temperature; the critical period is the third week of incubation, and higher temperatures (32-34 °C) produce females.

Two of the five species found in the coastal waters of southern Africa lay their eggs in the sandy beaches of Maputaland. Fully protected by the Kwazulu Government and Natal Parks Board (SA RDB, Vulnerable), their numbers are slowly increasing.

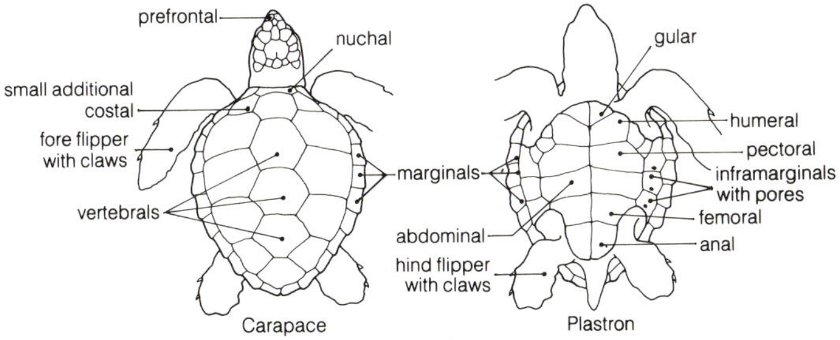

The shields of a sea turtle

LEATHERBACK TURTLES
Family Dermochelyidae

This group of large to giant sea turtles is characterized by various skull features (including the lack of nasal bones), extreme reduction of the bone the carapace and plastron, and a unique internal shell that is composed of small, polygonal bones.

Four fossil genera are known, dating back to the Eocene epoch. Only a single living representative survives.

LEATHERBACK TURTLE
Dermochelys

There is a single species in the genus, and it is peculiar in both anatomy an behaviour.

Leatherback Turtle ***Dermochelys coriacea*** (Pl. 9)

130-170 cm; max. 178 cm
A giant turtle (it can weigh up to 646 kg) with a deep, narrow, barrel-shape shell that lacks horny scutes, being covered instead with thick, smooth skir that resembles vulcanized rubber. The skin has 12 long ridges (five each o the carapace and the plastron, and one on each side). The shells of hatchli are covered with small, bead-like scales. The flippers are long and lack cla although these may be present for a short time in some hatchlings. The nec short and thick. The beak is bicuspid, sharp-edged and hooked. In adults, carapace and flippers are black, usually with scattered white spots. The plastron and lower surfaces of the head and flippers are white, suffused wi pink and grey-black. The carapace and flippers of juveniles are blue-grey when dry and blackish when wet, with white ridges on the shell and the trai edges of the flippers.

Biology and breeding: Leatherback turtles undertake long journeys an may temporarily enter cold currents to feed. Adults feed exclusively on jellyfish, but hatchlings may take other floating organisms. They have many internal adaptations to conserve heat in cold waters. The throat is coated w long, backwardly projecting spines that prevent slippery food from escapi Some leatherbacks have died after ingesting large sheets of clear plastic, presuming them to be edible. Nesting is restricted to tropical beaches, the most southerly of which is the small rookery on the Maputaland beaches. Gravid females haul themselves ashore at night in summer (November-January), favouring moonless nights with high tide around midnight. Each a deep, flask-shaped nest hole above the high-water mark. A thousand bill ball-sized eggs are laid in batches of 100-120 at nine- to 11-day intervals. Eggs have a high fertility (normally 70-75%, and up to 90%), and hatch in about 70 days. Females may return to nest every one to five years. Hatchlin emerge at night to avoid heavy predation by ghost crabs, sea gulls and fis Growth is phenomenally fast; the carapace grows about one centimetre a week, and weight increases a hundredfold in 29 weeks. Sexual maturity is reached in 3-4 years at approximately 140-cm carapace length.
Habitat: Surface waters of the temperate and tropical oceans.

Range: Worldwide. **Subspecies:** Indo-Pacific leatherbacks are sometir referred to a separate race (*Dermochelys coriacea schlegeli*), but this is questionable.

MODERN SEA TURTLES
Family Cheloniidae

These are advanced sea turtles that retain a hard shell and have strengthe limbs to increase their swimming efficiency.

Known from the Cretaceous period, there are 31 genera, 27 of these exti and four living. All are represented in southern Africa.

Key to the southern African genera in the Cheloniidae

1. Carapace with 4 pairs of costals, the first pair separated from the nuchal and never the smallest 2
   Carapace with 5 pairs of costals, the first being small and usually touching the nuchal 3

2. Snout not compressed; 2 prefrontals on head; shields not overlapping; 1 claw on each flipper .... *Chelonia* (Green and Flatback sea turtles, page 33)

Snout hooked; 4 prefrontals on head; shields strongly overlapping; 2 claws on each flipper .... *Eretmochelys* (Hawksbill turtle, page 34)

3. 5 pairs of costals; 3 inframarginals (without pores) across bridge; reddish colour .... *Caretta* (Loggerhead sea turtle, page 35)

6-9 pairs of costals; 4 inframarginals (often with pores) across bridge; olive colour .... *Lepidochelys* (Ridley sea turtles, page 34)

EEN AND ATBACK SEA RTLES
elonia

Widely distributed, three species are recognized at present. Extralimitally, the flatback sea turtle (*C. depressa*) occurs in Australasian waters, and is characterized by its flat shell and carnivorous habits; it lays only a few, large eggs. The Pacific green turtle (*C. agassizi*) occurs in the eastern Pacific, and is a small, dark turtle with a high, narrow carapace. Only the green turtle (*C. mydas*) occurs in southern African coastal waters.

Green Turtle *Chelonia mydas* (Pl. 8)

98-120 cm; max. 140 cm
A very large, hard-shelled turtle; nesting females on the Mozambique islands weighed 124-208 kg, and those from the Atlantic up to 300 kg. The shell is smooth, with thin, non-overlapping scutes, and a median keel in juveniles which disappears in adults. There are 12 pairs of marginals, the posterior ones being serrated in juveniles and smooth in adults. Four pairs of costals are present. The plastron is relatively large, with two long ridges in the young. The bridge is wide, with four inframarginals that lack pores. The head is compact and relatively small. The prefrontal scales on the nose are undivided and elongate. The front flippers each have a single claw (two in juveniles); there is only one claw on each of the hind flippers. Males have longer tails than females. Coloration is varied. Hatchlings have a black-brown carapace with bronze highlights on the vertebrals, and a white border and plastron. Juveniles have a carapace that is dark grey when approximately 20 cm long, and thereafter may have varied ground colour (pale red-brown to dark brown), streaked with dark brown, red-brown and yellow. The adult's carapace varies from greenish-brown to black; streaking may persist, break into spots or fade completely. The plastron is dirty white to yellow. The head shields are black to red-brown, and often white-edged. Females are usually darker in colour than males. **Biology and breeding:** During the first year of life, green turtles feed on jellyfish and other floating organisms, but then become predominantly herbivorous, grazing on sea grasses (eg. *Zostera* beds) in estuaries and shallow seas. Subadults and occasional adults are common along the eastern seaboard, with a few temporary residents in most open estuaries. It is not known whether these temporary residents return to northern waters to breed, or if they are non-breeding vagrants. In the west, adults come ashore on the Skeleton Coast, but are not known to breed and are probably basking out of cold waters. The green turtle does not breed on local beaches; the nearest rookery is Europa Island in the Mozambique Channel, where 4 000-9 000 turtles nest annually. Nesting occurs throughout the year, with a midsummer peak, and up to 750 females may come ashore each night. Mating takes place in shallow coastal waters in typical chelonian fashion: the males have enlarged

claws on their front flippers which they hook over the leading edge of the female's carapace. The eggs are spherical (41-47 mm dia., 38-58 g), and 115-197 are laid in each clutch. The female returns 2-3 (and up to six) time intervals of 10-20 days. The eggs hatch in approximately 56 days, the your emerging together, usually at night. Carapace length in hatchlings is 45-51 mm, and they weigh 18-29 g. Growth is slow, and sexual maturity is reached in 10-15 years. **Habitat:** Shallow waters with abundant submerge vegetation. **Range:** Tropical and subtropical seas, adults and subadults entering southern African coastal waters. **Subspecies:** None is recognize present; the E. Pacific *C. agassizi* is now treated as a full species.

HAWKSBILL TURTLE
Eretmochelys

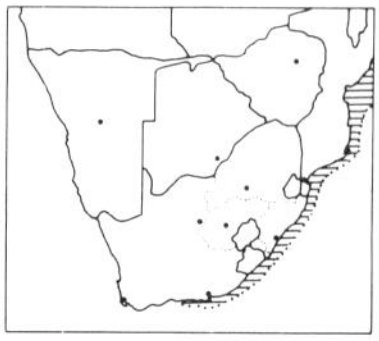

This is a medium-sized, highly distinctive sea turtle. Only a single species i recognized. It has no fossil history.

Hawksbill Turtle *Eretmochelys imbricata* (Pl. 8)

60-90 cm; max. 107 cm

A relatively small sea turtle that rarely weighs more than 50 kg, but has attained up to 139 kg. Its shell scales are thick and overlapping, and the he is narrow and anteriorly pointed, with a long, slightly bird-like beak (hence common name). It has two pairs of prefrontals, four pairs of costals and 12 pairs of marginals, the posterior ones markedly serrated. There are four poreless inframarginals. The forelimbs have extremely long digits, and ther are two claws on each limb. Males have long tails and narrower carapaces than females, and a small plastral concavity. In hatchlings, the carapace is uniform brown and the plastron dark, each scale having a large, dark spot. adults, the carapace shields are translucent amber, beautifully patterned v irregular, radiating streaks of light red-brown, black and yellow. The plastrc uniform yellow to orange-yellow, and the head is yellowish, with black-cent scales.

Biology and breeding: Young juveniles eat floating vegetation (eg. *Sargassum*), but adults are mainly carnivorous, feeding on hard-bodied, bottom-living marine invertebrates, including corals and urchins. They may become poisonous to man after feeding on toxic corals, and human deaths have resulted from eating the tainted flesh. Wild populations are endanger mainly by the curio trade. The beautiful scutes are made into the famous 'tortoiseshell' jewellery and other *objets d'art*. The volume of trade is horrendous; Indonesia and Japan imported 260 000 kg of raw shell for processing in 1978-9, while in Singapore and the Phillipines nearly 100 00 juveniles annually are stuffed, polished and sold as curios.

They do not breed locally; the nearest rookeries are in NE. Madagascar, Tromelin Island, and Primeiras and St. Brandon islands in Mauritius. Femal are very wary when emerging on to nesting beaches, which are usually of coarse sand. Depending on the region and the size of the female, 70-180 spherical eggs (40-42 mm dia.) are laid in a flask-shaped nest; 2-4 clutche are laid at 15- to 19-day intervals during the season. After an incubation pe of 58-64 days, hatchlings measuring approximately 40 mm emerge at nigh Growth is relatively rapid, and sexual maturity may be reached in 8-10 yea **Habitat:** Coral reefs of tropical waters. **Range:** Circumtropical; this is the second most common turtle in Mozambique coastal waters, but a relatively rare vagrant further south. **Subspecies:** A number of subspecies are recognized (eg. *E.i. imbricata* in the Atlantic, *E.i. bissa* and *E.i. squamata* ir the Indo-Pacific), but none seems well defined.

RIDLEY SEA TURTLES
Lepidochelys

These are the smallest of the sea turtles, with unusually broad shells. They were once famed for breeding in 'arribadas', when massive numbers of turt (up to 46 000) emerged to breed on 1 km of beach in a single day. Althoug this resulted in a tremendous wastage of eggs, as females often dug up the nests laid only minutes earlier by other females, it served to swamp predatc and resulted in overall survival of the young. Unfortunately, man is a most efficient predator, and these unique sights are no longer seen, the breeding females having been decimated.

Two species are recognized: extralimitally, Kemp's ridley (*L. kempi*) is a critically endangered species restricted to the Gulf of Mexico and the northern temperate areas of the Atlantic. The other species, *L. olivacea*, is a rare vagrant to the coastal waters of the subcontinent.

Olive Ridley Turtle *Lepidochelys olivacea* (Pl. 8)

50-65 cm; max. 75 cm
The smallest sea turtle, weighing up to 45 kg. It has a broad, flat-topped, smooth carapace, and numerous costal shields (5-9 on each side). Hatchlings and juveniles have three dorsal keels, and two on the plastron. There are 12-14 marginals, which are slightly serrated at the rear. There are four inframarginals, each with a pore on its posterior edge. The head is triangular, with two pairs of prefrontals. Each limb has two claws. Males have longer tails than females, and narrower, strongly tapered shells, with more intense pigmentation. In hatchlings, the carapace is uniform grey-black with light areas on the plastron, which in a few months become more extensive; adults are dark to light olive-green dorsally, with pale yellow, almost white plastrons. **Biology and breeding:** These turtles sometimes form large aggregations, migrating between nesting beaches and feeding grounds. They may sleep floating at the surface in large numbers, unlike other sea turtles, which usually sleep on the sea floor. They feed, sometimes at considerable depths, on bottom-living crustaceans, particularly prawns and shrimps, but also fish and squid. Perhaps the most common surviving sea turtle, it is still endangered by commercial trade in leather from the neck and limbs, and accidental capture in shrimp trawls. It does not breed locally, although a stray female once nested on Warner Beach in Natal; the nearest rookery is on the northern coast of Mozambique, where about 500-1 000 nests are laid each year. Local 'arribadas' are unknown; the nearest is in Orissa State in India. The female lays 2-3 clutches (at 17- to 29-day intervals) per season of 105-116 white, soft-shelled, spherical eggs (37-40 mm dia., 31-38 g), and may return to nest at one- to two-year intervals. Nest construction and egg-laying are rapid, and may be completed in one hour; the female tamps down the sand over the eggs with the thickened sides of her plastron. Incubation is short (42-62 days). Hatchlings measure 39-48 mm and weigh 14-18 g. Sexual maturity is possibly reached in 7-9 years. **Habitat:** Coastal mainland waters, often in major estuaries. **Range:** Circumglobal; present in tropical regions of Atlantic, Indian and Pacific oceans. Locally restricted to eastern coastal regions, very rarely entering Natal waters. **Subspecies:** The Caribbean *L. kempi* is sometimes treated as a localized race of the olive Ridley turtle.

GGERHEAD SEA
RTLE
etta

This genus contains a single living species, and a number of indeterminate fossil forms dating back to the late Cretaceous period of Europe.

Loggerhead Turtle *Caretta caretta* (Pl. 8)

70-100 cm; max. 107 cm
A large turtle, weighing 80-138 kg. Locally, a specimen has been recorded weighing 140 kg, while in the Atlantic, 159 kg has been recorded. (Reports of gigantic loggerheads weighing up to 400 kg are unconfirmed.) This sea turtle has a big head and an elongate shell that tapers at the rear and is smooth in adults, but has median keels on the costals and vertebrals in juveniles. There are five pairs of costals and 11-12 pairs of marginals which are sometimes bluntly serrated at the rear in juveniles. The plastron of hatchlings has two strong, long keels. There are three inframarginals without pores. The head is very broad (up to 25 cm across), and the jaw has extensive crushing surfaces.

There are two pairs of prefrontals. Each limb has two claws. Male loggerhe have longer tails than females, as well as an enlarged, strongly curved clav each front flipper, more massive heads, and smooth shell margins. In hatchlings, the carapace, plastron and head are uniform brown of varying shades, while the skin of the neck and flippers is darker. In both juveniles a adults, the carapace and the top of the head are uniform red-brown, with white-edged scales. The plastron is uniform yellowish, and the skin is light yellowish-grey.

Biology and breeding: After breeding in Maputaland, adult loggerhead move north to feed in the warm waters off Mozambique and E. Africa. Many hatchlings drift south in the Agulhas Current, some 'wrecking' on the E. and S. Cape coasts during onshore winds; others may continue in the southern Indian Ocean gyral, returning to the African coast three years later. For the three years of life, they drift in surface waters, eating bluebottles, comb jelli etc. Subsequently, they search for food in shallow coastal waters, becomin carnivorous. The extremely strong jaws are well adapted for feeding on cra molluscs and sea urchins, and can even crush giant clams.

They nest mainly in subtropical waters, including the sandy beaches of Maputaland in Natal, where 400-500 females breed each year. The largest rookery is on Masirah Island in Oman, where 30 000 females breed annual Mating takes places at the beginning of the nesting season. Females emer at night in late spring-summer (November-January in Maputaland) and lay about 500 eggs (40-42 mm dia.) per season in batches of 100-120 at 15-da intervals. Most return to nest after 2-3 years, but some may exceptionally ta up to eight years before breeding again, and others breed more frequently to six times in nine years). Incubation takes 47-66 days; hatchlings measur 39-49 mm. **Range:** Found worldwide in temperate and subtropical waters. More common on the east than on the west coast of the subcontinent. **Subspecies:** A number of subspecies have been proposed, including *C.c. gigas* for Indo-Pacific loggerheads, but none are recognized at prese

SOFT-SHELLED TERRAPINS
Family Trionychidae

These are unusual chelonians, surprisingly named not for their very odd, sc shells, but for having only three claws on each foot. Abundant in N. Americ and SE. Asia, there are a few African species. These terrapins are fully aquatic, with the shell flat and disc-like, often with a flexible edge. The horn shell is completely absent except for a few vestiges, and the underlying bo is also reduced. The neck is usually very long and extendable, with a snork like nose. Many grow to a large size. They are a shy, active species.

There are three genera containing five species in Africa; two species in separate genera just enter the subcontinent.

Key to the southern African genera in the Trionychidae

1. Plastron without femoral flaps .... *Trionyx* (Soft-shelled terrapins, page 36)

   Plastron with femoral flaps that cover retracted hind feet .... *Cycloderma* (African flapped soft-she terrapins, page 37)

SOFT-SHELLED TERRAPINS
Trionyx

This large genus is distributed in N. America (four species), Africa (one species) and Asia (10 species).

Nile Soft-shelled Terrapin *Trionyx triunguis* (Pl. 9)

40-60 cm; max. 95 cm
This terrapin attains a maximum weight of 40 kg. It is a very large soft-shell terrapin that lacks flexible flaps over the hind limbs. The head is elongate a flattened, with a snorkel-like snout. The forelimbs have three sharp-edged, crescentic skin folds. The young have indistinct median keels and wavy lin

of tubercles on the carapace. The carapace is dark brown or olive, usually covered with small white or yellow spots that may disappear in adults. The plastron is white, sometimes with dusky infusions at the front. The head and limbs are dark and profusely spotted.

Biology and breeding: Omnivorous, they eat molluscs, insects, frogs and fish, as well as palm nuts and fruits. The terrapins are eaten in many regions, and the flesh is reported to taste like rich and slightly oily veal. They move fast in water and on land. Usually vicious, they are ever-ready to bite. The very sharp and powerful jaws are efficient amputators of ill-placed fingers. They can survive seawater, and have apparently expanded their range by undertaking short marine excursions between river mouths. Long-lived, they have survived for longer than 42 years in captivity. The female lays 25-60 spherical, hard-shelled eggs (25-30 mm dia.) in nests buried 30-40 cm deep in exposed sand banks along rivers. Incubation is rapid, and the eggs hatch in 76-78 days.

Habitat: Deepish water in lakes, rivers and estuaries. **Range:** Throughout the Nile and most major river systems of W. and central Africa, but often absent upstream from major waterfalls (eg. absent from Lake Victoria). Just entering the subcontinent, occurring in Cunene River downstream of Ruacana Falls.

RICAN FLAPPED
FT-SHELLED
RRAPINS
cloderma and
clanorbis

These two genera of African soft-shelled terrapins are distinguished by the flexible flaps that cover the hind limbs when they are withdrawn. Each genus contains two species of medium-sized to large terrapins, distributed through the tropical regions of Africa. All are poorly known. In *Cycloderma*, the eyes are located closer to the snout than in *Cyclanorbis*, and there are other osteological differences. Only one species enters the region.

Zambezi Soft-shelled Terrapin *Cycloderma frenatum* (Pl. 9)

35-45 cm; max. 56 cm

This terrapin, attaining a weight of up to 14 kg, is a large soft-shelled species, with flexible flaps that cover the hind limbs. The head is elongate and flattened, with a snorkel-like snout. The forefeet have four or five sharp-edged, crescentic skin-flaps. The young terrapins have indistinct median keels and wavy lines of tubercles on the carapace. In colour, the carapace of adults is uniform pale to dark olive, sometimes faintly blotched; the young have a white edge to the carapace. The plastron is white to flesh-pink, sometimes with grey infusions in adults. The head and neck of the young are grey, with broad white stripes; these fade gradually in adult specimens, in which the head and neck are dark olive.

Biology and breeding: This terrapin digs with its forelegs in soft mud for snails and freshwater mussels, which it crushes with its strong jaws. It also eats fish, frogs and aquatic insects. Shy, the adults are rarely seen. When disturbed, they burrow into soft mud, and will bite and claw if captured. Enemies include crocodiles, otters and man. The eggs and terrapins are edible. Clutches of 15-22 hard-shelled, almost spherical eggs (30-35 mm dia.) are laid in December-March. Hatchlings (40-48 mm) emerge from nesting banks during the following rainy season.

Habitat: Rivers, lakes and stagnant ponds. **Range:** E. Africa, from Tanzania to Save River on the Mozambique floodplain, and Sabi-Lundi river confluence in Zimbabwe.

ND TERRAPINS
nily Emydidae

These are the characteristic terrapins of N. and central America and SE. Asia. The shell is hard, with some species having hinges in the plastron. The feet usually have some webbing between the toes.

There are two European genera (*Emys* and *Mauremys*) which are normally found in N. Africa. One N. American species, *Trachemys scripta*, has escaped from captivity and become established in some areas of South Africa.

SLIDERS
Trachemys

These terrapins have a hard carapace that is notched and serrated on the margin, and is usually keeled. The plastron lacks a hinge. The head is withdrawn backwards.

Five species are distributed throughout the swamps and slow-flowing riv of central America, with some extending into the USA. One species is bred commercially, and is distributed as a pet throughout the world. Escapees t seem to have become established have been discovered in Europe, Israel and South Africa.

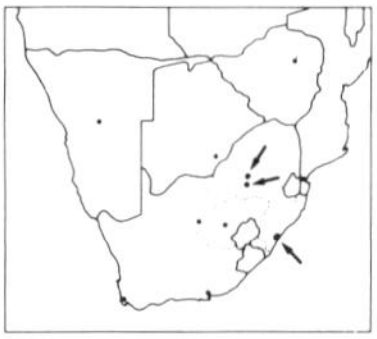

American Red-eared Terrapin *Trachemys scripta* (Pl. 10)

(Introduced) 15-25 cm; max. 28 cm
A medium-sized terrapin with a hard shell that lacks a hinge. The head is withdrawn straight back. The hind feet usually have slight webbing. The carapace is green, with irregular patterning that is darker and duller in adu The plastron is yellow with scattered patterning. The head has pale stripes a red ear spot in juveniles.

Biology and breeding: Omnivorous, this terrapin feeds on water insects snails, frogs, water weed, etc. It hibernates under water or in hollow logs during winter (when the water temperature drops below 10 °C). Although n well established in South Africa, this species has the potential to become a pest and it should not be released into the wild. Sexually mature males develop long claws on the front feet, with which they stroke the females in a elaborate courtship. Females lay 8-15 (up to 23) eggs in damp soil. Two or three clutches may be laid between August and February. The brightly coloured young hatch in two to three months.

Habitat: Quiet backwaters and ponds. **Range:** N. and central America; babies, imported with tropical fish, were released when they outgrew their homes and have been found in vleis and streams near Pretoria, Johannesburg, Silverton and Durban. **Subspecies:** Numerous races of thi widespread terrapin exist. Commercially farmed specimens are usually of Texas race, *T.s. elegans*.

Side-necked Terrapins

Suborder Pleurodira

'Pleurodira' means 'side-neck' and refers to the manner of head withdrawa side-necked terrapins: it is pulled to one side under the carapace, so that usually one eye can still be seen (some species have developed a hinge in plastron to further protect the head). The shell is flat and hard, and there is unpaired intergular at the front of the plastron. The feet are webbed. Some species have very long necks.

These are primitive terrapins, now restricted to the southern continents. Two families exist: the Chelidae in S. America and Australia, and the Pelomedusidae in Africa (including Madagascar) and S. America.

SIDE-NECKED TERRAPINS
Family
Pelomedusidae

The neck of these terrapins is completely retractable under the carapace edge, and the carapace lacks nuchal and supracaudal scutes. The plastrc has an intergular.

Two subfamilies exist: the Pelomedusinae (containing only two genera), which is restricted to Africa and is characterized by having five claws on th hind feet; and the Podocneminae (containing three genera), which is founc Madagascar and in S. America, with some species growing very large (to c one metre).

Fossils include the largest known chelonian, *Stupendemys geographicu* from Venezuela; it exceeded 2,5 metres in length and weighed more than 2 000 kg.

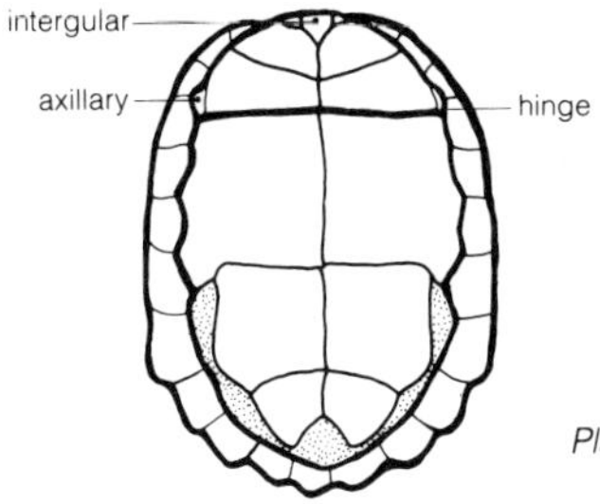

Plastron of Pelusios

y to the southern
'ican genera in the
lomedusidae

1. Plastron not hinged at front .... *Pelomedusa* (Marsh or Helmeted terrapin, page 39)

   Plastron hinged at front .... *Pelusios* (Hinged terrapins, page 40)

\RSH OR HEL-
:TED TERRAPIN
lomedusa

This genus contains a single species which is widely distributed in sub-Saharan Africa. The local name 'Cape water terrapin' is too parochial as this terrapin swims in the Nile and also occurs on Madagascar.

Marsh or Helmeted Terrapin *Pelomedusa subrufa* (Pl. 10)

20-30 cm; max. 32 cm
A medium-sized side-necked terrapin with a flat, hard, thin shell with no plastral hinge. This species has a large head with two small tentacles (of unknown function, but possibly used in mating) beneath the chin, and musk glands on the fourth to eighth marginals near the carapace edge. The hind feet have a webbed fringe. Males have longer tails and narrower shells than females, and grow to a larger size. The shell is uniform olive to dark brown above, sometimes with the shields black-edged and the marginals paler; below, it is either entirely black, pale-coloured, or has a symmetrical, pale-centred pattern. The head is dark on top with vermiculations, and pale below and on the jaws. **Biology and breeding:** This is a very common species, particularly in temporary pans, even in the central Karoo. Because of its thin shell, it is usually absent from permanent waters inhabited by crocodiles. Omnivorous, it eats almost anything, including water weed, insects and frogs. In some regions (eg. Etosha Pan) it acts 'crocodile' and ambushes, drowns and devours doves drinking at the water's edge. It emerges on to the bank or a log to bask, and aestivates during droughts by burrowing into moist soil. Many are killed on roads while migrating to new vleis after good rains. They have lived for over 16 years in captivity. Few people eat them, as they have an unpleasant, musky smell. They do not make the best pets; they are belligerent, long-necked and ever-willing to bite or to eject their cloacal contents. Mating occurs in water throughout summer. The male pursues the female, touching his snout against her vent and hindquarters. If she is receptive he grips the edge of her shell in all his feet and rubs the two short sensory tentacles which are situated under his chin on the back of her head; he also expels a stream of water from his nostrils over her face. Usually 10-30 (but up to and sometimes more than 40) soft-shelled, elongate eggs (30-40 x 18-28 mm, 10 g) are laid in a flask-shaped pit. The female digs this with her hind feet in soft, moist soil above the high-water mark. If the ground is too hard, she urinates to soften it. The young hatch in 90-110 days, emerging after the ground has been softened by rain. They measure 25-38 mm and weigh 8-10 g.
Habitat: Slow-moving and still water, including temporary pans.
Range: Found wherever water is present, even in the central Karoo and Etosha. Elsewhere, throughout most of sub-Saharan Africa; also on Madagascar.
Subspecies: Two races are recognized, both occurring in the region.

P.s. nigra has dull coloration; the plastron is black, and there are black triangles on the lower sutures of the marginals. It occurs in Natal, extending into adjacent OFS, Transkei and E. Cape. *P.s. subrufa* has large, off-white areas on the plastron, and occurs over the rest of the subcontinent.

HINGED TERRAPINS
Pelusios

These terrapins are immediately recognizable by the hinge at the front of th plastron, which can be closed, protecting the head and forelimbs. The shel usually domed and thick and in some species has a serrated top and/or posterior edge, especially in juveniles.

Distributed throughout most of sub-Saharan Africa, it is also present on Madagascar and the islands of the Indian Ocean. The number of species is debatable, but possibly amount to 15; five species are present on the subcontinent. The colour pattern of the plastron is an important local distinguishing feature.

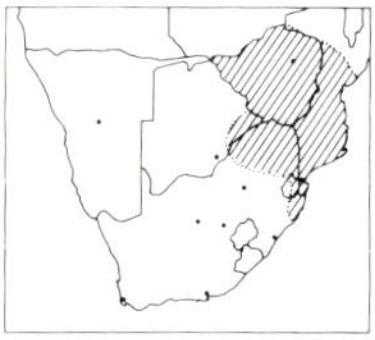

Serrated Hinged Terrapin *Pelusios sinuatus* (Pl. 11)

30-40 cm; max. 46,5 cm
The largest hinged terrapin. The posterior marginals are enlarged to give a serrated edge, and the vertebrals are keeled, particularly in juveniles. An axillary scale is present at the front junction of the carapace and the plastro The head is relatively small, with a weakly bicuspid beak and usually two longish tentacles under the chin. Females grow larger than males. The carapace and bridge are uniform black in colour. The plastron is yellow-centred, with a sharply defined, black, angular pattern around the edge. Th head is blackish-brown with yellow or brown vermiculations. The skin of the neck and limbs is pale olive-grey. **Biology and breeding:** Common in large water bodies, these terrapins are often seen basking on logs and rocks dur the day. Larger specimens eat mussels, while juveniles take invertebrates a frogs, etc. They scavenge at game killed by crocodiles, to which they also regularly fall prey. They take engorged ticks from the legs of drinking buffal and other large game. They may become a nuisance, taking baits set for fis They retreat into their shell when caught, and rarely bite. They have lived fo longer than 12 years in captivity. The female lays 7-13 eggs, up to 500 m frc the nearest water, in October-November; hatchlings appear in March-April **Habitat:** Perennial rivers and permanent lakes and pans. **Range:** Tropical E. Africa; along Zambezi River to Victoria Falls, and south to N. Zululand.

P.s. subniger

Pan Hinged Terrapin *Pelusios subniger* (Pl. 11)

13-18 cm; max. 20 cm
A small hinged terrapin with a rounded, smooth shell and a small plastral hinge. There is no axillary. The head is large, with a blunt snout and smooth beak; there are usually two tentacles under the chin. The carapace is unifor brown, and brown-grey when dry. The bridge is yellow and brown. The shie of the plastron have pale yellow centres. The head is uniform brown, not vermiculated and sometimes has black spots. The skin of the neck and limb is grey or black. **Biology and breeding:** Similar to the marsh terrapin, *Pelomedusa subrufa* (page 39), this terrapin often aestivates on land during droughts. Many have been found with their shells scarred by fire. They feed small frogs and invertebrates. In defence, they may discharge their cloacal contents. The females probably nest throughout summer; in captivity, eight eggs are laid in February-March. Hatchlings measuring 30 mm are found in March-April. **Habitat:** Pans and temporary water bodies. **Range:** Madagas and E. Africa, into Zimbabwe and N. Botswana, reaching S. Mozambique a N. Kruger National Park. An isolated race occurs on the Seychelles and possibly Mauritius. **Subspecies:** Two races are recognized. The typical rac *P.s. subniger*, occurs on the African mainland and Madagascar, and is replaced on the Indian Ocean islands by *P.s. parietalis*.

Okavango Hinged Terrapin *Pelusios bechuanicus* (Pl. 11)

25-30 cm; max. 33 cm
A large hinged terrapin with a heavy, domed, elongate shell with a small plastral hinge. There is no axillary. The vertebrals are slightly keeled in juveniles. The head is very large, with a smooth beak, and usually three tentacles under the chin. The carapace is uniform black. The plastron is largely or entirely black (but sometimes yellowish in the centre). The head is black, with symmetrical black and yellow markings that are prominent in juveniles. The skin of the neck and limbs is yellowish. **Biology and breeding:** This terrapin lives in deep, clear water and feeds on invertebrates and fish. The female lays large clutches of soft-shelled eggs (35-39 x 21-23 mm) in moist soil in early summer. **Habitat:** Clear waters of the Okavango Swamp and Kafue Flats. **Range:** Greater Okavango Basin, including Zambezi River above Victoria Falls.

Mashona Hinged Terrapin *Pelusios rhodesianus* (Pl. 11)

18-22 cm; max. 25 cm
This medium-sized hinged terrapin has a domed, elongate, smooth shell with a small plastral hinge. There is no axillary. The head is small, with a strongly bicuspid beak, and usually two tentacles under the chin. The carapace is uniform black. The plastron is black, with a diffuse yellow centre. The head is dark brown, and yellow on the sides. The skin of the neck and limbs is pale yellow. **Biology and breeding:** This terrapin feeds on aquatic insects, frogs and small fish. The female lays small clutches of small, soft-shelled eggs (33-37 x 20-23 mm). She may lay more than one clutch in a summer (nesting has been recorded in September-April). Hatchlings have been recorded in December-January. **Habitat:** Quiet, weed-choked backwaters on dams and vleis. **Range:** Central and SE. Africa, entering Okavango Swamp and central Zimbabwe, and with relict populations in N. Zululand and Durban.

Eastern Hinged Terrapin *Pelusios castanoides* (Pl. 11)

castanoides

18-20 cm; max. 22 cm
This terrapin is of medium size, with an elongate, smooth shell with a small plastral hinge. There is no axillary. The head is of moderate size, with a strongly bicuspid beak, and usually two tentacles under the chin. The carapace is olive, blackish-brown or yellowish. The plastron is yellow, usually with faint black markings on the front sutures. The head is blackish-brown, with fine yellow vermiculations. The skin of the neck and limbs is yellow. **Biology and breeding:** This terrapin frequents shallow water, burying itself in mud when the water dries up, and re-emerging with rains. Many specimens have shells scarred from dry-season fires. It feeds on aquatic insects and frogs, and also freshwater snails and floating vegetation. Two females were recorded laying 25 eggs (30-33 x 21-23 mm) each, at the end of September. **Habitat:** Still lakes and swamps at low altitudes. **Range:** E. Africa to central Mozambique, with isolated populations in N. Zululand and Madagascar. An isolated race occurs on the Seychelles. **Subspecies:** Two races are recognized. The typical race, *P.c. castanoides*, occurs in Africa and Madagascar, and is replaced by *P.c. intergularis* on the Seychelles.

SCALED REPTILES

Order Squamata

These reptiles are characterized by a scaly skin that is covered with a thin, horny layer that is periodically shed, either in bits or in one piece. They usua lack osteoderms, although these may be present on the head and body sca in some lizard families. The skull is diapsid, having openings in the upper a lower temporal regions. The bones of the palate do not form a continuous rc to the mouth. There are two nostrils. The cloacal aperture is transverse, and the paired male sexual organs (hemipenes) are stored in the tail base. They evert in use (so that the inner surface becomes the outer, rather like turning sock inside-out), and are usually adorned with spines and flounces. Only a single hemipenis is used at a time. Most species lay eggs, but viviparity has evolved on numerous occasions.

Squamates arose in the Triassic epoch, from eosuchian stock, although lizards did not become common until the Cretaceous period, and snakes a little later. Amphisbaenians are not known until the Eocene epoch, but must have evolved much earlier, probably from lizards. The squamates are now the peak of their radiation and represent the flowering of modern reptiles. T are distributed throughout the world's tropical and temperate regions, in all habitats, but with only one successful marine group (the sea snakes). The squamates constitute the largest group of living reptiles, containing more th 6 200 species, and exceeding in number all the world's mammal species. Three suborders are recognized. The lizards and snakes belong to large suborders, while amphisbaenians are much less diverse, and consist of on about 140 species.

Snake, amphisbaenian or legless lizard? Legless lizards are frequently confused with snakes. As a result, many are needlessly killed by people wh think they may be poisonous. Amphisbaenians look so like worms that they often called worm lizards, although they are neither worms nor lizards.

Legs are not always useful, and in some habitats can even get in the way A sinuous movement, much like that of fish, is used by some long-bodied reptiles to move in either long grass or loose, sandy soils; forward motion is achieved by pressing different parts of the body in waves against the grour or vegetation. Lizards in such habitats have evolved serpentine bodies. Movement in grass requires less strength than in sand, and is achieved by increased tail length; some legless grassland lizards have tails that are thre times the length of their body. Even in loose, sandy soils, burrowing require more effort, and fossorial reptiles have shorter tails and thicker, more muscu bodies. Snakes probably evolved from burrowing legless lizards. They have very unusual eye structure. This is most easily explained by assuming that i redeveloped from vestiges left in an ancestral blind legless lizard. Most of th primitive snakes are burrowing. More advanced snakes have radiated back on to the surface, and even into trees and the sea.

Amphisbaenians, which live permanently underground, probably evolved from very early lizards. Their bodies are usually pink, and are always covere with rings of squarish, non-overlapping scales, and the head often has a thi shield that resembles a thumbnail. While they may be confused with worms, they do not resemble snakes.

It is not always easy to tell snakes and legless lizards apart, particularly when the primitive burrowing snakes are compared with modern fossorial lizards. Most snakes have enlarged belly scales; these are never found in lizards, even in the burrowing species. Snakes' eyes (when present) are always covered with an immovable spectacle so they have an unblinking stare; many legless lizards retain movable eyelids, so you may presume that if it winks, it's safe. Unfortunately, many highly specialized fossorial lizards and primitive burrowing snakes lack external eyes, and all that can be seen are black dots under the head shields. Both groups also have small belly scales, so there is no simple rule for telling them apart. All blind snakes have extremely small tails, with a sharp spine at the tip. In addition, thread snakes and blind snakes have much rounder, blunter heads than any legless lizards. Confusion between these latter groups, however, is academic as no primitive snakes are venomous.

Snakes

Suborder Serpentes (Ophidia)

Snakes are legless squamates that lack eyelids, external ears and osteoderms. In some primitive families, vestiges of hind limbs are present. The left lung is reduced or absent. The tongue is retractile, into a sheath. Snakes have long backbones (some with more than 440 vertebrae), with many articulated ribs that are used for locomotion and to maintain the body shape.

All are carnivorous, and many are specialist feeders on particular prey. This may simply be engulfed alive, or first subdued, either by constriction or by the injection of a toxic venom. Teeth may be enlarged and used to grip prey, or modified for the injection of venom, which is derived from the salivary juices. All prey is swallowed whole, and in many species the lower jaw can be dislocated to allow large items to be engulfed. Some species have very toxic venoms which may be used defensively against predators. The tail can be shed in a few species, but it cannot be regenerated. All regularly shed their skin, usually in one piece, starting at the snout. Most lay soft-shelled eggs, but viviparity has evolved in some species, usually in cold climates.

The earliest fossils date from the Cretaceous period, about 80 million years ago. There are approximately 2 500 living species in 417 genera and at least 11 families, distributed throughout most of the temperate and tropical regions of the earth.

Snakes, like sharks, scare many people. Although most people realize that not all snakes are poisonous, few bother to learn which snakes are dangerous, and even fewer consider that even the venomous species may be beneficial.

Of the 130 species of snakes in southern Africa, 34 have fangs and venom glands; some of these have venoms that are less toxic than those of wasps and bees, and cannot be considered dangerous. Only 14 snakes have killed people on the subcontinent. These include the five species of cobra, the two species of mamba, the rinkhals, the Namibian coral snake, the puff adder and gaboon adder, the boomslang, the bird snake, and the rock python (these species are marked with a black skull-and-crossbones on the colour plates). The bites of several others (the shield-nose snake, the garter snakes, and the many-horned adder) can be serious, and may even have been fatal in exceptional circumstances. However, there have been no documented deaths from bites by these species (marked with a red skull-and-crossbones on the colour plates), or from the local burrowing asps, night adders, the small adders (including the berg adder and horned adder), the sea snake, skaapstekers, or any other back-fanged species. Obviously, the numbers and types of snakes differ from region to region, but no part of the Republic is fre of snakes, and at least one venomous species occurs in all regions. Even where there are large numbers of snakes, including many venomous specie (such as in Zululand), snakebite is rarely a significant risk. The following figures illustrate the point: in South Africa in 1986, over 10 000 people were killed in road accidents; 2 000 died from lung cancer and other diseases related to cigarette smoking; 220 were struck and killed by lightning; and fewer than 20, perhaps as few as 10, died from snakebite.

While a 'hug-a-snake' campaign is not necessary, greater tolerance and understanding of these efficient and unique predators is certainly called for The vast majority of our local snakes are harmless or clinically unimportant. In fact, they do a tremendous amount of good in controlling pests. Thailand provides an example: because many Asian people have a great belief in the healing properties of snake extracts, in 1985 Thailand exported 1,3 million snakes to Hong Kong, Korea, Japan and even the USA. With the crash in the country's snake population, rats proliferated and are estimated to have destroyed 400 000 hectares of rice fields. In southern Africa, snakes are as important as birds of prey in controlling dassies and cane-rats in farmlands.

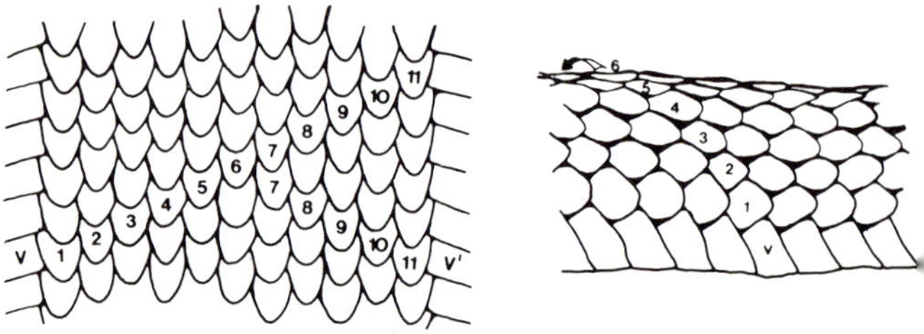

V = ventral scale

Counting midbody scale rows

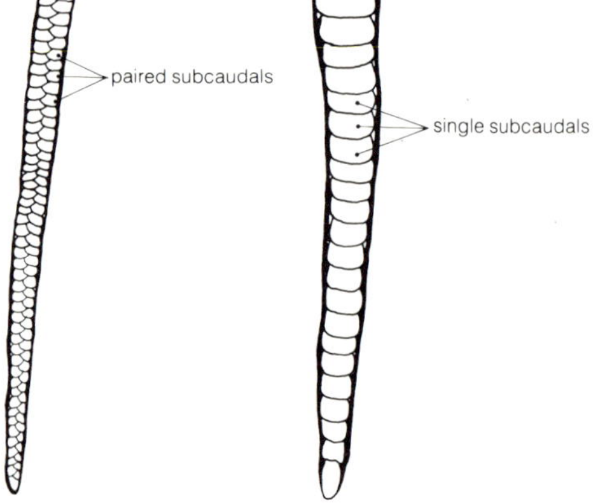

The ventral surface of a snake's tail

y to the Snake
nilies in southern
ica

1. Body worm-like, head hardly distinct from body, tail blunt; back and belly covered with small, similar-sized scales; eyes vestigial, buried under head shields 2
   Body not worm-like, head distinct from body, tail tapered; belly covered with transversely enlarged scales; eyes well developed and movable, with a transparent spectacle 3

2. Ocular shield not bordering lip; teeth present only in upper jaw; 20 or more scale rows at midbody; tail slightly longer than broad .... Typhlopidae (Blind snakes, page 45)
   Ocular shield bordering lip; teeth present only in lower jaw; 14 scale rows at midbody; tail at least three times longer than broad .... Leptotyphlopidae (Thread snakes, page 48)

3. More than 70 scale rows at midbody; some labials with deep, heat-sensitive pits; vestigial hind limbs present as a pair of claws bordering the vent .... Boidae (Boas and pythons, page 50)
   Fewer than 50 scale rows at midbody; labials without deep, heat-sensitive pits; vestigial hind limbs absent 4

4. Usually no enlarged poison fangs at front of upper jaw (but present in burrowing asps and harlequin snakes) .... Colubridae (Typical snakes, page 51)
   One or more pairs of enlarged, tubular poison fangs at front of upper jaw 5

5. Head covered with large, symmetrical shields; loreal absent; poison fangs small to moderate, fixed and not enclosed in membranous sheath .... Elapidae (Cobras, mambas and their relatives, page 88)
   Head usually covered with irregular, small, keeled scales; if with enlarged head shields, loreal present; poison fangs large, hinged and enclosed in membranous sheath .... Viperidae (Adders and vipers, page 96)

IND SNAKES
mily Typhlopidae

These are primitive snakes with a toothless lower jaw and internal vestiges of a pelvic girdle. They display many adaptations for burrowing, including having a cylindrical body and an indistinct head; polished, tightly fitting scales with ventrals that are not enlarged on the belly; reduced eyes under the head shields; and an exceptionally short tail with a terminal spine. All blind snakes are very similar in appearance and colour.

They feed mainly on termites but occasionally eat other invertebrates. They are commonly found under stones or in termite nests, and are sometimes forced to the surface by floods or exposed during ploughing. Oviparous and viviparous species are known. During mating the male coils tightly around the rear of the female. The skin is shed in compacted rings.

These snakes are distributed throughout the tropical regions of the world; there are about 160 species in two genera.

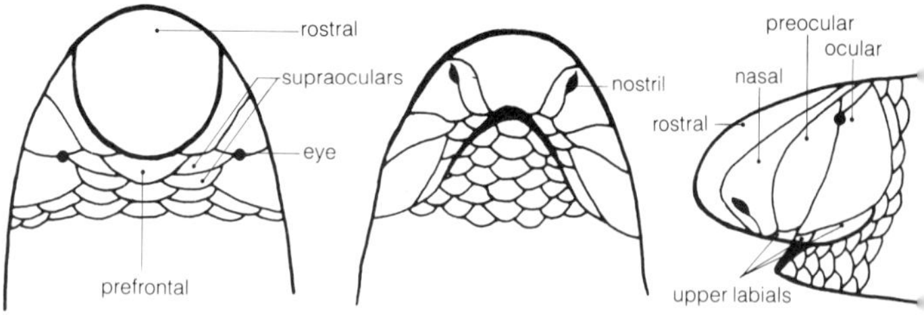

The head scales of a blind snake

Key to the Snake families in southern Africa

1. Nostril pierced laterally in a divided nasal; rostral narrow; 20 scales around midbody ..... *Ramphotyphlops* (Australasian blind snakes, page 46)

   Nostril pierced inferiorly in a semi-divided nasal; rostral wide; 22 or more scales around body ..... *Typhlops* (Blind snakes, page 46)

AUSTRALASIAN BLIND SNAKES
Ramphotyphlops

Australasian blind snakes are distinguished by a unique hemipenis that has a non-eversible tip. There are about 20 species; one introduced species occurs on the subcontinent.

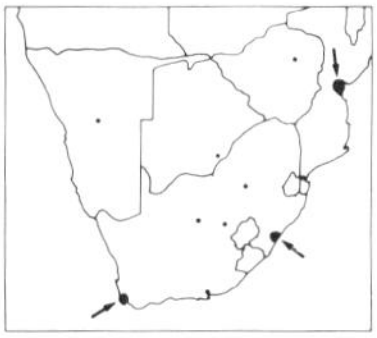

Flower-pot Snake *Ramphotyphlops braminus* (Pl. 40)

(Introduced) 14-16 cm; max. 17 cm
This very small, slender snake has 20 scales around its body, 300-350 dorsals, and a rounded snout. It is uniform grey to pale brown in colour, with paler belly. The snout, cloacal region and tail tip are cream. **Biology and breeding:** It feeds on ant and termite larvae. A parthenogenetic species, its unique reproduction has allowed the flower-pot snake to colonize many oceanic islands and most continents. It lays 2-6 minute eggs (2 x 6 mm). It is commonly transported in nursery plants, hence the common name. **Habitat:** Humic soil. **Range:** Introduced populations in Cape Town, Durban and Beira. Elsewhere, scattered along E. African coastal regions and throughout Indo-Pacific, and to Florida in USA.

BLIND SNAKES
Typhlops

The genus contains about 130 species, distributed throughout the tropics (except Australia). Forty-three species are found in subSaharan Africa. Of the seven occurring on the subcontinent, five are endemic. Some species that have a horizontal edge to the snout (eg. Delalande's and Schlegel's blind snakes) are sometimes placed in a separate genus (*Rhinotyphlops*).

Slender Blind Snake *Typhlops obtusus* (Pl. 39)

30-35 cm; max. 37 cm
This extremely slender snake has 24-26 scales around the body, more than 300 dorsals, and a smoothly rounded snout. Coloration is dark brown above and pale brown below. **Habitat:** Loose humic soil in forest. **Range:** Extreme E. Zimbabwe. Elsewhere, in S. Malawi and adjacent Mozambique.

Fornasini's Blind Snake *Typhlops fornasinii* (Pl. 40)

(Endemic) 14-16 cm; max. 18,5 cm
This very small snake has 22-27 scales around the body, fewer than 300 dorsals, and a rounded snout. In colour, it is uniform greyish to brown-black, with yellow blotches on the throat and anal region. **Habitat:** Coastal bush. **Range:** N. Zululand and adjacent Mozambique, to extreme SE. Zimbabwe.

Boyle's Blind Snake *Typhlops boylei*

(Endemic) 15-18 cm; max. 22 cm
This small, slender snake has 26-28 scales around the body, more than 300 dorsals, and an angular snout. The body is olive-brown above, with the scales yellow-edged; the flank scales are pale with dark brown centres. The belly is uniform pale yellow. **Habitat:** Sandveld. **Range:** Damaraland to W. Botswana.

Bibron's Blind Snake *Typhlops bibronii* (Pl. 39)

(Endemic) 35-38 cm; max. 46 cm
This stout species has 30 scales around the body, more than 300 dorsals, and an angular snout. Coloration is uniform brown to dark olive-brown above, with a paler belly; the young are paler than the adults. **Breeding:** The female lays 5-12 thin-walled eggs in late summer (February). Embryos are well developed, and the young (111-124 mm) hatch in 5-6 days. **Habitat:** Highveld and coastal grassland. **Range:** Transvaal and S. Natal to Albany district of E. Cape. A relict population occurs in E. Zimbabwe.

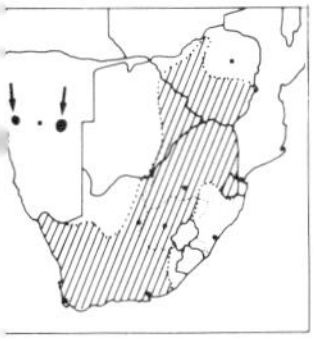

Delalande's Blind Snake *Typhlops lalandei* (Pl. 39)

(Endemic) 25-30 cm; max. 35 cm
This slender species has 26-30 scales around the body, more than 300 dorsals, and a prominent horizontal cutting edge to the snout. Coloration is uniform pinkish-slate to grey-brown above, with each scale pale-edged, giving a chequered effect. The belly is pale pink-grey. The young are flesh-coloured. **Breeding:** Females contain 2-4 eggs. **Habitat:** Varied: semi-desert, coastal bush, fynbos and savannah. **Range:** Cape, north to Transvaal, E. Botswana and Zimbabwe. Absent from Natal. Scattered old records in Namibia.

Beaked Blind Snake *Typhlops schinzi* (Pl. 39)

(Endemic) 16-19 cm; max. 23 cm
This small, slender snake has 22-26 scales around the body, more than 400 dorsals, and a prominent hooked snout. It is yellowish to flesh-coloured, with heavy blue-black to reddish-brown blotching that may form crossbars along the back. **Habitat:** Semi-desert and arid savannah. **Range:** Namibia and adjacent Botswana, south to Calvinia in the Cape.

T.s. schlegelii
T.s. mucroso
T.s. petersii

Schlegel's Blind Snake *Typhlops schlegelii* (Pl. 39)

60-70 cm; max. 95 cm
The largest typhlopid. It has 30-44 scales around the body (see Subspecies), over 300 (and up to 623) dorsals, and a prominent horizontal edge to the snout. The uniform colour phase is brown to black with a straw-yellow belly; the blotched phase has irregular dark brown to black blotches on the back, with the belly and flanks yellow to yellow-green; and in the striped phase, each scale is black-edged, forming dark lines that merge with age. Freshly sloughed snakes are blue-grey, changing to red-brown with time. **Biology and breeding:** Very large specimens are rarely seen; they live deep underground and lay down large fat stores for long fasts. The female lays

12-40 (up to 60 in very large females) eggs (20 x 10 mm) in late spring to summer. Embryos are well developed, and the young hatch in 5-6 weeks. **Habitat:** Varied; coastal bush to sandveld. **Range:** Northern regions of the subcontinent to Sudan. **Subspecies:** Four races are recognized. *T.s. brevi.* extralimital, occurring in NE. Africa. *T.s. schlegelii* has 36-44 scale rows; the is a uniform and a blotched phase. It occurs in extreme N. Zululand and S. Mozambique, through Transvaal to SE. Botswana. *T.s. mucroso* has 30-3 scale rows, and both striped and blotched phases occur (in a ratio of 2:1). It is found in E. Botswana, Zimbabwe and N. Mozambique, through to Keny *T.s. petersii* has 34-40 scale rows and occurs as a blotched phase only. It is found in Namibia, and N. Botswana to S. Angola.

THREAD SNAKES
Family
Leptotyphlopidae

These very small, thin, primitive, burrowing snakes have no teeth in the upp jaw, a single lung and oviduct, and internal vestiges of a pelvic girdle (some W. African species also have small external claws). The body is cylindrical, with a blunt head and a short tail. The scales are highly polished and not enlarged on the belly.

They are found throughout Africa (except in the Sahara) and adjacent Ne East, and in Amazonia and Middle America. There are 78 species in two genera. *Rhinoleptus* contains a single species and is restricted to W. Africa

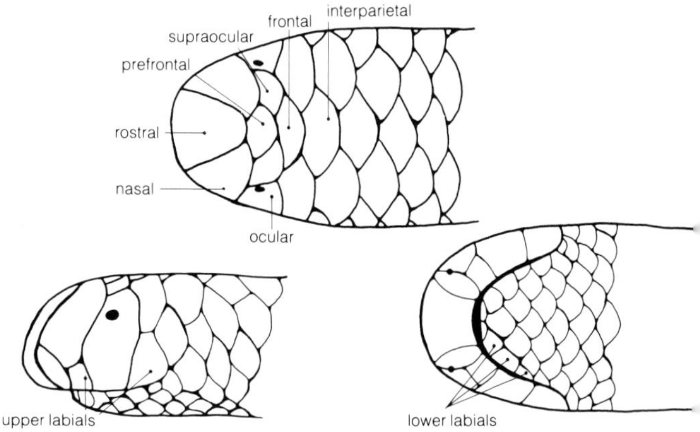

The head scales of a thread snake

THREAD SNAKES
Leptotyphlops

Thread snakes burrow underground, feeding almost exclusively on termites and swallowing only the soft abdomens of the larger species. They produce pheromones that prevent soldier termites from attacking them. Predators include snakes and large invertebrates, eg. scorpions and spiders. The females lay a few (1-7) elongate eggs. The Texas thread snake guards its eggs by coiling around them, but this is not known to occur in African speci Colour may change from pink-grey to black, depending on skin dryness.

There are 27 species in Africa, with nine species occurring on the subcontinent, five of which are endemic.

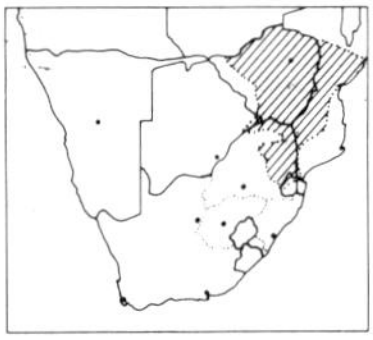

Long-tailed Thread Snake *Leptotyphlops longicaudus* (Pl. 40)

18-22 cm; max. 25 cm
This small, slender thread snake has a blunt, rounded head, a relatively long tail (34-58 subcaudals) and a discrete prefrontal. Coloration is uniform lilac dark grey above, tinged with pink; the belly is flesh-pink. **Breeding:** Two or more eggs (21 x 4 mm) are laid in summer. **Habitat:** Lowveld. **Range:** E. Transvaal, through Zimbabwe and Mozambique Plain to Kenya.

igricans

Black Thread Snake *Leptotyphlops nigricans*

13-17 cm; max. 20 cm
A very small, slender species with a short tail (19-33 subcaudals) and a discrete prefrontal. Coloration is uniform dark brown to black, with the scales sometimes pale-edged. **Habitat:** Varied. **Range:** Isolated populations in E. and S. Cape, N. Transvaal and Zambia to Sudan. **Subspecies:** Two races are recognized, the typical race occurring on the African mainland. *L.n. pembae* is endemic to Pemba Island off the Tanzanian coast.

Slender Thread Snake *Leptotyphlops gracilior* (Pl. 40)

(Endemic) 18-20 cm; max. 24 cm
This very slender snake has 300-350 dorsals, the prefrontal-frontal fused, 14 scale rows at midbody, and 10 scales around the tail. Coloration is uniform brown to black. **Habitat:** Karroid areas. **Range:** Inland W. Cape, with an isolated record in S. Namibia.

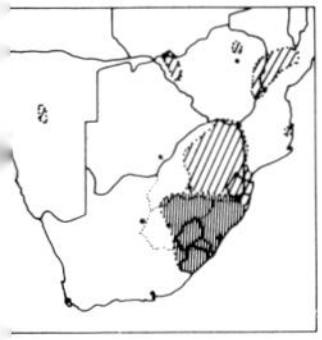

L.c. conjunctus
L.c. incognitus

Cape Thread Snake *Leptotyphlops conjunctus*

13-15 cm; max. 18 cm
This thread snake has fewer than 300 dorsals, the prefrontal-frontal fused, and 10 scales around the tail. It has a narrow rostral. It is uniform brown-black above and below; the preanal plate is sometimes white. **Breeding:** The eggs are very small (similar to rice grains). Hatchlings are 60 mm long. **Habitat:** Varied. **Range:** Eastern regions from East London to Malawi. **Subspecies:** Two races are recognized. *L.c. conjunctus* has more than 230 dorsals, and more than 27 subcaudals. It occurs from East London to S. Transvaal. *L.c. incognitus* has fewer than 230 dorsals, and fewer than 27 subcaudals. It occurs throughout much of Transvaal, with isolated populations in Namibia, Zimbabwe, Mozambique and Malawi.

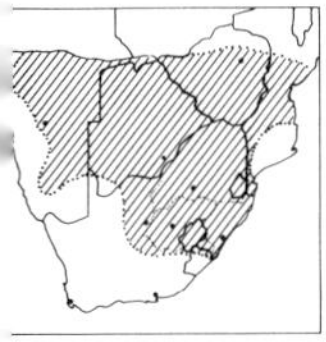

scutifrons

Peters's Thread Snake *Leptotyphlops scutifrons* (Pl. 40)

18-24 cm; max. 28 cm
This species is similar to the Cape thread snake, *L. conjunctus* (above), but has a wider rostral and a short tail (19-30 subcaudals). It is red-brown to black above, and sometimes paler below, with the scales often pale-edged. **Biology and breeding:** It may sham death when handled roughly. It lays 3-7 elongate eggs (3,5 x 14 mm) in November-December. The eggs may remain attached to each other like a string of sausages. **Habitat:** Varied; grassland, coastal bush, and mesic and arid savannah. **Range:** Throughout subcontinent, except most of Cape, S. and coastal Namibia and S. Mozambique Plain. **Subspecies:** Two races are recognized. The typical race occurs on the subcontinent and is replaced in N. Tanzania and Kenya by *L.s. merkeri.*

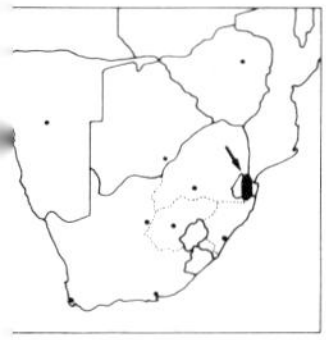

Tello's Thread Snake *Leptotyphlops telloi*

(Endemic) 13-15 cm; max. 18 cm
This species is similar to Distant's thread snake, *L. distanti* (below). The prefrontal-frontal is fused, and it has 12 scales around the tail. There are more than 300 dorsals, a narrow rostral and an undivided occipital. Uniform black, with white patches on the head. **Habitat:** Thornveld. **Range:** Lebombo Mtns.

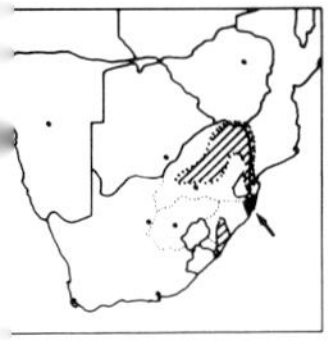

Distant's Thread Snake *Leptotyphlops distanti* (Pl. 40)

(Endemic) 13-17 cm; max. 23 cm
This species is similar to Tello's thread snake, *L. telloi* (above), except that the rostral is very broad and the occipital is divided. It is grey-black above and

paler below, with the scales often pale-edged. **Habitat:** Varied; coastal bu grassland and savannah. **Range:** N. and E. Transvaal, with scattered reco in Kwazulu and Natal.

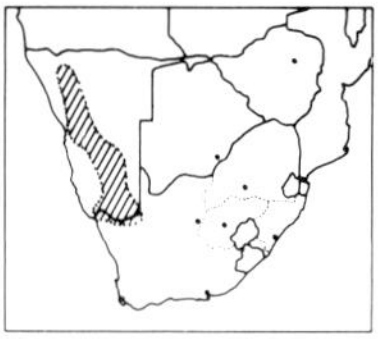

Western Thread Snake *Leptotyphlops occidentalis* (Pl. 40)

(Endemic) 18-20 cm; max. 28 cm
A very slender, small species that has its prefrontal-frontal fused, and 12 scales around the tail. There are fewer than 300 dorsals, and the anteric upper labial is fused with the supranasal. It is light grey-brown to purple-br in colour, with pale-edged scales giving a chequered effect; it is paler belc **Habitat:** Desert and arid savannah. **Range:** Namibia, from Kaokoveld to Li Namaqualand.

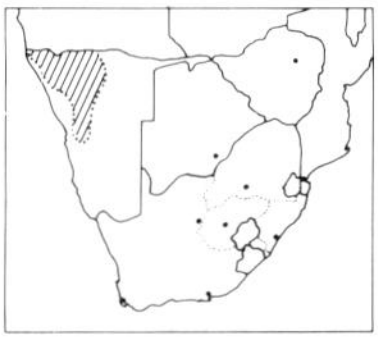

Damara Thread Snake *Leptotyphlops labialis*

(Endemic) 16-20 cm; max. 28 cm
A large, fairly slender species with 12 scales around the tail. It is locally uni in having the supraoculars fused with the oculars. Coloration is uniform grey-brown to brown above, and paler below, with the scales pale-edged. **Habitat:** Arid savannah. **Range:** N. Namibia.

BOAS AND PYTHONS
Family Boidae

This large, ancient family contains some of the largest snakes (eg. the anaconda and the reticulated python, which grow to 11 m). Internally, all ha minute limb bones and small spurs on either side of the cloaca that are vestiges of the pelvic girdles; these are slightly larger in males.

Fossils are known from the Upper Cretaceous period (100 million years ago). The family is distributed widely, with 65-70 species in 20 genera. It is usually split into two subfamilies: the egg-laying pythons (Pythoninae), and viviparous boas (Boinae). Only pythons are found on the subcontinent.

The fragmented head scales of the African rock python

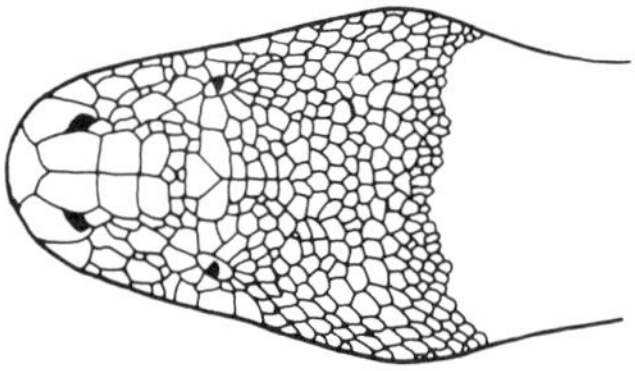

PYTHONS
Python

Pythons are medium to large snakes with small, smooth scales. The head h enlarged scales, but these are not symmetrically arranged or as large as in colubrid snakes. Pelvic spurs (vestiges of the hind limbs) are present on ei side of the anal shield.

All pythons lay eggs. The female coils tightly around the eggs, and some species shiver, generating extra body heat to incubate the embryos. The triangular pits on or between the lip scales detect infra-red radiation, enabl the snake to 'see' warm-blooded prey in the dark, or even when blind. All pythons constrict their prey.

Pythons are distributed throughout Africa and S. Asia to Australia. There 7-10 species (depending on the generic assignment), and three African species, with two on the subcontinent. The royal python (*P. regius*) from W. Africa is similar in size and habits to Anchieta's dwarf python.

Anchieta's Dwarf Python *Python anchietae* (Pl. 17)

100-120 cm; max. 180 cm
Stoutly built, Anchieta's dwarf python has a broad head and small, tubercu head shields. The upper lip has five heat-sensitive labial pits. The body sca

are small, smooth and in 57-61 rows. The head has a large, triangular, reddish-brown mark that is bordered by a white, black-edged band. The body is pale red-brown, with black-edged, white spots and bands. The belly is yellowish, with a few brown spots on the sides. **Biology and breeding:** A poorly known species, this small, gentle python is fully protected in Namibia. It is very rare in captivity. Wild specimens eat birds and gerbils, which they constrict. When captured, they roll into a tight, defensive ball and hide their heads. The female lays about five large eggs (62 x 37 mm) in November; these hatch in 60-70 days. The young measure 440-500 mm. Females may shiver during incubation. **Habitat:** Rugged, dry, rocky sandveld and riverine bush. **Range:** S. Angola and N. Namibia.

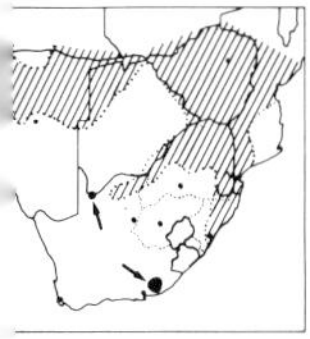

atalensis

African Rock Python *Python sebae* (Pl. 17)

300-500 cm; max. 560 cm

Africa's largest snake, the African rock python is very solid and stoutly built, with a triangular head with fragmented head shields. There are heat-sensitive pits on two of the upper labials and 4-6 of the lower labials. The body scales are very small, smooth and in 78-95 rows. There is a large, dark spearhead mark on the crown of the head, and dark and light bands radiating from the eye to the lip. The body is grey-green or grey-brown, with dark brown, black-edged bars and blotches on top, irregularly connected with sinuous dark brown bands that may form isolated blotches on the flanks. The belly is white with dark speckles. Juveniles are more brightly marked. **Biology and breeding:** They often bask, especially after feeding or when sloughing, and are fond of water, in which they may lie and hunt. Prey is ambushed and constricted, usually at dusk or after dark. Adults take small buck, monkeys, etc., although fish, leguaans and crocodiles are also eaten. They can swallow very large prey, but are vulnerable to attack by wild dogs and hyaenas when swollen with food. They may fast for long periods (two-and-a-half years has been recorded in captivity). Although they make good pets, some specimens never tame and the adults may grow too large to handle. They lunge and bite readily in defence. The teeth are very large and can inflict painful, ripping wounds. This is the only snake large enough to consider humans edible. but attacks on man are exceptionally rare, especially as most large pythons have been exterminated. The skins are used for fashion, but pythons are protected in southern Africa (SA RDB, Vulnerable). They are a valuable aid in controlling dassies, and cane-rats in Natal sugarcane fields. The female lays 30-50 (and more than 100 in very large females) large eggs (130-160 g each, about the size of an orange) in disused aardvark burrows, termite nests, caves, etc. The female coils around her clutch to protect the eggs, but does not incubate them by shivering. The young, measuring about 600 mm, hatch in 65-80 days. Sexual maturity is reached in 3-5 years, at 200-300 cm. **Habitat:** Usually open savannah regions, particularly rocky areas and riverine scrub. Absent only from true desert and dense rain forest. **Range:** Restricted mainly to the lowveld, reaching Natal south coast, and extending along Limpopo Valley to Lobatsi in Botswana and into N. Cape, with isolated records around Kalahari Gemsbok National Park. They also occur along Zambezi and Cunene valleys, extending into Okavango and N. Namibia. Extinct (since 1927) in E. Cape, but recently reintroduced. Elsewhere, throughout subSaharan Africa. **Subspecies:** Two races are recognized. Only the southern race (*P.s. natalensis*) is found on the subcontinent; this subspecies is replaced by the typical race (which has larger head shields and a different colour pattern) in W. Africa, to Uganda.

'ICAL SNAKES
ily Colubridae

This very large family contains some of the most successful and most common snakes. Most are medium sized. None has a functional left lung or pelvic vestiges. They usually have a 'typical' arrangement of enlarged head shields. Most lack fangs, but some groups have back fangs and a few have front fangs that may be hinged. The majority are harmless (even many of those specie with enlarged back fangs), but a few have powerful and unusual venoms th can cause death (eg. the boomslang, vine snakes and burrowing asps).

Colubrids occur on all continents (except Antarctica) and most islands. family includes more than 1 500 species in 292 genera. Many attempts hav been made to divide it into smaller groups, but no system has gained unive acceptance. Some authorities recognize as many as six separate families, with varying numbers of subfamilies and tribes. A conservative arrangeme has been adopted here, and a number of problematic local genera (five) h not been assigned to any subdivision. Four subfamilies occur on the subcontinent. Locally, the family contains 33 genera (five of which are endemic) and 85 species (29 of which are endemic).

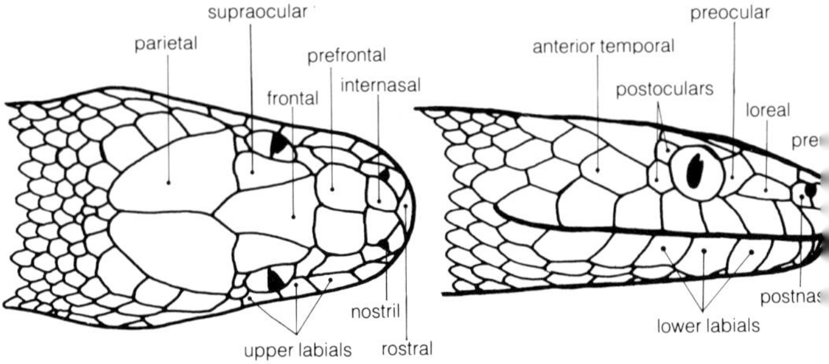

The head scales of a colubrid

Key to the southern African genera in the Colubridae

1. No enlarged grooved poison fangs in the upper jaw 2
   A pair of enlarged grooved poison fangs in the upper jaw (usually below the eye, but at the front of the jaw in *Homoroselaps* and *Atractaspis*) 14

2. Scales on back smooth (partly keeled only in *Prosymna janii*) 3
   Scales on back distinctly keeled 13

3. Nostril pierced between two nasal shields; tail moderate to long 4
   Nostril pierced in a single or semi-divided nasal shield; tail short 10

4. Anal entire; pupil vertical (occasionally almost round in *Lycodonomorphus*) 5
   Anal divided; pupil round or horizontal 6

5. Midbody scale rows numbering 19 or 21; 153-177 ventrals; belly usually yellow, orange or pink, often with a dark stripe below the tail and sometimes a dark stripe or scattered spots anteriorly; semi-aquatic .... *Lycodonomorphus* (Water snakes, page 57)

   Midbody scale rows usually numbering 23 or more (rarely 17-21); 165-230 ventrals; belly usually uniform white or dark grey; terrestrial .... *Lamprophis* (House snakes, page 58)

6. Snout rather pointed, prominent and with vertical sides; internasal shield entering nostril; scales in 25-31 rows at midbody .... *Pseudaspis* (Mole snake, page 63)

   Snout more or less rounded; internasal not entering the nostril; scales not exceeding 21 rows at midbody 7

7. Scales in not more than 15 rows at midbody; eye proportionately large; habit slender; tail 25-33% of total length; usually bright green, bluish-green or olive-green .... *Philothamnus* (Green and Bush snakes, page 82)

   Scales in 17-21 (exceptionally 15) rows at midbody; eye of moderate size; body moderately slender; tail 20-25% of total length; never bright green 8

8. A single, subtriangular internasal; parietal in contact with or narrowly separated from 6th upper labial; tail short, 15-20% of total length; a broad, pale, dorsolateral stripe on either side .... *Limnophis* (Striped swamp snake, page 64)

   Two internasals; parietal well separated from 6th upper labial by anterior temporals; tail moderate to long, 20-33% of total length 9

9. Ventrals not exceeding 150; scales in 17-19 (exceptionally 15) rows at midbody; mandibular teeth smallest in front .... *Natriciteres* (Marsh snakes, page 64)

   Ventrals exceeding 170; scales in 19-21 rows at midbody; mandibular teeth largest in front .... *Meizodon* (African smooth snakes, page 82)

10. Pupil vertical (or almost so); loreal shield present 11

    Pupil round; loreal shield normally absent (when present much reduced in size); nasal shield single; rostral small .... *Duberria* (Slug eaters, page 62)

11. Nostril pierced in a single nasal shield; snout rounded; rostral small; anterior maxillary teeth longest 12

    Nostril pierced in a semi-divided nasal; snout strongly depressed, projecting and with an angular, horizontal edge; rostral proportionately large; maxillary teeth very small, slightly larger posteriorly .... *Prosymna* (Shovel-snout snakes, page 66)

12. 8 upper labials, the 3rd, 4th and 5th entering the orbit; 8 lower labials .... *Lycophidion* (Wolf snakes, page 60)

Usually 6-7 upper labials, the 3rd and 4th entering the orbit; 6 (rarely 5 or 7) lower labials .... *Cryptolycus* (Dwarf wolf snake, page

13. Scales in 15-19 rows at midbody, without apical pits; scales along backbone enlarged and with 2 keels; loreal present; nostril large and pierced between 2 nasal shields; teeth normal and distinct .... *Mehelya* (File snakes, page 61)

Scales in 21-27 rows at midbody, with apical pits; scales along backbone not enlarged; loreal absent; nostril moderate and pierced in a semi-divided nasal shield; teeth few and rudimentary .... *Dasypeltis* (Egg eaters, page 84)

14. Eye moderate to large; loreal shield present; head more or less distinct from the neck; tail moderately long 15

Eye small to very small; loreal shield absent; head not distinct from the neck; tail very short 27

15. Subcaudals single; parietals broken up into small scales; pupil vertical .... *Pythonodipsas* (Western keeled snake, page 65)

Subcaudals paired; parietals entire 16

16. Pupil vertical; head much broader than neck 17

Pupil round or horizontal; head not or only moderately broader than neck 19

17. Anal usually divided (entire only in *T. beetzii*); loreal separated from the eye by a preocular; maxillary teeth smallest in front; body and tail conspicuously banded with black .... *Telescopus* (Tiger snakes, page 85)

Anal entire; body and tail not banded with black 18

18. Loreal entering the orbit; more than 70 subcaudals .... *Dipsadoboa* (Cat-eyed tree snakes, page 86)

Loreal separated from orbit by a preocular; less than 70 subcaudals .... *Crotaphopeltis* (Herald snakes, page 85

19. Eye large; pupil horizontal when fully dilated; normally keyhole- or dumbbell-shaped when partly dilated in daylight; habit very slender; scales in 19 rows at midbody .... *Thelotornis* (Bird snakes, page 87)

Pupil round; head short to moderately long; habit moderately slender 20

20. Scales keeled and in 19-21 rows at midbody; head very short and eye large .... *Dispholidus* (Boomslang, page 87)

Scales smooth and in 11-19 rows at midbody; head of moderate length and eye of moderate size 21

21. Rostral large and projecting; snout pointed and beak-like 22

Rostral of normal size; snout rounded and not beak-like 23

22. Snout strongly hooked in profile; loreal not longer than deep; preocular usually well separated from frontal; tail long, 53-125 subcaudals .... *Rhamphiophis* (Beaked snakes, page 68)

Snout not hooked in profile; loreal longer than deep; preocular in contact with frontal; tail short, 28-45 subcaudals .... *Dipsina* (Dwarf beaked snake, page 68)

23. Nostril pierced between at least 2 shields 24

Nostril pierced in a single, semi-divided nasal shield 26

24. Maxillary teeth interrupted below anterior part of eye by 2 much enlarged, fang-like teeth; nostril pierced between 2 or 3 nasal shields .... *Psammophis* (Sand and Grass snakes, page 70)

Maxillary teeth almost equal in size and continued without interruption to the gap separating them from the posterior pair of enlarged poison fangs 25

25. Nostril pierced between 2 nasal shields only; tail long, with 80+ subcaudals .... *Dromophis* (Olympic snakes, page 69)

Nostril pierced between 2 nasal shields and an internasal shield; tail short, subcaudals not exceeding 80 .... *Psammophylax* (Skaapstekers, page 69)

26. Mandibular teeth small and subequal in size; snout hollowed on either side, just anterior to the eye; anal entire .... *Amplorhinus* (Many-spotted snake, page 65)

	Mandibular teeth much enlarged in front; snout not hollowed on sides; anal scale divided	*Hemirhagerrhis* (Bark snakes, page 68)
27.	Subcaudals in pairs	28
	Subcaudals single	31
28.	Internasals absent; scales in 15 rows at midbody	29
	Internasals present; scales in 17-19 (exceptionally 21) rows at midbody	30
	Internasals present; scales in 15 rows at midbody; fangs in front of upper jaw	*Homoroselaps* (Harlequin snakes, page 79)
29.	A small preocular present; eye small, diameter being about equal to its distance from labial margin; head and neck black, and three well-marked longitudinal black stripes over back and tail	*Chilorhinophis* (Black and yellow burrowing snakes, page 79)
	Preocular absent; eye minute, its diameter being much less than its distance from the labial margin; uniformly dark above	*Amblyodipsas* (Purple-glossed snakes, page 77)
30.	Pseudo-preoculars (displaced prefrontals) present, snout strongly projecting, depressed and pointed; rostral very large, with an obtuse to sub-acute horizontal edge	*Xenocalamus* (Quill-snouted snakes, page 80)
	Preocular absent; prefrontals present; snout not strongly depressed or projecting; rostral of moderate size and with a rounded edge	*Amblyodipsas* (Purple-glossed snakes, page 77)
	Preoculars and prefrontals present, snout not strongly depressed or projecting, rostral of moderate size and with a rounded edge; hinged enlarged fangs in front of upper jaw	*Atractaspis* (Burrowing asps, page 7
31.	Scales in 23-27 rows at midbody; no preocular; body moderately large and stout	*Macrelaps* (Natal black snake, page 77)
	Scales in 15 rows at midbody; preocular present; body small and moderately slender	*Aparallactus* (Centipede eaters, page 76)

Scales in 19-23 rows at midbody; preocular present; body moderately long; tail terminating in a spine; hinged fangs in upper jaw .... *Atractaspis* (Burrowing asps, page 74)

.D WORLD
AKES
bfamily
aedontinae

This subfamily includes some of Africa's most typical snakes. They are characterized by vertebral and hemipenial features. They lack enlarged fangs, and none is venomous. Most are terrestrial, some are of burrowing habits, and others are either aquatic or semi-aquatic.

There are 15 genera with 44 species; seven genera (none of which is endemic) with 22 species (11 of which are endemic) occur in the region.

ATER SNAKES
codonomorphus

These medium-sized, aquatic snakes have a cylindrical body with a small head that is barely distinct from the neck. The eyes are moderately large, with round (sometimes elliptical) pupils. The scales are smooth, in 19-25 rows. The tail is moderately long, with paired subcaudals, and is longer in males, with an obvious hemipenial bulge. Females grow larger.

They are mainly nocturnal, although some species forage during the day. They are very common in suitable habitat. Prey consists mainly of frogs and tadpoles, and sometimes fish. These snakes capture their food underwater beneath stones, in crannies, etc. Large prey is constricted. They are oviparous. Harmless to man, they lack fangs and venom glands. They have a gentle disposition and very rarely bite.

Found throughout central and E. Africa, there are six species. Four of these, two of which are endemic, enter the subcontinent.

L.l. laevissimus
L.l. natalensis
L.l. fitzsimonsi

Dusky-bellied Water Snake
Lycodonomorphus laevissimus (Pl. 33)

(Endemic) 70-90 cm; max. 120 cm
This solid, smooth-scaled snake has a small, flattened head and small eyes with round pupils. The first upper labial has a backward projection. There are 60-85 subcaudals. The back is uniform dark olive-grey to brown-black, sometimes with a pale streak on the lower flank. The upper lip is spotted. The belly is cream to yellow, with a broad, central, dark band. **Biology and breeding:** Locally a common snake, it is aquatic, foraging mostly in water; small prey is swallowed while the snake is submerged. It lays 8-17 eggs in summer. **Habitat:** Pools in flowing, well-wooded streams. **Range:** E. Cape, through Transkei and Natal to SE. Transvaal. **Subspecies:** Three races are recognized: *L.l. laevissimus* (E. Cape to S. Natal) is poorly defined; it has 19 scale rows, and more than 170 ventrals; *L.l. natalensis* (Natal lowlands) has 19 scale rows and fewer than 170 ventrals; *L.l. fitzsimonsi* (Natal midlands and S.E. Transvaal) has 21 scale rows.

mlanjensis

Mulanje Water Snake *Lycodonomorphus leleupi* (Pl. 32)

70-85 cm; max. 93 cm
This snake is similar to the dusky-bellied water snake, *L. laevissimus* (above) but is slightly smaller, has a shorter tail (51-76 subcaudals), and lacks a dark belly band. The first upper labial has a backward projection. The back and upper lip are olive-black. The belly is dull orange with dark infusions. The tail has a dark stripe below. **Biology and breeding:** The Mulanje water snake is similar in habits and behaviour to the common brown water snake, *L. rufulus* (page 58), but is diurnal. It lays up to nine eggs (25 x 13 mm) in summer. **Habitat:** Small streams, pans, and vleis. **Range:** E. Zimbabwe and adjacent Mozambique, north through Malawi to Zaire. **Subspecies:** Two races are recognized, but only *L.l. mlanjensis* occurs on the subcontinent. The typical race of E. Zaire, *L.l. leleupi*, has a darker belly.

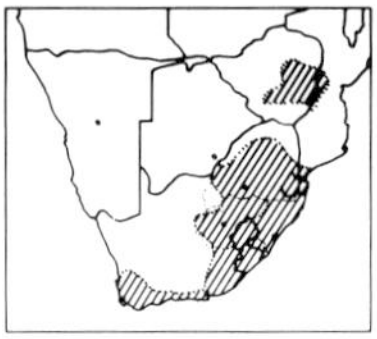

Common Brown Water Snake

Lycodonomorphus rufulus (Pl. 32)

(Endemic) 60-75 cm; max. 87 cm
This snake is similar to the Mulanje water snake, *L. leleupi* (page 57), but is slightly smaller, has vertical pupils, and lacks the backward projection to the first upper labial. Subcaudals number 53-86. The back is uniform olive-brow the upper lip and belly pale yellow-pink, and the tail darker below. **Biology and breeding:** These nocturnal water snakes are common under cover around water margins, and feed mainly on frogs. They are powerful constrictors. They tame easily and make gentle pets. The female lays 6-10 eggs in midsummer. The young measure about 150 mm. **Habitat:** Small streams, pans and vleis. **Range:** Eastern half of the subcontinent, extendin along the coast to W. Cape.

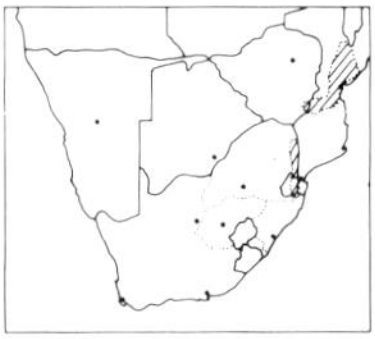

L.w. obscuriventris

Whyte's Water Snake *Lycodonomorphus whytii* (Pl. 33)

50-60 cm; max. 66 cm
This snake is similar in habits and appearance to the Mulanje water snake, *L. leleupi* (page 57), but is much smaller and has a short tail (37-52 subcaudals). The back is uniform dark olive to blackish. The upper lip has a pale stripe. The belly is orange-yellow, and sometimes faintly spotted, and t tail is darker. **Habitat:** Small streams, pans and vleis. **Range:** Mozambique floodplain, into adjacent regions. **Subspecies:** Two races are recognized, with only *L.w. obscuriventris* occurring on the subcontinent. The typical rac *L.w. whytii* (occurring in Malawi and Tanzania), has fewer ventrals and a longer tail.

HOUSE SNAKES
Lamprophis

These characteristic African snakes are small to medium-sized; only a few exceed a metre in length. The eyes are small, with vertical pupils (as is usua nocturnal snakes). The scales are smooth, with the anal undivided and the subcaudals paired.

Most house snakes are terrestrial, although some live underground (in termite nests, etc.) and others forage in rock cracks. These harmless constrictors are attracted to houses by food (usually rodents, but some also take reptiles). Oviparous, they lay white, oval eggs.

Thirteen species are found throughout the continent, with isolated populations in Arabia (*L. arabicus*) and even on the Seychelles (*L. geometricus*). There are seven species on the subcontinent, six of which are endemic.

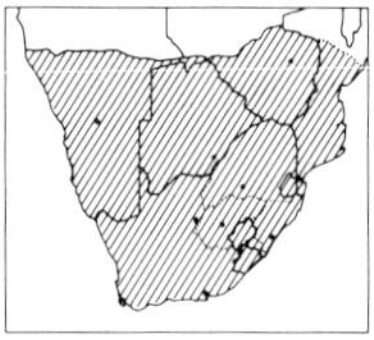

Brown House Snake *Lamprophis fuliginosus* (Pl. 28)

60-90 cm; max. 150 cm
A large house snake with an obvious head and small body scales (27-29 rows). It is uniform red-brown in colour (snakes from arid areas are light orange). Large, old snakes are darker, almost black. There are two pale yel streaks on the side of the head, which sometimes extend on to the front half the body. The belly is off-white. The young sometimes have indistinct, pale lateral spots that may persist in adults. Xanthic specimens are known. Male have a prominent hemipeneal bulge.

Biology and breeding: Terrestrial, these snakes forage for rodents and other small vertebrates (including bats) at night. In arid regions they freque eat lizards. They bite readily at first, but settle down easily in captivity. Up to eggs (30 x 15 mm) are laid in summer, and these take 60-90 days to develo Hatchlings measure 190-260 mm. In captivity, the female may lay clutches every 1-2 months during the breeding season. **Habitat:** Common in highvel grassland and arid karroid regions, but found everywhere, and tolerant of urban sprawl. **Range:** Throughout southern Africa and most of the continen

Olive House Snake *Lamprophis inornatus* (Pl. 32)

(Endemic) 60-80 cm; max. 130 cm
This large house snake has 21-25 scale rows and small eyes. It is uniform olive-green in colour, and occasionally almost dark brown; the belly is light grey-green. **Biology and breeding:** This snake is similar to the brown house snake, *L. fuliginosus* (page 58), but prefers moister habitats. It often eats other snakes. The female lays 5-15 eggs (38 x 25 mm) in October-December; these hatch in 70-90 days. The young measure 210-240 mm. **Habitat:** Moist coastal bushveld and fynbos, extending into grassveld. **Range:** Coastal belt from SW. Cape to East London, extending through Natal lowlands on to Transvaal highveld.

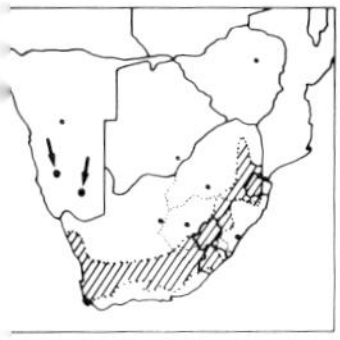

Spotted House Snake *Lamprophis guttatus* (Pl. 17)

(Endemic) 30-50 cm; max. 68 cm
A small, slender house snake with 21-25 scale rows and relatively large eyes. Its coloration is blotched, and regionally variable; there is a series of blotches on the back, arranged in alternating or adjacent pairs, that may merge to form a zigzag. In the north, the body is pinkish-grey or brown, distinctly marked with dark brown, often dark-edged, spots. The colour is duller in E. Cape, becoming light brown or tan in W. Cape, with only diffuse spots on the front of the body; the belly is yellowish-white. **Biology and breeding:** This snake is common in rocky habitats, sheltering in cracks and under flakes during the day, and feeding at night on geckos and sleeping skinks. It is a shy snake, easily stressed in captivity. The female lays 3-6 elongate eggs (38 x 20 mm) in midsummer. **Habitat:** Arid areas. **Range:** Inland mountains of Cape and Cape fold mountains, extending into S. Namibia and through Natal to E. Transvaal.

Aurora House Snake *Lamprophis aurora* (Pls. 25 and 32)

(Endemic) 50-60 cm; max. 90 cm
This beautiful, short, stocky house snake has 21-23 scale rows and a short tail (35-58 subcaudals). The body is green (rich olive to citrine) above, with a prominent orange-yellow stripe along the backbone, and the belly is white, with adjacent scale rows usually yellow. Adults are drabber than the sparkling juveniles, which are speckled with a pale bar on each scale and have a black-spotted head. **Biology and breeding:** Unfortunately, this snake is not common; it is a secretive, terrestrial species, shy and rarely attempting to bite. Its staple diet is nestling rodents. The female lays 8-12 eggs (35 x 20 mm) in summer. **Habitat:** Grassland, entering coastal bush and fynbos.
Range: Highveld of Transvaal and OFS, extending into Natal and E. Cape. Isolated records from S. Cape and Cape escarpment.

Fisk's House Snake *Lamprophis fiskii* (Pl. 19)

(Endemic) 25-35 cm; max. 40 cm
A small house snake with an abrupt head and a short tail. The scales are smooth and in 21-23 rows, the loreal as long as it is deep, and the subcaudals number 28-34. The head is lemon-yellow in colour, with symmetrical dark blotches. The body is lemon to dirty yellow, with a double row of alternating, dark brown blotches (sometimes fused on the forebody to form a zigzag pattern), and the belly is creamy white.

Biology and breeding: Little is known of this snake's habits and behaviour; a few specimens have been collected on roads at night (SA RDB, Rare). It lives underground, and constricts and eats small lizards. When threatened, it hisses and tightly coils and uncoils the front and rear of its body. A female laid eight eggs in summer. **Habitat:** Karroid sandy veld. **Range:** Few specimens from widely scattered localities around Great Karoo.

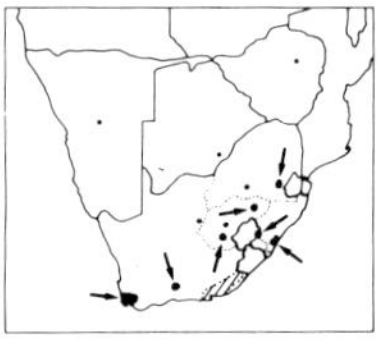

Yellow-bellied House Snake *Lamprophis fuscus* (Pl. 32)

(Endemic) 40-50 cm; max. 65 cm
This small, slender house snake has a small head and 19 body scale rows. It is uniform olive-brown to light olive-green above; the upper lip and scale rows bordering the belly are yellow-green, and the belly is light yellow. **Biology:** These snakes are generally rare, and are usually found in old term nests. They rarely bite, and do not settle easily in captivity. The diet is mainly lizards. **Habitat:** Grassveld and fynbos. **Range:** Cape Town to E. Transvaal.

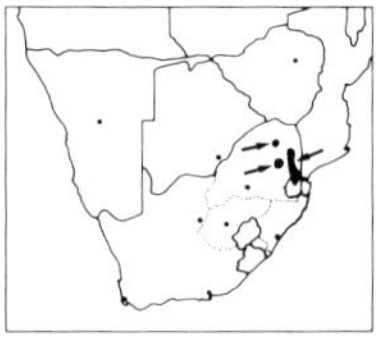

Swazi Rock Snake *Lamprophis swazicus* (Pl. 28)

(Endemic) 50-60 cm; max. 73 cm
This very slender snake has an obvious head, bulging eyes and 17 scale ro It is dark red-brown to light beige; the belly is creamy white. The scales may be dark-edged. **Biology:** The Swazi rock snake is similar in habits and behaviour to the spotted house snake, *L. guttatus* (page 59), being nocturn sheltering during the day in rock cracks, and feeding on small lizards. **Habitat:** Savannah. **Range:** E. Transvaal highlands and W. Swaziland.

WOLF SNAKES
Lycophidion

These small snakes are peculiar to Africa. The body is cylindrical, with the flattened head barely distinct from the neck. The eyes are small, with vertica pupils. The front teeth in the upper and lower jaw are long and recurved (hence 'wolf' snake), but these snakes lack venom glands and are harmless man. The scales are smooth, in 17-19 rows. The anal is undivided. The tail is short, with the subcaudals paired. The tail is longer in males.

Nocturnal, they shelter during the day under cover, emerging to hunt at dusk. Prey consists mostly of diurnal lizards (skinks, lacertids, etc.) which ar caught sleeping in their retreats. They are gripped by the large teeth of the wolf snake and constricted. Snakes are sometimes taken. Slow-moving and relatively docile, these snakes will bite if provoked. They are oviparous.

Twelve species are found throughout sub-Saharan Africa, with four occurring on the subcontinent, one of which is endemic.

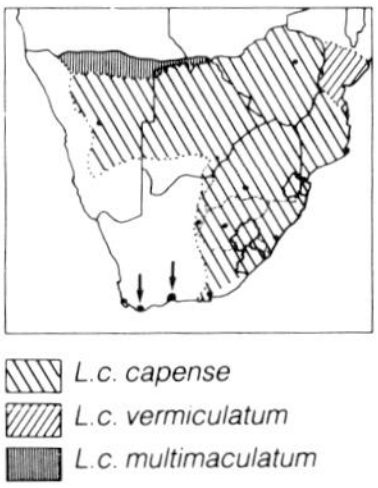

Cape Wolf Snake *Lycophidion capense* (Pl. 36)

30-40 cm; max. 64 cm
A small snake that grows larger in the southern part of its range. It has a flattened head. The postnasal touches the first upper labial. The ventrals number 159-200. Coloration is varied (see Subspecies). **Biology and breeding:** This common terrestrial species prefers damp situations. The female lays 3-9 eggs (22 x 10 mm) in early summer. Hatchlings measure 120 mm. **Habitat:** Grassland and savannah, entering coastal bush and fynb in the Cape. **Range:** Throughout most of the subcontinent, except Namib Desert and most of Cape. Isolated records from S. Cape and S. Namib Dese Elsewhere, through E. Africa to Egypt. **Subspecies:** Six races are recognize three of which occur on the subcontinent. *L.c. capense* has a uniform dark brown to black back, sometimes with each scale white-tipped, and a white c black-speckled belly. It occurs over most of the subcontinent, north to Zamb *L.c. vermiculatum* is uniform dark brown to black on the back and belly. It is found on Mozambique Plain to Save River. *L.c. multimaculatum* has a red-brown back, stippled with white, sometimes with a series of paired, dark blotches. It occurs in Caprivi Strip, north to Zaïre.

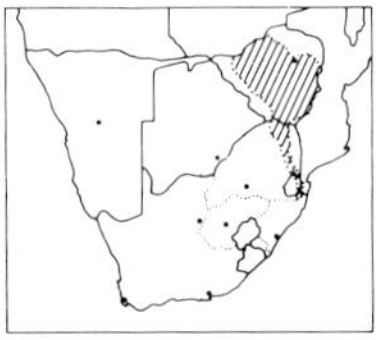

Variegated Wolf Snake *Lycophidion variegatum* (Pl. 29)

(Endemic) 30-35 cm; max. 43 cm
A small, slender snake. The postnasal is separate from the first upper labial There are 185-204 ventrals. Its speckled coloration is similar to that of some

Cape wolf snakes, but the white markings are more extensive; the belly is dark, sometimes with pale marks. **Biology and breeding:** It prefers rocky areas, hunting for sleeping skinks and geckos in cracks. Three eggs have been recorded in the largest female. **Habitat:** Savannah. **Range:** Lebombo Range in N. Zululand, through Transvaal escarpment to Zimbabwe.

Hellmich's Wolf Snake *Lycophidion hellmichi* (Pl. 25)

30-35 cm; max. 47 cm
This small, slender snake is similar to the Cape wolf snake, *L. capense* (page 60), but has higher ventral (197-214) and subcaudal (32-40) counts. It has a wide, pale brown band covering the back, and paired, dark brown scales (which may form a zigzag or bars) on either side of the backbone. The flanks are pale brown. The belly is dark, with the ventrals pale-edged. **Habitat:** Rocky arid savannah. **Range:** Central and N. Namibia, into Angola.

Eastern Wolf Snake *Lycophidion semiannule* (Pl. 36)

20-25 cm; max. 28 cm
This small species lacks a postnasal and has few ventrals (139-157). Coloration is varied. The back is greyish-blue to purple-brown (rarely with black crossbars), with the scales faintly pale-edged. The head is pale-edged, with a dark arrowhead on the crown. The belly is dark, with the ventrals pale-edged. **Biology:** This terrestrial species is rare and secretive, living in grass tussocks, etc. **Habitat:** Coastal grassland. **Range:** Mozambique Plain to N. Zululand.

VARF WOLF
IAKE
yptolycus

This snake is similar to the wolf snakes, *Lycophidion* (page 60), but is small and has fewer upper and lower labials and different dentition.

The genus consists of a single species, which is endemic to the subcontinent.

Dwarf Wolf Snake *Cryptolycus nanus* (Pl. 36)

(Endemic) 20-23 cm; max. 27 cm
This small snake has a blunt, flattened head, small eyes and vertical pupils. Its scalation is similar to that of the eastern wolf snake, *Lycophidion semiannule* (above), but the dwarf wolf snake has only 6-7 upper labials, six lower labials and scales in 17 rows to the tail. The back and belly are uniform blue-black, with a stippled white band around the snout. **Biology and breeding:** This secretive species is a specialist feeder on amphisbaenians (*Chirindia*). The female lays two elongate eggs (26 x 6 mm). **Habitat:** Floodplain with miombo woodland. **Range:** Central Mozambique Plain.

LE SNAKES
helya

These peculiar African snakes are closely related to the wolf snakes, *Lycophidion* (page 60). The body is almost triangular in cross-section, with the broad, flat head very distinct from the neck. They have large nostrils, and smallish eyes with vertical pupils. The body scales are strongly keeled and almost conical, in 15-19 rows, often non-overlapping; the scales along the backbone are enlarged and have two keels. The anal is undivided. The tail is longish, with paired subcaudals. The common name derives from the similarity of the body in shape and texture to a three-cornered file.

Nocturnal and terrestrial, file snakes are very secretive, and are rarely seen in spite of being widely distributed. They are constrictors, feeding on small vertebrates; some are specialist feeders on snakes. The oviparous females are larger than the males. Harmless to man, they lack fangs and venom glands. They are very docile, never attempting to bite, but are restless in the hand.

Ten species are distributed throughout most of sub-Saharan Africa, with three on the subcontinent, none of which is endemic.

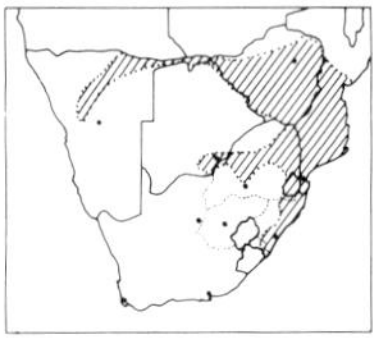

M.c. capensis

Cape File Snake *Mehelya capensis* (Pl. 25)

100-120 cm; max. 165 cm
This large snake has a very flat head and a thickset, triangular body. Two labials enter each eye. The scales are strongly keeled, in 15 rows, with thos on the spine fused to the backbone. There is extensive bare skin between th scales. There are 193-244 ventrals and 44-58 subcaudals. The back is grey grey-brown, and sometimes dark olive to purple-brown, with a white to yello vertebral stripe. The skin between the scales is pink-purple. The belly is ivor white to cream, extending on to the flanks.

Biology and breeding: The Cape file snake is a formidable predator on other snakes (even venomous species, to whose venom it appears immune) and also eats other small vertebrates. It never bites when captured, but may empty its bowels. A few (5-13) large eggs (47-55 x 20-31 mm) are laid in lea litter. The young measure 390-420 mm. In captivity, eggs hatch in 90-100 days. Females may lay two clutches in a summer.

Habitat: Mainly savannah, but entering coastal forest and arid regions. **Range:** In the east from Natal, through Transvaal, Zimbabwe and Caprivi St to N. Namibia. Elsewhere, to Cameroon and Somalia. **Subspecies:** Three races are recognized, with only the typical race occurring in the southern African subcontinent.

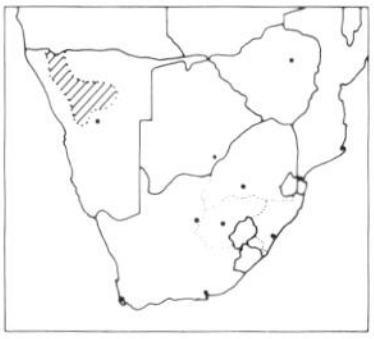

Angola File Snake *Mehelya vernayi* (Pl. 36)

90-110 cm; max. 134 cm
This species is similar in appearance to the Cape file snake, *M. capensis* (above), but has three labials entering each eye. The scales are in 19 rows and the ventrals number 256-268. It lacks the white vertebral stripe of the Cape file snake, and the back is dark red-brown on the forebody, with the scales increasingly pale-centred towards the tail, giving it a speckled appearance. The belly is cream-yellow.

Biology and breeding: This snake is similar in behaviour and habits to the Cape file snake; toads and lizards have been found in the gut contents. A few eggs (37 x 15 mm) are laid. **Habitat:** Rocky mountainous country. **Range:** N. Namibia to W. Angola.

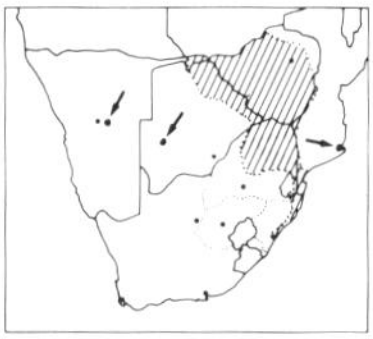

Black File Snake *Mehelya nyassae* (Pl. 36)

45-50 cm; max. 65 cm
This small species has two labials entering each eye, the scales in 15 rows, ventrals numbering 165-184, and 51-77 subcaudals. The back is uniform da brown to purple-brown above. The skin between the scales is pink. The bell dark olive to black, or white to cream-olive. **Biology and breeding:** This sna is similar in behaviour and habits to the Cape file snake, *M. capensis* (above but eats mainly lizards, particularly skinks. The female lays up to six eggs. T young measure 200-220 mm. **Habitat:** Savannah, entering coastal forest. **Range:** Eastern regions from Natal to Kenya, and extending through Botswana to central Namibia.

SLUG EATERS
Duberria

These stout-bodied little snakes have a small head that is hardly distinct fro the neck. The small eyes have round pupils, and each nostril pierces a singl nasal. The scales are smooth, in 15 rows. The anal is undivided. The tail is short, with paired subcaudals.

These gentle, slow-moving, shy snakes forage among grass roots and rotting timber in damp situations. The diet consists exclusively of slugs and land snails; the slug eater follows a slime trail and simply swallows the prey i finds at the end of it. The females are larger than the males.

There are two species, both of which enter the subcontinent, and one of which is endemic.

D.l. lutrix
D.l. rhodesiana

Common Slug Eater *Duberria lutrix* (Pl. 28)

30-35 cm; max. 43 cm
A short, small-headed snake, with 116-142 ventrals. The back is brick-red to pale brown above, sometimes with a broken black line along the backbone. The flanks are paler, being grey to light brown. The belly is cream, edged with a dark, dotted line.

Biology and breeding: This is a common species in suitable habitat, and a boon to gardeners and farmers. Inoffensive, it rolls into a tight spiral when alarmed (hence the Afrikaans name 'tabakrolletjie'). Six to eight (max. 12) babies are born in late summer (January-February); they measure 80-90 mm. **Habitat:** Savannah, entering coastal bush and fynbos. **Range:** Cape fold mountains, through E. Cape to Natal, Transvaal and Zimbabwe. Elsewhere, through E. Africa to Ethiopia. **Subspecies:** There are six races, two of which occur on the subcontinent. *D.l. lutrix* has two postoculars, and broken black dorsolateral lines; it occurs in Transvaal, south to the Cape. *D.l. rhodesiana* has one postocular, and lacks dorsolateral lines; it is found in Zimbabwe.

Variegated or Spotted Slug Eater *Duberria variegata* (Pl. 16)

(Endemic) 25-30 cm; max. 39 cm
This small, stout snake is similar to the common slug eater, *D. lutrix* (above), but has a more prominent snout and 116-142 ventrals. The back is brown to olive or dark brown, with three rows of blackish blotches that may form crossbars and are sometimes obscured by pale speckling. The belly is dirty yellow with dark reticulation, particularly at the rear. **Biology and breeding:** Similar in most aspects of its behaviour and habits to the common slug eater, this species has not been observed to form a coiled defensive posture. Seven to 20 young (larger litters in larger snakes) are born in late November-January; they measure about 90 mm. **Habitat:** Coastal dune forest. **Range:** N. Zululand and adjacent Mozambique.

LE SNAKE
eudaspis

This genus contains a single, widely distributed species.

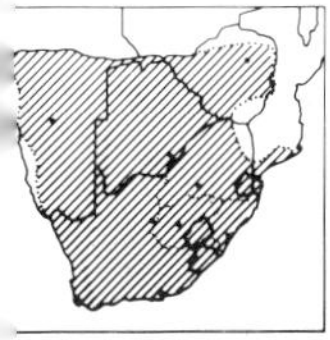

Mole Snake *Pseudaspis cana* (Pls. 18, 28 and 35)

100-150 cm; max. 210 cm
This is a large, thick, solid snake with a slightly hooked nose. The body scales are smooth (but sometimes keeled in black snakes from W. Cape), in 25-31 rows. Each nostril is pierced between two nasals. The anal is divided. The eyes have round pupils. Coloration is variable. The young are blotched, the body being light red-brown with four rows of dark, pale-edged spots (the centre pair may fuse to form a zigzag pattern); the adults are plain (the juvenile pattern is sometimes retained), usually light to red-brown, but occasionally olive, grey, dark brown or black (common in W. Cape). Males have thicker, longer tails.

Biology and breeding: Mole snakes are extremely useful, harmless constrictors that live underground in abandoned animal burrows and feed on moles, rodents and other small mammals. Juveniles also eat lizards. Some mole snakes eat eggs, which they swallow whole. They are aggressive when first caught, and often bite and twist, leaving deep gashes that may require stitches. They calm down in captivity and make good pets, although they are prone to long fasts. Males fight in mating season and bite each other severely. They mate in late spring (October), and 25-40 (up to 95) young are born during March-April, each measuring about 200 mm. **Habitat:** Sandy scrubland in SW. Cape, highveld grassland, and mountainous and desert regions. **Range:** Throughout southern Africa (and on Robben Island). Elsewhere, north to Angola and Kenya.

SNAKES WHOSE RELATIONSHIPS REMAIN UNRESOLVED

The following species are unusual African snakes whose relationships to ot snakes have not been resolved. Various theories concerning their affinities have been proposed, but these are not uniformly accepted. Many are endemic to the subcontinent.

MARSH SNAKES
Natriciteres

These small, inconspicuous snakes, along with the striped swamp snake, *Limnophis bicolor* (page 65), may be related to natricine snakes (which are mainly Palaearctic). The body is cylindrical, with a small head that is barely discernible from the neck. The eyes are largish, with round pupils. The internasals are paired. The scales are smooth, in 15-19 rows. The anal is sometimes divided. The tail is moderately long, with paired subcaudals.

Diurnal and terrestrial, they are found sheltering under cover in damp situations. They feed mainly on fish and frogs. The females are oviparous, a grow larger than the males. They are very docile, rarely biting. If grabbed b the tail, they spin wildly, breaking it off. This caudal autotomy is common in lizards, but rare in snakes, and may be an adaptation to predation by fish a wading birds. Harmless to man, they lack fangs and venom glands.

Three species are distributed throughout tropical Africa, with two enterin the subcontinent, neither of which is endemic.

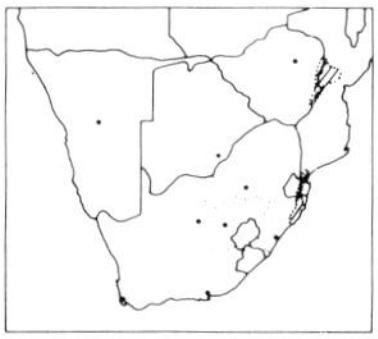

N.v. sylvatica

Forest Marsh Snake *Natriciteres variegata* (Pl. 33)

30-35 cm; max. 45 cm

A small, thickset snake with smooth scales in 17 rows at the front and 15 ro at the rear. There are usually eight lower labials, the first four touching the anterior chin shields. Ventrals number 125-143, and subcaudals 60-84. Th back is dark olive to chestnut-black above (sometimes with a faint yellow collar), with a broad, darker band down the backbone, usually bordered b row of minute white dots. The top lip is barred in yellow and black. The bell yellow to orange with dark grey edges. **Biology and breeding:** This snake shelters under cover at forest fringes, feeding on frogs and fish-eating spid The female lays 5-6 eggs. **Habitat:** Montane and lowland evergreen forest. **Range:** Isolated populations in N. Zululand and adjacent Mozambique anc eastern escarpment of Zimbabwe. Elsewhere, through E. and central Afric Sierra Leone. **Subspecies:** Four races are recognized, with only *N.v. sylva* occurring on the subcontinent. This subspecies is distinguished from the typical race (which occurs from Sierra Leone to Zaïre, and has 15 scale ro around its body); from *N.v. bipostocularis* (which occurs in Zaire, Angola a Zambia) by having three postoculars; and from *N.v. pembana* (occurring o Pemba Island, off Tanzania) by its higher ventral and subcaudal counts.

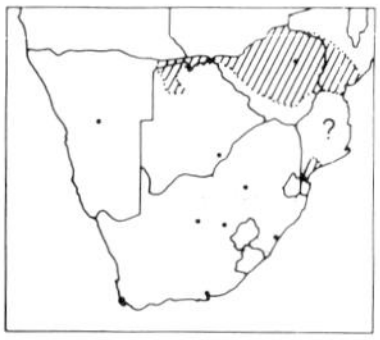

Olive Marsh Snake *Natriciteres olivacea* (Pl. 33)

35-40 cm; max. 52 cm

This small snake is similar to the forest marsh snake, *N. variegata* (above), has 19 scale rows at the front and 17 at the rear; there are nine lower labial the first five touching the anterior chin shields. In colour, it is similar to the forest marsh snake, but it lacks the yellow collar and is more often an olive colour. **Biology and breeding:** Similar in habits and behaviour to the forest marsh snake, the olive marsh snake inhabits vleis and pans, and has been observed to eat winged termites. It tames well and is very docile in captivit The female lays 6-8 eggs (22 x 9 mm) in early summer. **Habitat:** Savannah. **Range:** S. Mozambique, Zimbabwe and N. Botswana, north to Sudan and Africa.

STRIPED SWAMP SNAKE
Limnophis

There is a single species in the genus. It is a small, inconspicuous snake w cylindrical body, and is distinguishable from the marsh snakes, *Natriciteres* (above), by having a single, triangular internasal.

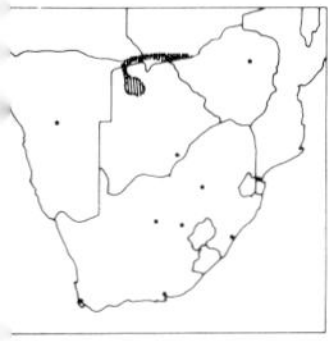

bangweolicus

Striped Swamp Snake *Limnophis bicolor* (Pl. 25)

45-50 cm; max. 62 cm
This snake has a small head that is barely distinct from the neck. The eyes are largish, with round pupils. The sixth upper labial is very large, often touching the parietal. The scales are smooth, in 19 rows. The anal is divided. The tail is short, with paired subcaudals. The back is dark olive-brown, flanked by a paler stripe. The sides are pale, with 3-4 black-edged scale rows simulating thin stripes. The belly is yellow to brick-red. **Biology and breeding:** A poorly known snake that feeds mainly on fish and amphibians. It lays eggs. **Habitat:** Marshy areas. **Range:** Okavango and Zambezi drainage basins. **Subspecies:** Two races are recognized, with only *L.b. bangweolicus* (which has a narrower snout) being found on the subcontinent.

ESTERN KEELED
NAKE
thonodipsas

This unusual snake is easily recognized by its small, irregular head shields, a condition usually found only in pythons and adders.

There is a single species in the genus.

Western Keeled Snake *Pythonodipsas carinata* (Pl. 16)

45-55 cm; max. 62 cm
This snake's body is cylindrical, with smooth (or faintly keeled) scales, in 21 rows. The head is very flat and distinct from the neck. The eyes are largish, with vertical pupils. The nostrils are situated on top of the snout. The anal is entire, and the tail longish with undivided subcaudals. The back is yellow-olive, pale buff, sand or grey above, with a double series of dark-edged, brown to grey-brown blotches that may form a zigzag or crossbars. The flanks have smaller, less distinct bars. The head is variegated in the body colours. The belly is white, and sometimes spotted on the sides. **Biology and breeding:** Nocturnal and terrestrial, this snake feeds on small lizards and rodents. It is very viper-like in appearance, and may mimic the horned adder, *Bitis caudalis* (page 100). It has large back fangs and bites readily, but is not venomous. It settles well in captivity, but should be fed sparingly. **Habitat:** Rocky desert. **Range:** W. Namibia into SW. Angola.

NY-SPOTTED
NAKE
nplorhinus

This is an unusual snake of nondescript appearance. There is a single species in the genus, which is endemic to the subcontinent.

Many-spotted Snake *Amplorhinus multimaculatus* (Pl. 18)

(Endemic) 45-55 cm; max. 63 cm
The body of this snake is cylindrical, with a small head that is just distinct from the neck. The eyes are of moderate size, with round pupils. The scales are smooth (except over the posterior back and tail base), in 17 rows. The anal is undivided. The tail is of moderate length, with divided subcaudals. It has prominent, grooved back fangs, but its venom is considered harmless. The back is green or olive-green to brown above, usually with a series of dark brown to black blotches, and sometimes with a pale dorsolateral stripe; scattered scales are often pale-edged, giving a flecked effect. The belly is dull green to blue-grey. **Biology and breeding:** Secretive, but locally common, this snake forages for frogs and lizards in reed beds and waterside vegetation. It bites readily when caught and may coil into a tight spring, similar to the common slug eater, *Duberria lutrix* (page 63). Its venom has caused local pain, inflammation and swelling that resolved in a short time; the bite also bled freely but severe haemorrhage did not develop. Antivenom is unnecessary and probably ineffective for bites by this species. Usually 4-8 (max. 12) young, measuring 125-200 mm, are born in late summer. **Habitat:** Mountain streams and vleis. **Range:** Cape fold mountains, Natal and Transvaal Drakensberg, with an isolated population on Zimbabwe's eastern escarpment.

SHOVEL-SNOUT SNAKES
Prosymna

These small, strange African snakes are not obviously related to any other group. The body is cylindrical, with the head not distinct from the neck. The snout is characteristically depressed, forming an angular horizontal 'shovel' that is sometimes upturned. The scales are usually smooth, in 15-21 rows. T anal is undivided. The tail is short, ending in a spine.

They burrow in loose soil, feeding almost exclusively on reptile eggs. The shell is punctured by the maxillary teeth, and the egg, which may contain a well-developed embryo, is swallowed whole. They lay a few (3-6) elongate eggs. Some species have an unusual defensive display, forming a tight 'watchspring', usually with the head hidden beneath the coils. When touche they wildly coil and uncoil. Harmless to man, they lack fangs or venom glan

There are 12 species distributed throughout sub-Saharan Africa, with sev occurring on the subcontinent, three of which are endemic.

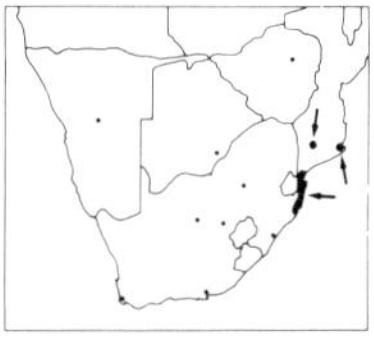

Mozambique Shovel-snout *Prosymna janii* (Pl. 16)

(Endemic) 22-26 cm; max. 31 cm
A small snake with a rounded, angular snout, and eyes with round pupils. It has few ventrals (107-129), and keeled scales in 15-17 rows. The body is yellow to pale red-brown above, with a series of paired, large, dark brown to black blotches that are larger on the forebody. The belly is uniform cream-white. **Biology and breeding:** These snakes are rarely found, except in spr when males come to the surface to search for females. In defensive display the front of the body is inflated and raised, and moved slowly to and fro with the mouth open. Up to five eggs are laid in December. **Habitat:** Coastal dur forest. **Range:** S. Mozambique and N. Zululand.

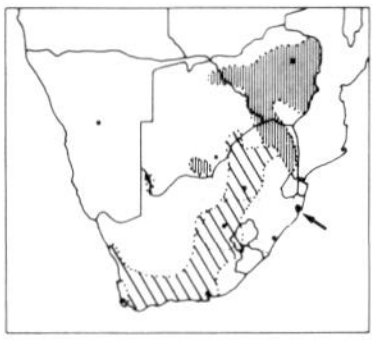

P.s. sundevallii
P.s. lineata

Sundevall's Shovel-snout *Prosymna sundevallii* (Pls. 16 and 2

(Endemic) 24-26 cm; max. 36 cm
This small snake has smooth scales in 15 rows. The snout is upturned. The pupils are round. The paired internasals touch each other (or are just apart) There are 131-170 ventrals. The body is pale to dark brown above, with numerous light and dark mottlings, and sometimes with a series of dark vertebral blotches. The belly is white. **Breeding:** The female lays 3-5 elonga eggs (28 x 9 mm). Hatchlings measure 100-110 mm. **Habitat:** Dry areas, including savannah woodlands, highveld and karroid areas, entering fynbo and valley bushveld in the Cape. **Range:** Eastern half of the subcontinent. **Subspecies:** Two races are recognized. *P.s. sundevallii* has its internasals separated, paired spots, and a longer tail; it occurs from S. Transvaal to the Cape, with a few records from N. Cape/S. Botswana border. *P.s. lineata* has internasals touching, single spots, and a shorter tail; it occurs in E. Transvaa and Zimbabwe, with an isolated record from N. Zululand.

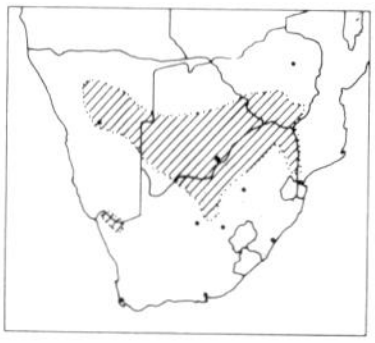

Two-striped Shovel-snout *Prosymna bivittata* (Pl. 19)

(Endemic) 26-28 cm; max. 35 cm
This slender snake has smooth scales in 15 rows. The snout is upturned. Th pupils are round. The paired internasals are well separated. Ventrals numbe 154-180. The back is purple-brown to red-brown, with a broken orange strip on the backbone. The belly is white. **Breeding:** The female lays up to 4 eggs (27 x 7 mm). **Habitat:** Acacia savannan, entering sandveld. **Range:** Transva through S. Zimbabwe and Botswana to Namibia and Namaqualand.

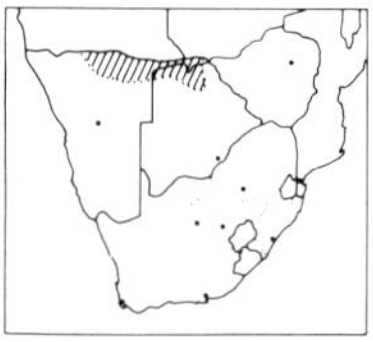

Angola Shovel-snout *Prosymna angolensis*

24-28 cm; max. 36 cm
A small snake with an angular snout. It has smooth scales in 15 rows. There a single bandlike internasal. Ventrals number 121-163. The tail is very short

(16-26 subcaudals). The back is light yellow-brown, with a paired series of black spots (which may be absent); the scales are sometimes dark-edged with a pale centre. The belly and flanks are yellow-white. **Habitat:** Savannah. **Range:** N. South West Africa/Namibia and Caprivi Strip, north to Angola and Zambia.

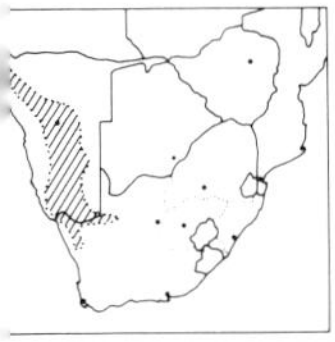

South-western Shovel-snout *Prosymna frontalis* (Pl. 26)

30-35 cm; max. 44 cm
This small, slender snake has an angular snout. The scales are smooth, in 15 rows. There is a single bandlike internasal. The eyes have vertical pupils. Ventrals number 153-199. The tail is long (32-54 subcaudals). The body is light brown to chestnut above; the scales are dark-edged, giving a striped or mosaic effect. It has a broad, dark brown to black collar (which is followed by fainter crossbars in juveniles). The belly is uniform white. **Habitat:** Rocky areas in arid regions. **Range:** NW. Cape, through Namibia to S. Angola.

stuhlmannii

East African Shovel-snout *Prosymna ambigua* (Pl. 29)

24-26 cm; max. 30 cm
A small snake with an angular snout. Its smooth scales are in 15 rows. It has a single bandlike internasal. There are 124-164 ventrals. The tail is long (17-39 subcaudals). The back is dark brown to metallic blue-black, sometimes with the scales pale-centred; a paired series of small white spots flanking the backbone may be present. The belly is white, and sometimes brown-black. **Biology and breeding:** The defensive 'watchspring' display has not been recorded in this species. The female lays 3-4 eggs (19-30 x 6-8 mm). **Habitat:** Savannah, extending into wooded hills. **Range:** Central and E. Africa, south to Zululand. **Subspecies:** Three races are recognized, with one entering the region. *P.a. stuhlmannii* has a more rounded snout, dorsal spots, and a longer tail; it occurs in N. Zululand, E. Transvaal, Zimbabwe and Mozambique, and elsewhere to S. Somalia.

Visser's Shovel-snout *Prosymna visseri*

28-31 cm; max. 35 cm
This small snake has a rounded snout, a single internasal, and a very large prefrontal touching the first upper labial. The scales are smooth, in 15 rows. The eyes have round pupils. Ventrals number 189-208, and subcaudals 37-57. The back is dark brown, with a pale yellowish vertebral stripe (which may be broken). The belly is white. **Biology:** It differs from other shovel-snout snakes in that it lives in deep granite cracks in rocky areas, where it feeds on gecko eggs. **Habitat:** Mopane tree savannah. **Range:** S. Angola; one specimen recorded recently from Kamanyab in N. Namibia.

ND SNAKES AND
IEIR RELATIVES
bfamily
ammophinae

These snakes share certain vertebral and dental features and are unusual in having a vestigial hemipenis.

Varying from very small to large in size, most are diurnal, fast, active terrestrial snakes, while some are partly arboreal (living under tree bark), and others burrow in loose sand. They are rare in forest or thick bush, and are usually found in savannah or semi-desert, with some entering montane grassland. Egg-laying is usual, but there is one viviparous species. All possess large back fangs, and some have toxic saliva, although only a very few species are clinically dangerous. Serious envenomation has been reported for the Mediterranean *Malpolon*, and mild envenomation for some local *Psammophis*.

The subfamily is distributed throughout Africa, with some entering Europe and the near East; one species is present on Madagascar. There are eight genera, with six on the subcontinent, one of which is endemic.

BARK SNAKES
Hemirhagerrhis

These small, arboreal snakes have a short maxillary with 9-10 teeth, followe after a gap by two curved fangs.

There are two species in the genus, one of which occurs in the region. It is replaced by *H. kelleri* in the dry savannahs of Ethiopia and Somalia.

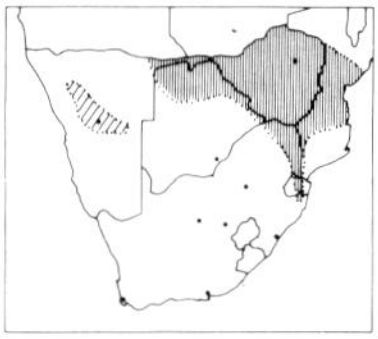

H.n. nototaenia
H.n. viperinus

Bark or Mopane Snake *Hemirhagerrhis nototaenia* (Pl. 16)

25-35 cm; max. 43 cm
A small, slender snake with a flattened head, an obvious neck, small eyes w vertical pupils, and a semi-divided nasal. The body is cylindrical, with smoo scales in 17 rows. The anal is divided. The tail is longish, with paired subcaudals. The back is grey or grey-brown, with a dark vertebral stripe tha flanked by, and often fused with, a series of black spots, forming crossbars a zigzag. The top of the head is black. The belly is dirty white, mottled with grey. The tail is paler, with the last third often yellow or orange to salmon pin **Biology and breeding:** They shelter under loose bark, feeding on day geckos, small skinks, and occasionally small frogs; these are eaten while th snake hangs head-down. They may also eat gecko eggs. They never bite. T female lays 2-8 elongate eggs (24 x 6 mm). **Habitat:** Savannah woodlands. **Range:** Extreme N. Natal and E. Transvaal lowveld, into Zimbabwe and Mozambique, and through N. Botswana to N. Namibia and S. Angola. Elsewhere, to tropical E. Africa. **Subspecies:** Two races are recognized. *H.n. nototaenia* has its blotches often connected by a vertebral stripe, and a long tail (68-98 subcaudals); it occurs in the east. *H.n. viperinus* usually lac the vertebral stripe, and has a short tail (52-75 subcaudals); it occurs as scattered records in central and N. Namibia and S. Angola.

BEAKED SNAKES
Rhamphiophis

These are large, stout snakes with a shortened skull and a hooked snout. Th are terrestrial, searching underground in rodent burrows for their food.

There are three species, one of which occurs in the region.

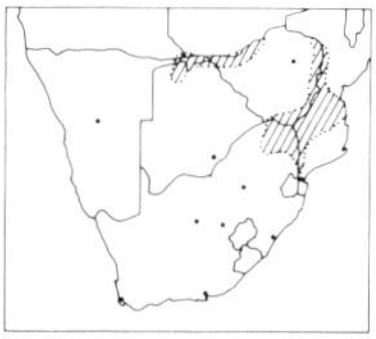

R.o. rostratus

Rufous Beaked Snake *Rhamphiophis oxyrhynchus* (Pl. 34)

120-140 cm; max. 157 cm
A large, stout-bodied snake. The head is distinct from the body, and has a prominent, hooked snout. The eyes are largish, with round pupils. The body cylindrical and stout, with smooth scales in 17-19 rows. The anal is divided. The tail is long, with paired subcaudals. The back is uniform yellowish-brow to red-brown, with the scales sometimes pale-centred. The head has a dark stripe on the sides. The belly is cream to yellow-white. **Biology and breeding:** Slow-moving and diurnal, they shelter in mammal burrows and termitaria. They eat a variety of small vertebrates, including other snakes; juveniles also take insects. They rarely bite, and although they hiss and strik when first captured, they tame well. The female lays 8-17 large, cylindrical eggs (34-40 x 20-22 mm), often staggered over a few days. **Habitat:** Sandy thornveld or bushveld. **Range:** E. Transvaal, N. Botswana, Zimbabwe (whe is absent from the central plateau) and Mozambique floodplain, and north t Sudan. **Subspecies:** Two races are recognized, with only the eastern race, *R.o. rostratus*, occurring in the region; the typical race in W. Africa (reachin Uganda) lacks the head stripe, and has 14 scale rows and a single preocul

DWARF BEAKED SNAKE
Dipsina

The genus is endemic to the subcontinent, and contains a single species.

Dwarf Beaked Snake *Dipsina multimaculata* (Pls. 15 and 17)

(Endemic) 30-32 cm; max. 45 cm
This small, slender snake has a distinct head with a prominent, hooked snou The eyes are largish, with round pupils. The body is cylindrical and slender,

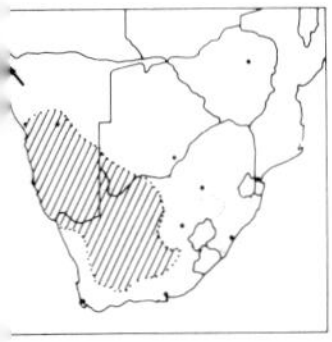

with smooth scales in 17 rows. The anal is divided. The tail is short, with paired subcaudals. The back may be various shades of buff, grey and pink-brown, with 3-5 rows of darker blotches that may be pale-centred and/or fuse to form crossbars. The back of the neck has a dark V-shape. The head has an eye stripe. The belly is white to pink-cream, with dark lateral spots. **Biology and breeding:** This snake hides beneath stones or in loose sand at the base of bushes, ambushing small lizards, particularly barking geckos and small lacertids. It is very docile, but when threatened may adopt a coiled posture that mimics the horned adder, *Bitis caudalis* (page 100). The female lays 2-4 eggs. **Habitat:** Rocky, sandy areas. **Range:** Cape karroid areas, through SW. Botswana to S. and W. Namibia.

.YMPIC SNAKES
omophis

These slender, medium-sized snakes have the maxillary teeth longest in the middle, and the anterior mandibular teeth the largest.

There are two species in the genus, one of which occurs in the region. *D. praeornatus* is restricted to W. Africa, from Senegal to the Central African Republic.

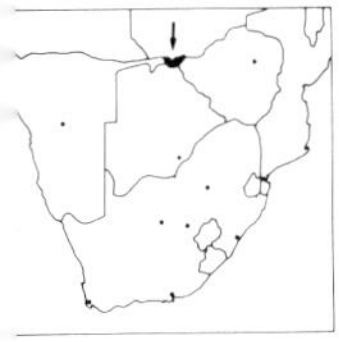

Lined Olympic Snake *Dromophis lineatus* (Pl. 37)

70-90 cm; max. 121 cm
A slender snake with a cylindrical body and a distinct head. The eyes are medium-sized, with round pupils. The scales are smooth, in 15-17 rows. The anal is divided. The tail is long, with paired subcaudals. The back is olive in colour, with three thin, greenish-yellow stripes. The scales are black-edged. The belly is greenish-yellow to pale green, usually with distinctive black bars on the sides of the ventrals. **Biology and breeding:** Diurnal, these snakes forage around rivers and swamps for frogs and small mammals. They are gentle, and rarely bite. The female lays 6-9 eggs (23-27 x 12-18 mm). **Habitat:** Low, waterside vegetation. **Range:** Extreme W. Zimbabwe and Caprivi Strip. Elsewhere, through tropical Africa to Sudan.

KAAPSTEKERS
ammophylax

These solid-bodied snakes have a distinct head, and moderately sized eyes with round pupils. The maxillary teeth are the same size to slightly larger in the rear; the mandibular teeth are largest at the front. The body is cylindrical, with smooth scales in 17 rows. The anal is divided. The tail is of moderate length, with paired subcaudals.

These are terrestrial grassland snakes that forage during the day and evening. They feed on a wide variety of small vertebrates which they bite, then grip until the prey has been subdued by the venom. They may bite if they are restrained. The venom is clinically unimportant, and no serious symptoms have been reported; experimentally, it is very toxic, but yields are minute. The common name is ludicrous: it is doubtful if they produce enough venom to kill a rat, let alone a sheep. Reproduction is varied, the species showing interesting maternal care in cold climates.

There are three species in the genus, all occurring in the region. None is endemic.

rhombeatus

Spotted or Rhombic Skaapsteker
Psammophylax rhombeatus (Pls. 17 and 22)

80-90 cm; max. 146 cm
A medium-sized snake that has a smallish head with a rounded snout. The rostral is usually deeper than it is broad, and sometimes separates the internasals. There are usually two anterior temporals. Ventrals number 143-177, and subcaudals 60-84. The back is yellowish-brown to pale olive, with three (rarely four) rows of dark-edged blotches down the back, that may fuse to form a zigzag or irregular stripes. The upper lip is dark-spotted. The belly is yellowish with dark blotches. **Biology and breeding:** Common in moist

grasslands, this species actively pursues its prey. The diet includes lizards, frogs, rodents, birds, and even other snakes. It may bite, but only if provoke It tames well. Up to 30 eggs (20-35 x 12-18 mm) are laid in a hole; the female guards the eggs by coiling around them. They take 35-45 days to hatch, and the young measure 180-220 mm. **Habitat:** Highveld grasslands and fynbos, entering karroid areas. **Range:** Widespread in the highveld, entering coasta Natal and S. Cape, with scattered records in Little Namaqualand and Namib **Subspecies:** Two races are recognized, with only the typical race occurring on the subcontinent; it is replaced in S. Angola by *P.r. ocellatus*, which has a different colour pattern and more than 172 ventrals.

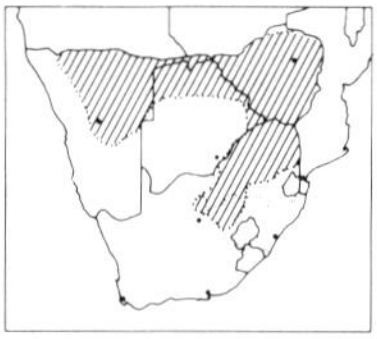

Striped Skaapsteker *Psammophylax tritaeniatus* (Pl. 22)

60-70 cm; max. 89 cm
This snake is of medium size. The head is small, with a pointed snout. The rostral is usually broader than it is deep. The eyes are small. The tail is shorti (49-69 subcaudals). The back is grey to pale grey-olive, with three distinct, black-edged, dark brown stripes, one of these forming a vertebral line that may be divided by a fine yellowish line. The lateral stripes pass through the eyes. The upper lip and belly are uniform white to cream. **Biology and breeding:** These gentle, beautiful and inoffensive snakes forage in grasslar for small mammals (the young take lizards and small frogs). When disturbed they make a dash for cover and then freeze; they are very well camouflaged and easily overlooked. They wriggle wildly in the hand, but never bite. The female lays 5-18 eggs (23-28 x 11-15 mm, 2-3 g) in a hole, but is not known t guard the eggs. Hatchlings measure 130-220 mm. **Habitat:** Open grasslanc **Range:** Central part of the region, from S. OFS through Transvaal, Zimbabw N. Botswana and Namibia, north to Angola and Tanzania.

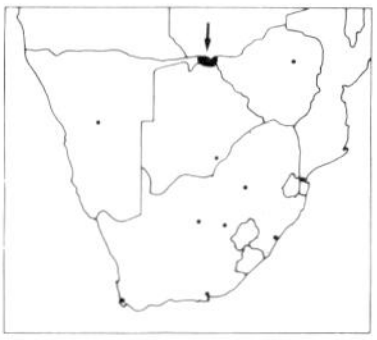

P.v. variabilis

Grey-bellied Grass Snake *Psammophylax variabilis* (Pl. 22)

80-90 cm; max. 100 cm
A medium-sized snake. The head is small, with a rounded snout. The rostral always broader than it is deep. The eyes are small. There is one anterior temporal. The tail is shortish (49-61 subcaudals). The back is grey or olive-brown, sometimes with three thin (one scale wide), dark stripes which may be white-flecked. The upper lip is grey-white. The belly is uniform grey. **Biology and breeding:** These snakes eat a wide variety of small mammals, well as fish. They are viviparous; four fully formed babies, measuring 151-155 mm, are born in December. **Habitat:** Floodplain grasslands (elsewhe in montane grassland). **Range:** Restricted locally to Chobe floodplain. Elsewhere, through central and E. Africa to Ethiopia. **Subspecies:** Three races are recognized, with only the typical race occurring in the region.

SAND AND GRASS SNAKES
Psammophis

These small to large snakes have the head distinct from the neck, and moderate to large eyes with round pupils. There are 10-15 maxillary teeth, th largest at the front of the eye, and two large, grooved fangs at the back of th eye. The anterior mandibular teeth are distinctly enlarged. Usually the first fc lower labials touch the anterior chin shields. The body is cylindrical, with smooth scales in 11-19 rows. The anal is sometimes divided. The tail is long, with paired subcaudals.

These fast, active, diurnal snakes inhabit savannah or arid scrubland, but are sometimes restricted to montane grassland. Some species (*P. sibilans* and *P. schokari*) 'polish' themselves with a nasal gland secretion, which reduces skin water-loss. Prey (small vertebrates) is pursued, grabbed and chewed until the venom takes effect, then swallowed head first. These snake usually struggle wildly when first caught, and the larger species bite readily. When they are caught by the tail, they spin wildly, causing it to break off. The tail tip cone may be regenerated, resulting in false low subcaudal counts.

Truncated tails are more common in some species (63% in *P. biseriatus*) than in others (8% in *P. crucifer*). They tame well, but require frequent food. The venom is usually harmless, but a bite from the Kenyan *P. biseriatus* caused mild haemorrhage; pain, swelling and nausea has been noted in some *P. sibilans* and *P. phillipsii* bites. They are oviparous.

This is a large genus, containing more than 20 species which are found throughout Africa, with three species entering the near East. Nine occur in the region, one of which is endemic.

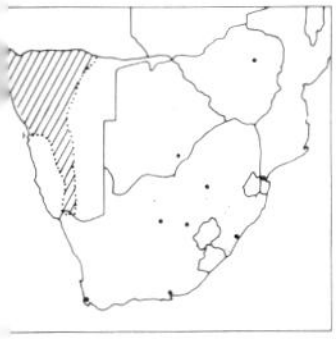

Western Sand Snake *Psammophis trigrammus* (Pl. 34)

90-110 cm; max. 118 cm
A very slender snake, with 17 scale rows. There are nine upper labials (numbers five and six enter the eye). The first five lower labials touch the anterior chin shields. Ventrals number 183-197. The anal is divided. The tail is long (132-155 subcaudals). The body is pale olive to grey-brown anteriorly, and reddish to yellowish behind; sometimes black-edged scales along the back form a stripe that may be flanked by a yellowish-white stripe. The belly is off-white, with a light grey to olive median band. **Biology:** This very fast snake chases skinks and lacertids during the heat of the day. **Habitat:** Arid scrubland. **Range:** Namaqualand, through Namibia to S. Angola.

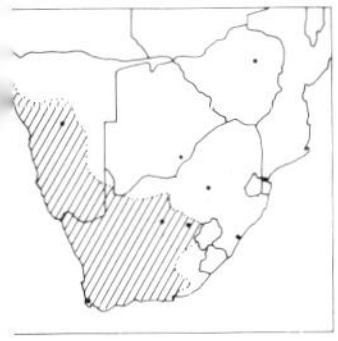

Karoo Sand Snake or Whip Snake
Psammophis notostictus (Pl. 23)

75-90 cm; max. 100 cm
This slender snake has 17 scale rows and eight upper labials (numbers four and five enter the eye). There are two preoculars. Ventrals number 155-183. The anal is undivided. The tail is long (80-107 subcaudals). Coloration may be uniform or striped. The back is light grey to dark brown, with paler flanks; sometimes there is a pale stripe or spots along the backbone and a pale stripe on the side. The edge of the belly has white and grey stripes, the rest being off-white with grey blushes. **Biology and breeding:** This snake is a very fast predator on skinks and lacertids, which it chases during the heat of the day. It is the most common snake in karroid areas; many specimens are killed while crossing roads. A gravid female contained three eggs (28 x 6 mm) in October. **Habitat:** Arid scrubland and karroid regions. **Range:** Cape and S. OFS, through Namibia to S. Angola.

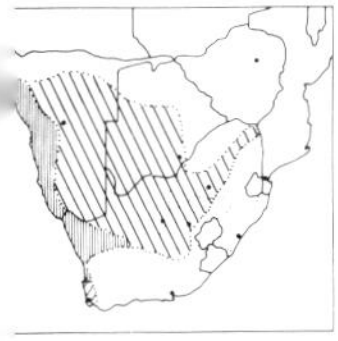

P.l. leightoni
P.l. trinasalis
P.l. namibensis

Cape, Namib and Fork-marked Sand Snakes
Psammophis leightoni (Pl. 23)

80-100 cm; max. 136 cm
Slender snakes with 17 scale rows and eight upper labials (numbers four and five enter the eye). There is one preocular. The posterior nasal is divided, as is the anal. Coloration may be striped or blotched (see Subspecies). **Biology and breeding:** These snakes chase and eat small vertebrates, mainly rodents and lizards, but also take other snakes. A female, gravid in October, contained eight eggs (25 x 9 mm). The hatchlings measure 220-240 mm. **Habitat:** Coastal fynbos, desert and semi-desert, entering savannah. **Range:** Western regions, from Cape Town to N. Transvaal and S. Angola. **Subspecies:** There are three races, all of which occur in the region. The Cape sand snake, *P.l. leightoni*, has the top of the head spotted or barred, 155-161 ventrals, and 92-97 subcaudals; it is found in SW. Cape. The fork-marked sand snake, *P.l. trinasalis*, has the top of the head striped, 150-172 ventrals, and 84-102 subcaudals; it is found in E. Namibia, Botswana, OFS and Transvaal. The Namib sand snake, *P.l. namibensis*, has the top of the head spotted or barred, 167-187 ventrals, and 94-112 subcaudals; it is found in Namaqualand, extending through Namib Desert, to S. Angola.

Jalla's Sand Snake *Psammophis jallae* (Pl. 23)

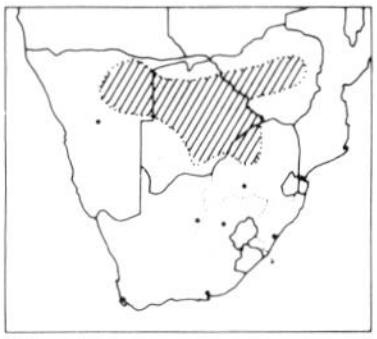

80-90 cm; max. 109 cm
This snake has 15 scale rows, and seven upper labials (numbers three and four enter the eye). There is one preocular. The posterior nasal is divided, as the anal. Ventrals number 154-175. The tail is long (84-112 subcaudals). The back is light grey to olive-brown, usually with a broad black-edged dark ban (sometimes with yellow vertebral spots), bordered by a narrow white-yellow streak. The flanks are buff to red-brown. The belly has a broad olive-yellow central stripe which is sometimes dark-edged, with the outer ventral edges white to cream. **Biology:** A rare snake. The diet consists mainly of skinks an lacertids. **Habitat:** Grassland and savannah woodlands. **Range:** Transvaal, Zimbabwe, Botswana and N. Namibia, north to SE. Angola and W. Zambia.

Stripe-bellied Sand Snake *Psammophis subtaeniatus* (Pl. 22)

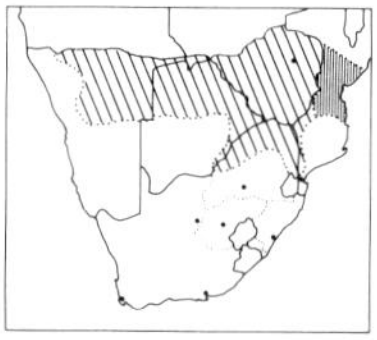

P.s. subtaeniatus
P.s. orientalis

90-110 cm; max. 137 cm
This is a slender snake with 17 scale rows. There is one preocular. The posterior nasal is divided, as is the anal. The back is grey-olive to brown (an paler at the rear) with a broad, black-edged dorsal stripe, flanked by a crear to yellowish stripe, then a dark lateral stripe and a black stripe. The head ha pale, dark-edged blotches. The belly has a bright yellow, black-edged strip bordered at the edge by a white stripe. **Biology and breeding:** This snake readily climbs into low bush, and will eat small birds in addition to lizards an rodents. It is very fast and difficult to catch, although many are killed by bird of prey. The female lays 4-10 eggs (32 x 12 mm). **Habitat:** Open dry savannah, thornveld and bushveld. **Range:** Northern regions from coast to coast. Elsewhere, to E. Africa. **Subspecies:** Two races are recognized, and both occur in the region. *P.s. subtaeniatus* has nine upper labials, and 106-132 subcaudals; it is found over most of the region, except in Mozambique *P.s. orientalis* has eight upper labials, 94-116 subcaudals, and its head markings and pale dorsolateral stripes are faint; it is found in Mozambique.

Dwarf Sand Snake *Psammophis angolensis* (Pl. 22)

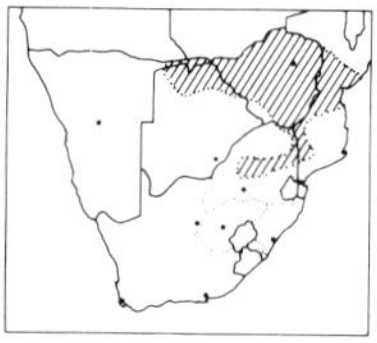

25-35 cm; max. 50 cm
This is a very small, slender snake with only 11 scale rows. It has eight uppe labials (numbers four and five enter the eye). There is one preocular. The posterior nasal is usually undivided, while the anal is divided. Ventrals numb 135-156. The tail is short (57-82 subcaudals). The back is grey to yellow-bronze, with a broad brown-black dorsal stripe, and a faint, broken black stripe on the lower flank. The head is dark brown, with three narrow, pale crossbars and 1-2 dark neck collars. The lips, throat and belly are yellow-white. **Biology and breeding:** This small, secretive snake forages in grass tussocks, under fallen logs, etc., feeding on small lizards and frogs. It never bites, but is difficult to tame. The female lays 3-5 elongate, small eggs (15-18 x 5-6 mm). **Habitat:** Moist and dry wooded savannah. **Range:** N. Transvaal, through Zimbabwe, N. Botswana and Namibia, north t Zaire and Tanzania.

Leopard and Short-snouted Grass Snakes
Psammophis sibilans (Pl. 34)

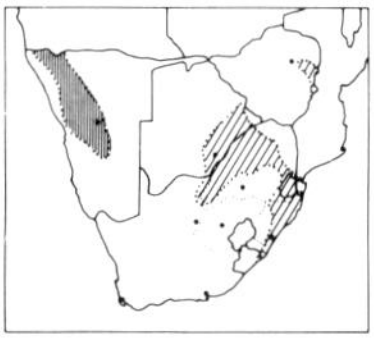

P.s. leopardinus
P.s. brevirostris

90-110 cm; max. 135 cm
These are slender snakes with 17 scale rows, and eight upper labials (numbers four and five enter the eye). There is one preocular. The posterior nasal is usually undivided, while the anal is divided. Ventrals number 146-17 The tail is long (72-109 subcaudals). Coloration is usually striped (see Subspecies). **Biology and breeding:** These alert, active snakes are

frequently seen crossing roads. They eat small vertebrates. They bite readily when first caught. The females lay 4-15 eggs (28 x 15 mm) in summer; they may lay two clutches per season. The young are 190-250 mm long. **Habitat:** Varied; eastern race in highveld and montane grassland, western race in rocky arid savannah. **Range:** Southern races have a patchy distribution on the subcontinent (see Subspecies). Elsewhere, the typical race occurs through central and E. Africa to Morocco and Egypt, and to Senegal in the west. **Subspecies:** Three races are recognized, with two occurring in the region. *P.s. leopardinus* has variable coloration, the back being light red-brown to light olive, usually with a pale buff chain pattern on the forebody and pale yellow stripes on the rear. There are pale bars on the back of the head. The chin and throat have grey blotches. The belly is off-white, sometimes with grey bands. It is found in Namibia and S. Angola. In *P.s. brevirostris*, the back is olive-brown, with a broad dorsal stripe of black-edged scales, flanked by pale dorsolateral stripes. There is sometimes a narrow, broken white stripe down the backbone. The belly is white. This race occurs in the highveld and adjacent regions, extending to Natal coast, and with a relict population in E. Zimbabwe.

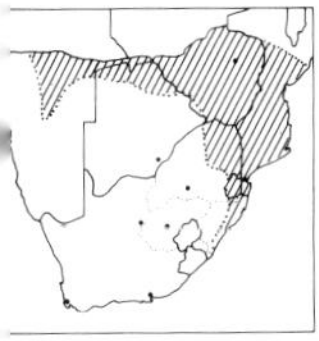

Olive Grass Snake *Psammophis phillipsii* (Pl. 23)

110-130 cm; max. 174 cm
A large, robust snake with 17 scale rows, and eight upper labials (numbers four and five enter the eye). There is one preocular. The posterior nasal is usually undivided, while the anal is divided. The snout is not flattened. Ventrals number 151-183. The tail is long (82-110 subcaudals). The back is olive-brown (paler towards the tail), sometimes with the scales black-edged, forming thin black lines, or with scattered black flecks on the forebody. The belly is white-yellow, sometimes with black streaks. **Biology and breeding:** The olive grass snake is similar to the short-snouted grass snake, *P. sibilans brevirostris* (page 72), but forages more readily in marshy areas. It occasionally eats other snakes, even young black mambas. It should be treated with respect: it bites readily, and its venom may cause nausea and pain; antivenom is ineffective in the treatment of this snakebite. The female lays 10-30 eggs (28-40 x 10-20 mm) in dead leaves, etc., in midsummer. The young measure 275-300 mm. **Habitat:** Moist savannah and low-lying grasslands. **Range:** Northern part of the region, extending south along Natal coast. Elsewhere, to Kenya and Senegal.

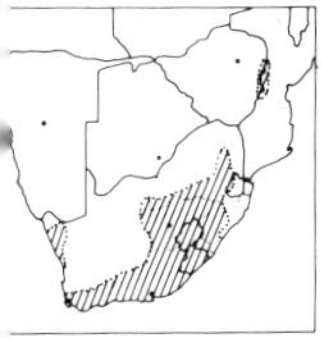

Cross-marked or Montane Grass Snake
Psammophis crucifer (Pls. 23 and 34)

(Endemic) 40-50 cm; max. 70 cm
This small, robust snake has 15 scale rows, and eight upper labials (numbers four and five enter the eye). There is one preocular. The posterior nasal is usually undivided, while the anal is divided. The snout is not flattened. Ventrals number 136-165. The tail is short (61-81 subcaudals). The back is silver-grey to olive-brown, with a broad, dorsal, black-edged, brown stripe; there is a similar stripe with a white lower border on the flank. The head has dark-edged, cream-yellow crossbars. The belly is yellow-orange, often with grey lateral streaks. Occasional specimens are uniform olive-grey with an off-white belly. **Biology and breeding:** This species occurs in the temperate zones, and is active on mountain plateaus and in moist grasslands. It eats mainly lizards, but also takes frogs. It wriggles wildly when caught, but rarely bites, and it tames well. The female lays 5-13 eggs (18-21 x 10 mm, exceptionally up to 32 mm long) in midsummer. The eggs may hatch in 45 days. **Habitat:** Highveld and montane grassland, entering fynbos. **Range:** Cape fold mountains and Natal midlands, on to the highveld of the Transvaal and OFS, with relict populations along the old escarpment in the Cape, Transvaal and E. Zimbabwe.

AFRICAN BURROWING SNAKES
Subfamily
Atractaspidinae

These are unusual African burrowing snakes that show a wide range of fang types and associated glands. Usually back-fanged, some have non-erectile front fangs, while others have long fangs that can rotate forwards. The head small and not distinct from the neck. The eyes are small to minute, with round pupils. The snout is round to very pointed. The head shields are often fused and thus reduced in number. The loreal is always absent. The body is cylindrical, and sometimes very long and thin. The scales are smooth and la apical pits. Many of these snakes are uniform grey to black, but some speci are very brightly coloured. They all have similar skulls, vertebrae and hemipenes. Most are harmless, but a few (those previously placed with the vipers and cobras) may be clinically dangerous.

All live underground, some utilizing cracks or insect and mammal tunnels others pushing through loose sand or leaf litter. Many have specialized diet With one exception, all are oviparous.

The taxonomy is confused; at present, there are 12 genera and about 60- species. They are distributed throughout Africa, with one genus (*Micrelaps*) entering the Near East. Seven genera (2 endemic) occur in subcontinent.

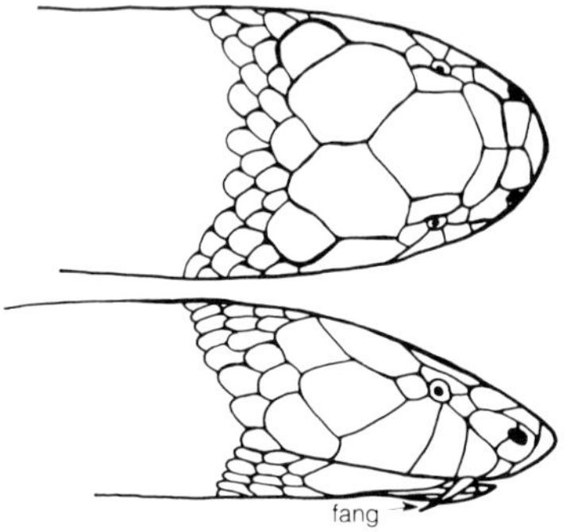

The head scales and fangs of a burrowing asp

BURROWING ASPS
Atractaspis

These unusual snakes have a small head that is not distinct from the neck, a is covered with symmetrical head shields. The eyes are very small, with rour pupils. The body is cylindrical, with smooth, shiny scales in 17-37 rows. The anal is undivided. The tail is short, with paired or single subcaudals. Burrow asps are distinguished by the large, hollow poison fangs that are situated at the front of the mouth on an otherwise toothless maxilla. In colour, they are usually uniform purple-brown to black, with a white, grey or black belly. The are easily confused with some harmless species, eg. purple-glossed snake *Amblyodipsas* (page 77), and wolf snakes, *Lycophidion* (page 60). Previou called 'mole vipers' or 'burrowing adders' because of their erectile front fang these similarities are now known to be due to convergent evolution (similar evolution to meet a similar need). Other common names include 'stilleto snakes' and 'side-stabbing snakes'.

These burrowing snakes may emerge on the surface at night, particularly after rain. They are very common in suitable habitat, feeding on small vertebrates, particularly other burrowing reptiles; baby rodents, sleeping skinks and lacertids, legless lizards and other snakes are all eaten. They ma excavate a chamber, using the side of the head, beneath a sun-warmed stone. They have a peculiar 'aromatic' smell, the function of which is unknow Oviparous, they lay a few (usually 4-7, but up to 11) elongate eggs in leaf litt The neck is unusually flexed on a hard surface, with the nose pointing down vertically. They bite with a peculiar sideways 'stab', which is why they cannc be safely held behind the head. The fangs have limited rotation and are hooked into the prey without the snake fully opening its mouth; usually only a single fang is injected, but multiple bites may be delivered. The venom glan are large, and in some species extend into the neck. Bites from most species are mild, causing only pain and local swelling, but deaths have been caused by three W. and N. African species. Polyvalent antivenom is ineffective in the treatment of their bites.

They are distributed throughout most of subSaharan Africa, with an isolated species in the Jordan Valley. The taxonomy is confused; there are about 12 species, with only three occurring in our region, one of which is endemic.

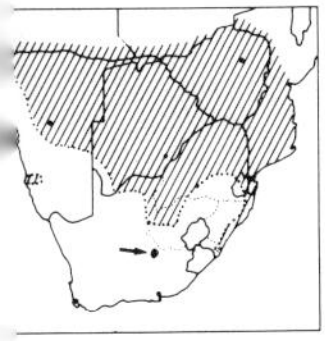

Southern or Bibron's Burrowing Asp
Atractaspis bibronii (Pl. 38)

40-50 cm; max. 63 cm
This short, stocky snake has smooth, close-fitting scales in 21-23 rows. The head shields are symmetrical. It has enlarged, erectile front fangs. Ventrals number 196-260. The anal is undivided. The tail is short (18-28 single subcaudals), with a terminal spine. In colour, the back is uniform purple-brown to black. The belly colour is variable; it is either uniform white to cream, sometimes with scattered dark blotches, or uniform dark grey. **Biology and breeding:** Usually found under cover (in old termitaria, under stones, etc.), these snakes emerge on the surface on warm, wet summer nights. The diet is varied; mainly other burrowing reptiles are taken, but it also eats small mammals and frogs. Irascible and ever-willing to bite, these snakes are best left undisturbed. The glands yield minute amounts of venom (1,3-7,4 mg), which is straw-yellow in colour. The venom causes immediate pain and local swelling; mild neurotoxic symptoms (eg. nausea, dry throat and vertigo) may be present in the early stages, but necrosis is rare, and is usually a result of bad treatment. Bites are common in Zululand and E. Transvaal lowveld, but no fatalities have been recorded. Polyvalent antivenom is ineffective in the treatment of this snakebite. The female lays 3-7 eggs (27-36 x 10-12 mm) in summer. The young measure 150 mm. **Habitat:** Varied; ranging from highveld grassland and semi-desert to coastal bush. **Range:** Throughout the northern regions, to Natal south coast, and with scattered inland records. Elsewhere, north to Kenya.

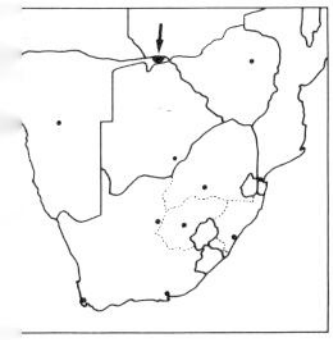

orientalis

Eastern Congo Burrowing Asp *Atractaspis congica*

40-50 cm; max. 53 cm
This short, stocky snake has smooth, close-fitting scales in 19-21 rows. The head shields are symmetrical. It has enlarged, erectile front fangs. Ventrals number 193-208. The anal is undivided. The tail is short (18-25 paired subcaudals), with a terminal spine. In colour, the back and belly are uniform purple-brown to black. **Biology and breeding:** This species' venom has not been studied; there have been no bites reported. Three elongate eggs (62 x 12 mm) have been recorded in a large female. **Habitat:** Moist savannah. **Range:** Caprivi Strip, north to SE. Zaire, E. Angola and Zambia. **Subspecies:** Three races are recognized, with the eastern race (*A.c. orientalis*) just entering the region.

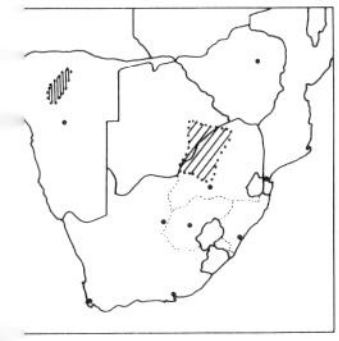

A.d. duerdeni
A. duerdeni subsp. nov.

Duerden's Burrowing Asp *Atractaspis duerdeni* (Pl. 38)

(Endemic) 30-40 cm; max. 48 cm
This short, stocky snake's abrupt, rounded snout has a sharp horizontal angle at its tip. The smooth, close-fitting scales are in 23-25 rows. It has symmetrical head shields, and enlarged, erectile front fangs. Ventrals number 195-228 (see Subspecies). The anal is undivided, and the tail is short (21-25 unpaired subcaudals), with a terminal spine. In colour, the body is uniform blackish-brown or grey above, and uniform white or cream-pink on the belly, extending on to the lower flanks and lips. **Biology:** The behaviour and habits of this snake are poorly known; it appears to be less aggressive than Bibron's

burrowing asp, *A. bibronii* (page 75). **Habitat:** Sandy soil in thornbush savannah. **Range:** Isolated populations in central Namibia and SE. Botswa and adjacent Transvaal. **Subspecies:** Two races are recognized. *A.d. duerdeni* has a ventral count of 195-202 in males, and 209-216 in females; it occurs in SE. Botswana and adjacent Transvaal. A western race (being described at present) has a ventral count of 200-209 in males, and 217-228 in females; it occurs in central Namibia.

CENTIPEDE EATERS *Aparallactus*

These are small, unusual snakes with a small head that is barely distinct fro the neck, and very small eyes with round pupils. There are usually two large fangs (absent in *A. modetus*), which are sometimes grooved, on the maxilla below the eye. The body is cylindrical, with smooth scales in 15 rows. The a is undivided, and the tail is short to moderately long, with unpaired subcaudals

Burrowers, these snakes live in loose soil, old termitaria, under rocks, or i rotting logs. They feed almost exclusively on centipedes, which they grab a chew; the prey quickly succumbs to the venom, and is swallowed head firs Most are oviparous, but the Kenyan *A. jacksonii* gives birth to a few live you

Centipede eaters are distributed throughout subSaharan Africa, reaching Ethiopia in the east and Guinea in the west, but are absent from most of the Cape. There are 11 species, with four in the region, one of which is endemic

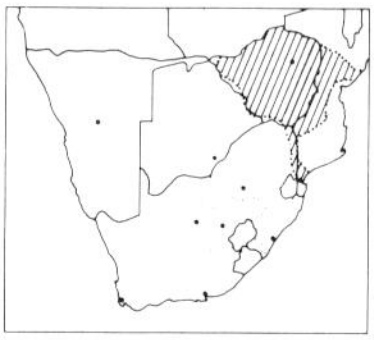

A.l. lunulatus

Reticulated Centipede Eater *Aparallactus lunulatus* (Pl. 26)

30-40 cm; max. 53 cm
This small snake has its first lower labials in contact behind the mental. It ha six lower labials, 144-176 ventrals, and 48-65 subcaudals. The back is pale grey to olive or dark brown, often with the scales dark-edged, giving a reticulated effect; juveniles have a black collar and up to 12 blotches on the forebody that fade in adults. The belly is greenish-white, sometimes suffuse with dark grey. **Biology and breeding:** It is found under logs and stones at low altitudes, and eats scorpions as well as centipedes. The female lays 3-4 elongate eggs (30 x 7 mm) in summer. **Habitat:** Sandy lowveld. **Range:** Zimbabwe and adjacent Mozambique, south into Kruger National Park. Elsewhere, to Sudan and Zaire. **Subspecies:** Two races are recogniz The typical race occurs in the region and throughout most of the range; it is replaced in Somalia by *A.l. scortecci*.

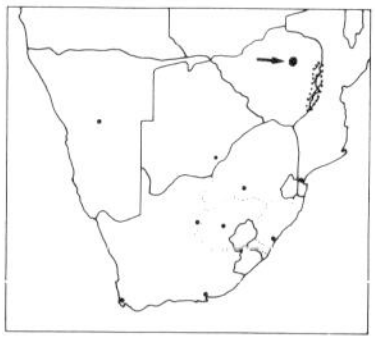

Black Centipede Eater *Aparallactus guentheri* (Pl. 26)

30-40 cm; max. 47 cm
This small, slender snake has its first lower labials separated by the mental. The nasal is divided. There are five lower labials, 150-173 ventrals, and 49- subcaudals. The head and back are blue-grey to black, with two narrow ye collars on the neck; the chin and belly are off-white. **Biology:** This snake is similar in habits and behaviour to the reticulated centipede eater, *A. lunulat* (above). **Habitat:** Areas of high rainfall supporting evergreen forest. **Range:** E. Zimbabwe, north to Kenya.

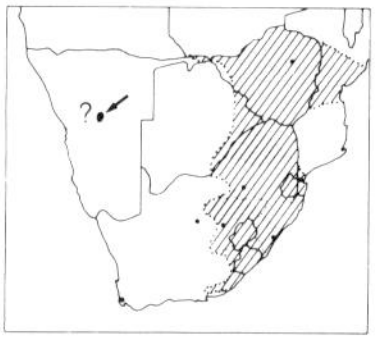

Cape Centipede Eater *Aparallactus capensis* (Pl. 26)

25-30 cm; max. 41 cm
This small, slender snake has its first lower labials separated by the mental. Each nostril pierces the undivided nasal. There are five lower labials, 126-1 ventrals, and 29-63 subcaudals. The back is red-brown to grey-buff, with a black collar and a brownish head. The belly is grey-white. **Biology and breeding:** They are common in old termitaria, which offer shelter, warmth a food. When caught, they struggle wildly and may attempt to bite, but their minute teeth are harmless. They lay 2-4 very elongate eggs (32 x 4-5 mm). young are 90-120 mm long. **Habitat:** Varied; including highveld and monta

grassland, savannah and coastal bush. **Range:** Eastern regions, from Port Elizabeth to Zimbabwe, N. Botswana and Caprivi Strip. There is an old record from central Namibia. It is absent from S. Mozambique plain.

Mozambique Centipede Eater *Aparallactus nigriceps*

(Endemic) 20-30 cm
This small snake is similar to the Cape centipede eater, *A. capensis* (page 76), but it has fewer ventrals (108-123) and subcaudals (20-35). Its coloration is also similar to that of the Cape centipede eater, except that the nape collar is twice as broad. **Biology and breeding:** Nothing is known about this rare and localized snake, but it is presumed to have similar habits and behaviour to the Cape centipede eater. **Habitat:** Coastal bush. **Range:** Area around Inhambane in S. Mozambique.

TAL BLACK
AKE
crelaps

This unusual genus is endemic to South Africa, and contains a single species.

Natal Black Snake *Macrelaps microlepidotus* (Pl. 37)

(Endemic) 70-90 cm; max. 108 cm
A thick-bodied snake with a blunt head that is not distinct from the neck. The minute eyes have round pupils. There is no loreal or preocular. The prefrontal enters the eye. The maxillary is short, with four teeth and two very large, grooved fangs under the eye. The body is cylindrical, with smooth scales (but feebly keeled in the anal region) in 25-27 rows. Ventrals number 158-172. The anal is undivided, and the tail is short (35-50 single subcaudals). This snake may be confused with the purple-glossed snakes, *Amblyodipsas* (below), which differ by having paired subcaudals, and the burrowing asps, *Atractaspis* (page 74), which can be distinguished by their erectile front fangs and preocular. In colour, the back is ashy-black to jet-black, and sometimes paler on the belly. **Biology and breeding:** This snake burrows in moist leaf litter and humic soil, coming to the surface on warm, damp nights. It feeds on frogs (*Breviceps*), small mammals, legless lizards (particularly the giant legless skink) and other snakes. The prey is seized, restrained in the body coils and chewed to introduce venom. The Natal black snake is a good swimmer. Docile, it rarely bites. The venom is poorly known; bites are reported to cause loss of consciousness and collapse for up to 30 minutes, but are very rare. Antivenom is ineffective in treating this snakebite. The female lays 3-10 large eggs (38-56 x 23-31 mm) in December-January in leaf litter. The young, which hatch in March, measure 200-290 mm. **Habitat:** Coastal bush. **Range:** Eastern coastal regions from East London to Zululand.

RPLE-GLOSSED
AKES
blyodipsas

These burrowing snakes are widely distributed through subSaharan Africa. They are very similar in appearance to the Natal black snake, *Macrelaps microlepidotus* (above), and the burrowing asps, *Atractaspis* (page 74). The head is small and not distinct from the neck. The eyes are very small, with round pupils. There is no loreal, preocular or anterior temporal. The maxillary is very short, with 3-5 teeth and two large, grooved fangs under the eye. The body is cylindrical and solid, with smooth scales in 15-21 rows. The anal is divided, and the tail is very short, with paired subcaudals.

Purple-glossed snakes are found in various habitats, from coastal bush and moist forest to dry savannah, and in the Kalahari. They burrow in loose soil and feed on other burrowing reptiles, small mammals and amphibians. Prey is subdued by venom while it is held in the body coils (this may take up to four hours). Although rarely seen, they may surface on warm, damp nights. Docile, they rarely bite. They are usually oviparous, the exception being *A. concolor*. Males are often much smaller than females, and have longer tails.

There are nine species, with four in the region, two of which are endemic.

Natal Purple-glossed Snake *Amblyodipsas concolor* (Pl. 38)

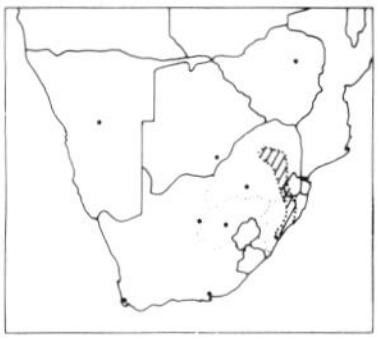

(Endemic) 40-50 cm; max. 75 cm
A thick-bodied snake with a blunt snout. It has seven upper labials, the fifth which is the largest. There are two pairs of chin shields. The scales are smooth, in 17 rows. There are 133-157 ventrals. The tail is short (28-39 subcaudals), tapering to a point. The back is uniform glossy dark brown to black with a purple sheen; the belly is paler. **Biology and breeding:** This ra species burrows in humic soil in forested areas; when hungry, it may lie just below the surface with its head partially exposed. Adults readily eat other snakes, while juveniles take small lizards. Very docile, they rarely bite. Ther some confusion about this snake's breeding habits: 10 hatchlings and two eggs (30 x 18 mm) were found in February in moist soil (in Natal); 11 eggs (26-32 x 15-17 mm) were laid by a female in December (in Transvaal); and female (also in Transvaal) gave birth to 12 babies in March. **Habitat:** Moist forested areas. **Range:** Natal lowlands to NE. Transvaal.

Common Purple-glossed Snake
Amblyodipsas polylepis (Pl. 38)

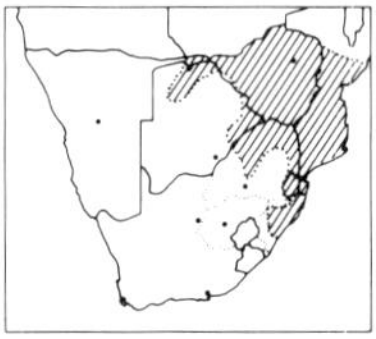

A.p. polylepis

50-80 cm; max. 111 cm
This stocky snake has a blunt snout, and six upper labials, the fifth of which the largest. There are seven lower labials, the first four touching the single p of chin shields. The internasals are not fused with the prefrontals. The scale are smooth, in 19-21 rows. Ventrals number 154-215 (more than 180 in mal less than 185 in females), and subcaudals 15-31 (more than 24 in females, less than 25 in males). The back and belly are glossy dark brown to black, w a purple sheen. **Biology:** It feeds on burrowing reptiles, including blind snakes; it captures sleeping lizards in their burrows at night. It is docile, rar biting. **Habitat:** Savannah, entering dry forest. **Range:** Natal, through Transvaal, Caprivi Strip, N. Botswana, Zimbabwe and Mozambique. Elsewhe to Angola, Zaire and Zambia. **Subspecies:** Two races are recognized, with or the typical race occurring in the region. It is replaced in coastal Kenya and Tanzania by *A.p. hildebrandtii*, which has only 17-19 scale rows.

Eastern Purple-glossed or White-lipped Snake
Amblyodipsas microphthalma (Pl. 24)

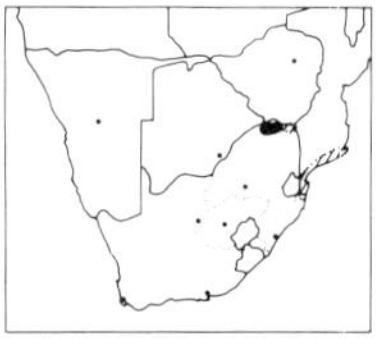

A.m. microphthalma
A.m. nigra

(Endemic) 25-30 cm; max. 35 cm
This is a small snake with a bluntish snout. There are five upper labials, the fourth of which is the largest, and six lower labials, the first four touching the single pair of chin shields. The internasals are fused with the prefrontals. Th scales are smooth, in 15 rows. The back is uniform dark brown to black, wit purple sheen. The upper lip, throat and belly are white or yellow, the belly either having a dark median stripe, or being uniform black. **Biology:** The typical race burrows in deep sandy soil, while *A.m. nigra* hides under stone rocky terrain. Legless skinks are eaten. **Habitat:** Deep alluvial soil or rocky thornveld. **Range:** St Lucia in Natal and adjacent Mozambique Plain, to NE Transvaal. **Subspecies:** Two races are recognized, both occurring in the region. *A.m. microphthalma* has a white-yellow belly, and 120-153 ventrals is found in Natal, Mozambique and N. Kruger National Park. *A.m. nigra* has black belly, and 146-168 ventrals; it is found in NE. Transvaal.

Kalahari Purple-glossed Snake
Amblyodipsas ventrimaculata (Pl. 24)

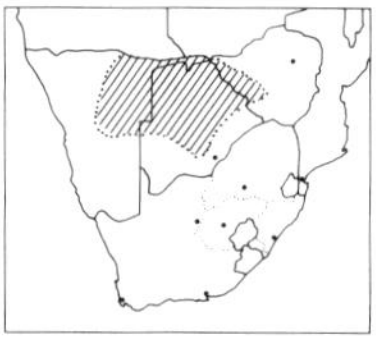

30-35 cm; max. 48 cm
This small snake has a bluntish snout. There are five upper labials, the fourt which is the largest, and five lower labials, the first three touching the single

pair of chin shields. The internasals are not fused with the prefrontals. The scales are smooth, in 15 rows. Ventrals number 172-205. The tail is short (18-29 subcaudals). The back has a broad purple-brown to black dorsal stripe; the scales may be tipped with yellow. The flanks, upper lip and subcaudals are yellow. The belly is uniform white, rarely with scattered black blotches. **Biology and breeding:** Found in sandy soil in moist regions of the Kalahari, this snake feeds on burrowing reptiles, including amphisbaenids (*Zygaspis*), legless skinks (*Typhlacontias gracilis*) and garter snakes. Three eggs have been found in a female. **Habitat:** Kalahari sand. **Range:** W. Botswana and adjacent Namibia and NW. Zimbabwe, north to Zambia.

ACK AND
LLOW
RROWING
AKES
ilorhinophis

These very small, elongate snakes are found in E. and central Africa. The head is small, not distinct from the neck, and has a rounded snout. The eyes are small, with round (sometimes elliptical) pupils. There is no internasal or loreal. The prefrontals are very large, covering the snout. Each nostril pierces a single nasal. The maxillary is very short, with 3-4 teeth and two large, grooved fangs. The body is slender and cylindrical, with smooth scales in15 rows. The anal is divided, and the tail is short, with paired subcaudals.

All these snakes are brightly coloured, with the tail the same shape and colour as the head. In defense, they draw themselves into a loose coil, hide the head and raise the tail, waving it slowly to deflect danger away from the vital organs. They are oviparous.

There are three species, one of which just enters the region.

gerardi

Gerard's Black and Yellow Burrowing Snake

Chilorhinophis gerardi (Pl. 24)

30-40 cm; max. 51 cm

This is a very slender, small snake with a rounded snout. The eyes have round pupils. A small preocular is present. There are four upper labials, the third entering the eye, and five lower labials. Ventrals number 244-294, and subcaudals 19-31. The back is chrome to pale greenish-yellow, with three black stripes. The head and tail tip are black, the head with pink-yellow blotches. The belly is white. The anterior of the tail is orange-yellow, followed by a black band, with the rest being light blue. **Biology and breeding:** This snake burrows in loose soil, feeding on small amphisbaenids and snakes. The tail display is a dramatic bluff, and it may even give a mock 'strike'. The female lays six elongate eggs (30 x 6 mm) in summer. **Habitat:** Savannah. **Range:** N. Zimbabwe, north to S. Zaire. **Subspecies:** Two races are recognized. The typical race occurs in the region, and is replaced by *C.g. tanganyikae* in N. Zambia, SE. Zaire and W. Tanzania.

RLEQUIN
AKES
moroselaps

These small, brightly coloured snakes have the head barely distinct from the neck, and small eyes with round pupils. Each nostril pierces a single nasal. The maxillary is extended, bearing two large, hollow front fangs. The body is cylindrical, with smooth scales in 15 rows. The anal is divided. The tail is short, with paired subcaudals.

These snakes burrow in loose soil and forage underground in tunnels and cracks. They are often exposed in old termitaria or under stones. They feed on other burrowing reptiles, which are seized, and killed by venom. They are oviparous. Docile, they rarely bite.

The taxonomic relationships of these snakes has long been confused. They were previously called dwarf garter snakes (*Elaps*) and were placed with cobras and mambas in the Elapidae because they have fixed front fangs. They are now known to be related to the burrowing asps, *Atractaspis* (page 74), and their scientific name was changed to avoid having to change the family name of cobras and mambas (Elapidae).

There are two species, and both are endemic to South Africa.

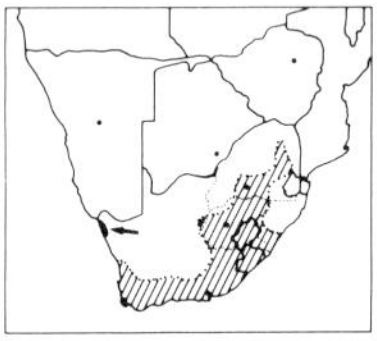

Spotted Harlequin Snake *Homoroselaps lacteus* (Pl. 19)

(Endemic) 30-40 cm; max. 55 cm
This small, slender snake has six upper and six lower labials, 160-209 ventrals, and 24-43 subcaudals. Varied colour phases occur: either the bac is yellowish-white with numerous broken black bands, and usually with an orange-yellow vertebral stripe; or it is black with irregular yellow-white crossbars and a vertebral series of red to orange spots; or it is black, each scale with a yellow dot, and with a bright orange-yellow vertebral streak.

Biology and breeding: This snake is common in certain regions, living ir old termite nests, under stones, etc. It is known to eat legless skinks, blind snakes and other snakes, but rarely feeds in captivity. Its venom has not be studied. Bites are very rare because of its small gape and reluctance to bit Two known cases suffered mild to severe swelling of the limb with mild haemorrhage, swollen lymph vessels and painful glands that resolved in 3- days; the only neurological symptom was a persistent headache that lastec 24 hours. Polyvalent antivenom is ineffective in the treatment of this snakeb Up to six eggs are laid in December.

Habitat: Varied; semi-desert to savannah and coastal bush.
Range: Southern part of the region, from Namaqualand to Transvaal.

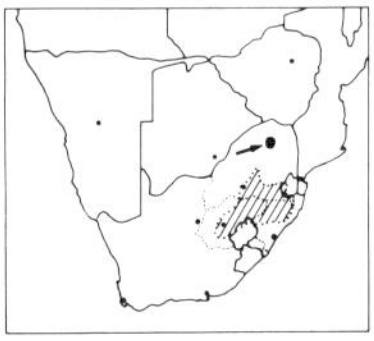

Striped Harlequin Snake *Homoroselaps dorsalis* (Pl. 24)

(Endemic) 20-30 cm; max. 31 cm
This small, very slender snake has six upper and five lower labials, 210-23 ventrals, and 22-33 subcaudals. The body is black above, with a conspicu yellow stripe along the backbone. The lips, belly and lower flanks are yellow-white. **Biology:** Very little is known about this very rare, minute snak is probably similar to the spotted harlequin snake, *H. lacteus* (above). Its venom has not been studied as yields are minute. There have been no recorded bites by this species. **Habitat:** Grassland. **Range:** Highveld, extending to Natal midlands.

QUILL-SNOUTED SNAKES
Xenocalamus

These unusual African snakes have a small, elongate (quill-shaped) head which is not or is hardly distinct from the neck. The snout is almost hooked, with an underslung mouth. The minute eyes have round pupils. The elonga 'preocular' is actually a displaced prefrontal. The maxillary is short, with 4-6 teeth and two large fangs under the eye. The body is cylindrical, with smoo scales in 17 rows. The anal is divided. The tail is short and blunt, with paire subcaudals.

The very thin body, pointed snout and underslung mouth of these snake are adaptations for burrowing in sandy soils. They feed exclusively on amphisbaenians. They are slow-moving, and struggle when captured, but never bite.

There are five species, which are distributed through central and southe Africa, with four in the region, two of which are endemic.

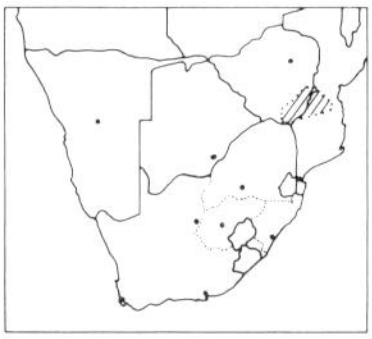

Sabi Quill-snouted Snake *Xenocalamus sabiensis*

(Endemic) 35-40 cm; max. 50 cm
This small, slender snake has its snout depressed and prominent, but not hooked. There are 5-6 upper labials and six lower labials. The nasal is usua divided. Ventrals number 187-218, and subcaudals 22-33. In colour, the b is uniform black, the lower flanks chrome-yellow, and the belly off-white (sometimes with scattered brown blotches).

Biology and breeding: These snakes are rarely seen unless they are forced to the surface by rain. The female lays approximately three large, elongate eggs. **Habitat:** Alluvial sands. **Range:** SE. Zimbabwe and adjace Mozambique.

Transvaal Quill-snouted Snake
Xenocalamus transvaalensis **(Pl. 29)**

(Endemic) 30-40 cm; max. 41 cm
This small, slender snake has its snout depressed and prominent. There are five upper and five lower labials. Each nostril pierces a single large nasal. There are 183-192 ventrals and 23-32 subcaudals. The back is black, sometimes with pale-centred scales, giving a speckled appearance. The belly and lower flanks are white-yellow (sometimes black-blotched). **Breeding:** The female lays two eggs (28 x 6 mm). (SA RDB, Rare.) **Habitat:** Kalahari and alluvial sand. **Range:** N. Zululand and adjacent S. Mozambique, and N. Transvaal.

X.b. bicolor
X.b. lineatus
X.b. australis

Bicoloured Quill-snouted Snake
Xenocalamus bicolor **(Pls. 18 and 24)**

45-55 cm; max. 71 cm
A slender snake with a prominent, depressed and hooked snout. It has 5-6 upper labials and five lower labials. The nasal is divided. There are 186-250 ventrals and 20-37 subcaudals. Coloration is varied, with four basic colour phases. The striped phase is black to purple-brown above, with the lower flanks off-white to chrome-yellow and the belly white (sometimes with dark blotches). The spotted phase is yellow above, with two rows of purple-brown spots, and a white belly. The reticulated phase has the back brown or grey, with each scale pale-edged, and the lower flanks and belly white. The melanistic phase is uniform black above and below. **Biology and breeding:** These snakes live in deep sand. The female lays 3-4 elongate eggs (40-47 x 15 mm) in December. The young are 200 mm long. **Habitat:** Kalahari and alluvial sands. **Range:** Northern parts of the region, reaching N. Zululand, and with an isolated population in OFS. Elsewhere, north to Angola and Zaire. **Subspecies:** There are four races, with three occurring in the region. *X.b. bicolor* (the bicoloured quill-snouted snake) occurs in all colour phases, and has six upper labials and high ventral counts; it is found over most of the range. *X.b. lineatus* (the striped quill-snouted snake) is striped, and has six upper labials, a very strongly compressed head, and high ventral counts; it is found in SE. Zimbabwe, south to N. Zululand. *X.b. australis* (the Waterberg quill-snouted snake) is striped, and has five upper labials and low ventral counts; it is found in the Waterberg range in Transvaal. Extralimitally, *X.b. machadoi* occurs in Angola and Zaire.

. inornatus

Elongate Quill-snouted Snake ***Xenocalamus mechowii*** **(Pl. 18)**

50-60 cm; max. 84 cm
This very elongate snake has a prominent, depressed snout. There are six upper labials and five lower labials. The nasal is divided. The frontal enters the orbit. The ventral count is 247-296, and subcaudals number 22-32. The back is light purple-brown to lemon-yellow, with two irregular rows of dark blotches. The belly is white to dirty yellow, and sometimes faintly blotched. Specimens that are uniform black are rarely found. **Biology and breeding:** This secretive species burrows for amphisbaenians in regions of Kalahari sand. The female lays up to four large eggs. **Habitat:** Kalahari sand. **Range:** Northern regions, from NE. Namibia to NW. Zimbabwe. Elsewhere, to N. Angola and Zaire. **Subspecies:** Two races are recognized, with only *X.m. inornatus* occurring in the region. The typical race is found in N. Angola and SW. Zaire.

LUBRINE
AKES
bfamily Colubrinae

This subfamily contains the common snakes of the northern hemisphere, including the racers, rat snakes, king snakes and their allies. Although relatively poorly represented in Africa, it does include a number of clinically important species. They are distinguished by an asymmetrical hemipenis, with a simple sulcus. Back fangs have developed independently in a number of lineages, while in one group (the egg eaters) the teeth are greatly reduced. Included here are a number of groups sometimes placed in separate tribes even elevated to distinct subfamilies (eg. boigines, philothamnines, etc).

They are distributed worldwide, but occur particularly in the northern hemisphere. There are 70-80 genera with over 400 species; eight genera wi 16 species occur on the subcontinent. Three species are endemic.

AFRICAN SMOOTH SNAKES
Meizodon

This is an endemic African genus (extending just into Arabia) of small, secretive snakes, related to European smooth snakes and rat snakes. They are oviparous. There are three species, with one entering the subcontinent.

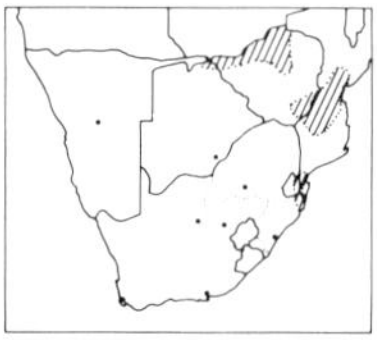

Semiornate Snake *Meizodon semiornatus* (Pl. 26)

40-50 cm; max. 66 cm
A small, slender snake with a flat head, and eyes with round pupils. The sca are smooth, in 21 rows at midbody. The tail is moderately long. The back is grey to olive-brown, with irregular black crossbars on the front of the body. T head is black, with the eye partially ringed with white. The throat is white, an the belly is grey. **Biology and breeding:** This shy, diurnal snake forages in thick vegetation along river courses for skinks, day geckos and small frogs. rarely bites, and is harmless to man. (SA RDB, Peripheral.) The female lays a few (2-3) large, elongate eggs (35 x 10 mm). **Habitat:** Arid and mesic savannah. **Range:** N. Zululand and Swaziland, through Mozambique Plain and Zimbabwe to E. Africa and Yemen.

GREEN AND BUSH SNAKES
Philothamnus

These agile, diurnal African snakes have very slender, usually green bodies an obvious head, and large eyes with round pupils. The body scales are smooth and usually in 15 rows at midbody. Harmless to man, they lack fang and venom glands, but are often mistaken for green mambas or boomslang and are needlessly killed.

Most species are terrestrial, but they readily climb low vegetation. Inhabit damp areas, they feed on small vertebrates, particularly amphibians, which they grab but do not constrict.

There are 18 species distributed throughout subSaharan Africa, with five the subcontinent, one of which is endemic.

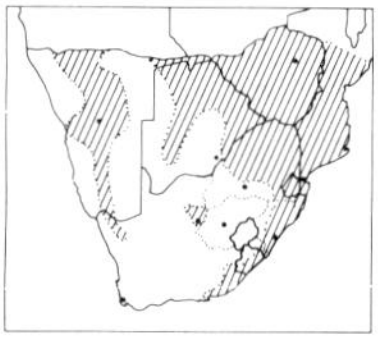

Spotted Bush Snake *Philothamnus semivariegatus* (Pl. 30)

70-100 cm; max. 126 cm
A very slender snake with a flat, distinct head and a long tail. There are two pairs of temporals on each side of the head, and three upper labials enter each eye. The ventrals and subcaudals are strongly keeled and laterally notched. The body is bright green to olive, usually with dark spots and bars the forebody (these are sometimes absent in specimens from E. Transvaal), becoming grey-bronze towards the rear. The head is green or blue-green, a the eyes have golden irises. The belly is greenish-white to yellowish. **Biolog and breeding:** This beautiful, graceful snake hunts among shrubs and bush on rocky ridges or along river courses for geckos, chamaeleons and tree frogs. An expert and speedy climber, it is difficult to detect in foliage. Its visi is excellent, and prey and danger are quickly spotted. It is often seen with th head and neck undulating sideways while the body remains motionless. In threat display, the neck is inflated, revealing the bright blue skin, and it may strike. It takes time to settle in captivity. The female lays 3-12 elongate eggs (28-41 x 8-12 mm) in midsummer. Hatchlings measure 230-260 mm.
Habitat: Open forest or savannah, extending into arid regions.
Range: Northern regions, extending into Kalahari and N. Cape, reaching Namaqualand in the west and Port Elizabeth in the east. Elsewhere, through central Africa to Sudan and Guinea.

Ornate Green Snake *Philothamnus ornatus* (Pl. 25)

50-60 cm; max. 78 cm
A slender snake with a rounded head, two temporals on each side of the head, and three labials entering each eye. It has a longish tail (86-100 smooth subcaudals). The back is emerald to olive-green, with a yellow-edged, red-brown dorsal stripe, and a white to bronze-cream belly. **Biology:** This snake lives in reed beds alongside streams, and eats small amphibians. It is shy and rarely bites. **Habitat:** Mesic savannah. **Range:** It has a patchy distribution; E. Zimbabwe and Okavango Swamp, and elsewhere to Angola and Lake Malawi.

Western Green Snake *Philothamnus angolensis* (Pl. 30)

80-100 cm; max. 116 cm
A slender snake with two temporals on each side of the head, and three labials entering each eye. The tail is long (87-120 smooth subcaudals). The body is uniform bright emerald to olive-green, often with scattered blue-white spots. The skin between the scales is black, and the belly is pale green to yellowish-green. **Biology and breeding:** This active snake climbs into reed beds along river courses and vleis and feeds on small reed birds and lizards, as well as frogs. In threat display, it inflates the throat, showing the black skin (similar to the boomslang), and bites readily. (SA RDB, Peripheral.) The female lays 5-8 (max. 16) eggs in summer (December-February); these vary in size (25-43 x 9-18 mm). They hatch in about two months. The young measure 220-260 mm. **Habitat:** Mesic savannah. **Range:** It has a patchy distribution; N. Zululand and adjacent Mozambique, E. and central Zimbabwe, Okavango Swamp and Caprivi Strip, and central Namibia. Elsewhere, north to Cameroon.

Green Water Snake *Philothamnus hoplogaster* (Pl. 30)

50-70 cm; max. 96 cm
This small, slender species has two temporals on each side of the head, and two labials entering each eye. There are 73-106 smooth subcaudals. The body is uniform bright emerald to blue-green, sometimes with black bars on the forebody. The skin between the scales is black. The belly is bluish-white to yellow. **Biology and breeding:** This alert, active snake hunts in reed beds alongside vleis and streams. It is an expert swimmer. The diet consists mainly of small frogs and fish, but lizards may also be taken. The young also eat grasshoppers. The neck is not inflated in threat display. The female lays 3-8 elongate eggs (25-34 x 8-12 mm) in early summer. The young measure 150-200 mm, and are darker than the adults. **Habitat:** Varied; coastal bush, fynbos, and arid and mesic savannah. **Range:** E. Cape coast, through Natal, Transvaal and S. Mozambique to Zimbabwe. Elsewhere, to Kenya.

P.n. natalensis
P.n. occidentalis

Natal Green Snake *Philothamnus natalensis* (Pl. 30)

(Endemic) 70-90 cm; max. 112 cm
This slender snake may easily be confused with the green water snake, *P. hoplogaster* (above). It differs in being larger, having keeled ventrals and subcaudals (see Subspecies) and two pairs of temporals on each side of the head. The body is uniform bright green above (juveniles sometimes have dark bars on the forebody), and pale green below. **Biology and breeding:** This snake is similar in habits and behaviour to the green water snake.
Habitat: Varied; wet montane and dry forest, and miombo woodland.
Range: S. Cape (to Riversdale), through Natal and Transvaal to Zimbabwe and S. Mozambique. **Subspecies:** Two races are recognized. *P.n. natalensis* has keeled subcaudals and nine upper labials; it occurs in N. Zululand to S. Mozambique. *P.n. occidentalis* has smooth subcaudals and eight upper labials; it occurs over the rest of the range.

EGG EATERS
Dasypeltis

These unusual African snakes have numerous adaptations for feeding exclusively on birds' eggs. The mouth is almost toothless, with minute teeth embedded in the thick gums, and the skin of the neck and lower jaw is very elastic. The head is small, the moderately sized eyes have vertical pupils, a there is no loreal scale. The body scales are strongly keeled, particularly th lateral rows.

Egg eaters are capable of swallowing eggs with a diameter three times th of their head. The egg is taken into the throat and moved back and forth. Special 'gular' teeth (which are actually projections into the gullet from the backbone) saw through the shell. The liquid content is then swallowed, and the collapsed shell discarded. Only fresh eggs are eaten. As a defence against predators (jackals, mongooses, etc.), they mimic venomous adders they have similar colour patterns, and can 'hiss' by forming a nested horseshoe shape and rubbing their serrated lateral scales together. They strike readily, with the mouth agape; while this is effective defence, these snakes are quite harmless. They are oviparous. Nocturnal and mainly terrestrial, they will readily climb trees and rock faces in search of birds' nes

There are six species, with three occurring on the subcontinent, one of which is endemic.

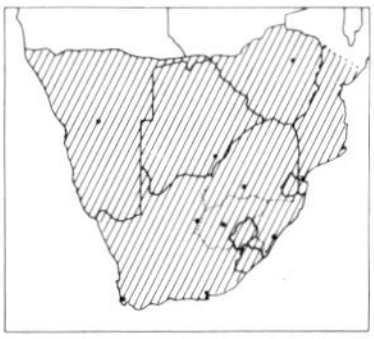

Common or Rhombic Egg Eater *Dasypeltis scabra* (Pl. 15)

50-70 cm; max. 116 cm
A slender, solid snake with a small, rounded head. The lateral body scales are small and have serrated keels. The tail is short (38-78 subcaudals); males hav longer tails. The back is slate-grey, light brown or olive-brown, with a median series of dark, squarish blotches flanked by narrow, dark bars. The top of the head has two narrow, V-shaped marks followed by a similar prominent mark o the nape. Lining of the mouth is black. Belly is white, and sometimes flecked. T occasional red-brown, patternless specimen may be confused with the southe brown egg eater (below). **Biology and breeding:** This snake is very common, but is rarely seen. It feeds gluttonously in spring and summer, laying down fat the winter fast. Colour patterns vary to match the local soil colour and mimic venomous species; the brown-red specimens in the west mimic the horned adder (page 100), and the grey-black specimens along the east coast mimic t night adder (page 97). It lays 6-25 eggs (36 x 18 mm) in summer; these take 8 90 days to hatch. The young measure 210-240 mm. In captivity, two clutches may be laid per season. **Habitat:** Absent only from true desert and closed-canopy forest. **Range:** Throughout the subcontinent. Elsewhere, to Sudan in th north and Gambia in the west.

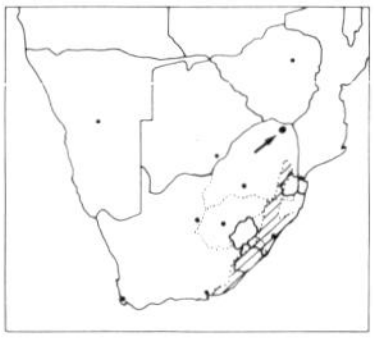

Southern Brown Egg Eater *Dasypeltis inornata* (Pl. 28)

(Endemic) 60-80 cm; max. 97 cm
This species is similar to the common egg eater, *D. scabra* (above), but has longer tail (69-92 subcaudals) and slightly smaller, unserrated lateral scale Its uniform coloration is reddish-brown, and sometimes yellow-olive to dark brown, with dark skin between the scales. The belly is white to dirty yellow. **Biology and breeding:** This snake is similar in behaviour and habits to the common egg eater. The female lays 8-16 eggs (33 x 20 mm) in summer. Th egg surface is pimpled. The young measure 230-280 mm. **Habitat:** Open coastal woodland. **Range:** Alexandria in E. Cape, through Natal and S. Zululand, to SE. Transvaal and Swaziland.

East African Egg Eater *Dasypeltis medici* (Pl. 15)

50-60 cm; max. 90 cm
This species is similar to the common egg eater, *D. scabra* (above), but has longer, more slender tail (71-109 subcaudals), and reduced and serrated

lateral scales. The back is pinkish to red-brown, sometimes with a darker brown vertebral stripe interrupted by white patches. The flanks have narrow vertical bars. The head and neck have about five faint V-shaped marks. The lining of the mouth is pink. The belly is cream with brown stippling. **Biology and breeding:** This snake's habits and behaviour are poorly known. (SA RDB, Peripheral.) A female laid six eggs (24 x 8 mm). **Habitat:** Lowland evergreen forest. **Range:** N. Zululand (Sodwana), through Mozambique Plain and eastern escarpment of Zimbabwe, to Kenya. **Subspecies:** Two races are recognized, with only the typical race occurring on the subcontinent. *D.m. lamuensis* has an unpatterned body and fewer ventrals, and occurs from Somalia to Kenya.

ı. medici

ERALD SNAKES
otaphopeltis

These nocturnal, terrestrial African snakes have a short, flattened head and big eyes with vertical pupils. The scales are mainly smooth and in 17-21 rows. They lack large venom glands.

Herald snakes inhabit marshy areas, feeding on amphibians. They are oviparous.

There are six species, with two occurring on the subcontinent.

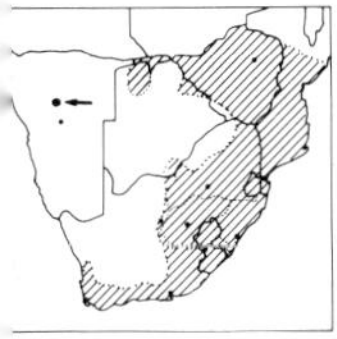

Herald or Red-lipped Snake
Crotaphopeltis hotamboeia (Pl. 33)

60-75 cm; max. 81 cm
A small snake with a broad, obvious head and a short tail. The scales are in 19 rows at midbody and are dull, but the head is iridescent when the skin is freshly shed. The back is olive to green-black above, sometimes with white dots. The head is iridescent blue-black, and the upper lip is orange-red (but white or blackish in the north). The belly is uniform white.

Biology and breeding: The presence of this snake in South Africa was first noted in the Eastern Province Herald newspaper, hence its common name. Living in marshy areas, it feeds at night on amphibians; the prey is grabbed and held until it has been immobilized by the venom. The Herald snake has large, blade-like back fangs. It is belligerent, and when threatened, it flattens its head and flares its lips, which results in a viper-like appearance. It bites readily, and the wound bleeds freely, but no toxic symptoms have been recorded in a bite by this species. It tames well. The female lays 6-12 eggs (26-32 x 10-13 mm) in leaf litter in early summer; these hatch in 61-64 days. The young measure 130-180 mm. **Habitat:** Savannah and open woodland. **Range:** Eastern half of the region, from SW. Cape to Zimbabwe. Elsewhere, to tropical Africa.

Barotse Water Snake *Crotaphopeltis barotseensis* (Pl. 33)

40-50 cm; max. 62 cm
This snake looks similar to the Herald snake, *C. hotamboeia* (above), but has a more elongate body and smooth, glossy body scales in 17 rows. The back is uniform light grey-brown, with dark-edged scales. The head is not dark. The belly is pale brown. **Biology and breeding:** This aquatic species feeds on frogs. It has a gentle disposition and rarely bites. The female lays 6-8 eggs in February. **Habitat:** Papyrus swamp. **Range:** Okavango Swamp, along Chobe River to upper Zambezi River.

GER SNAKES
lescopus

These nocturnal, very slender snakes are found throughout Africa, extending into Europe and the near East. Elsewhere, they are called cat snakes because of their large eyes and vertical pupils.

They are usually found in dry, rocky regions or open savannah. The diet consists mainly of birds, lizards and occasionally bats.

There are six species in the genus, with two in southern Africa, one of which is endemic.

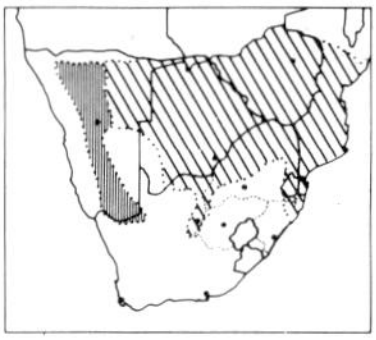

T.s. semiannulatus
T.s. polystictus

Eastern Tiger Snake *Telescopus semiannulatus* (Pl. 19)

55-80 cm; max. 105 cm
A thin-bodied snake with a distinct head and large eyes with vertical pupils. The scales are smooth, in 19 rows at midbody. The anal is divided. The hea uniform orange, with a dark nape band and brown-orange eyes. The body i orange-pink to dull salmon, with 22-50 dark blotches that are larger on the forebody. The belly is uniform yellowish to orange-pink. **Biology and breeding:** Mainly terrestrial, these snakes regularly climb dead trees, old thatched huts, etc. They are slow-moving and unpredictable. When first encountered, they will strike readily and often. In captivity, they become calmer, but may still bite. Small roosting birds, bats, lizards and rodents are eaten. The venom is mild and innocuous to man. The female lays 6-20 elongate eggs (30 x 16 mm) in moist leaf litter in summer; these hatch in 71- days. The hatchlings measure 170-230 mm. In captivity, females may retair sperm and lay eggs every two months during summer. **Habitat:** Savannah a sandveld. **Range:** Northern bushveld (reaching Pretoria), eastern lowveld (reaching mid-Natal) and extending through Kalahari bushveld into Namibi and N. Cape. Elsewhere, to Kenya and Zaire. **Subspecies:** Two races occu *T.s. semiannulatus* has 20-50 dark blotches on the back; it is found in Kalah and eastern regions. *T.s. polystictus* has 52-75 dark blotches on the back; i occurs in highveld regions of Namibia.

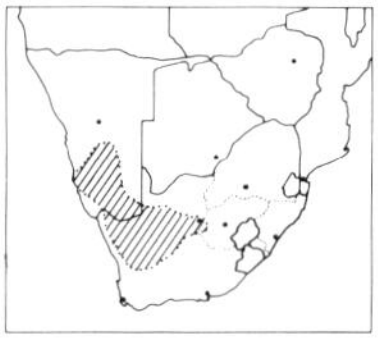

Namib Tiger Snake *Telescopus beetzii* (Pl. 19)

(Endemic) 40-50 cm; max. 67 cm
This snake has a slender build, 21 scale rows at midbody, and an undivided anal. It is sandy-buff on the body, with a series of dark round blotches on the back (30-39 on the body, and 12-20 on the tail). The belly is pinkish-tan. **Biology and breeding:** This rare species lives on rock outcrops, sheltering cracks during the day and emerging at night to feed on lizards. It is usually only found dead on roads. The female lays 3-5 elongate eggs (10-14 x 33-55 mm) in December; these hatch after 80-90 days. The young measure 170-190 mm. **Habitat:** Rocky, arid regions. **Range:** S. Namibia, extending through the Karoo to Victoria West and SW. OFS. Absent from Namib Deser despite its common name.

CAT-EYED TREE SNAKES *Dipsadoboa*

Endemic, nocturnal, African tree snakes, the cat-eyed tree snakes have a long, slender body, a flattened, distinct head, and eyes with vertical pupils. They are closely related to the Herald snakes, *Crotaphopeltis* (page 85), bu differ in having well-developed venom glands and posterior gular shields.

Arboreal, they shelter during the day under bark or in hollow stumps, emerging at night to feed on tree frogs. They are oviparous.

There are six species in the genus, with one entering the subcontinent.

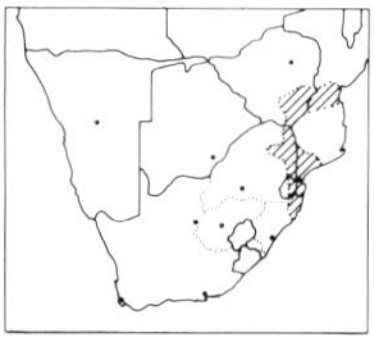

D.a. aulica

Cross-barred or Marbled Tree Snake *Dipsadoboa aulica* (Pl. 31)

55-70 cm; max. 85 cm
This small, slender snake has smooth scales in 17 rows at midbody, an obvious head with the loreal entering the eye, the anal undivided, and a longish tail. The back is brown to light brown, with 38-50 faint white crossba The head is finely marbled with white, and the tongue is white. The belly is o white, laterally flecked with red-brown. **Biology and breeding:** This snake shelters during the day in hollow logs, under bark, in thatched roofs, etc., emerging at dusk to feed on geckos and tree frogs. When disturbed, it adop an open coiled posture with the head well raised. It bites readily but is not dangerous. The female lays 7-8 small eggs (23-28 x 10-13 mm) in midsummer. **Habitat:** Lowveld riverine forest. **Range:** N. Zululand, through

Swaziland, E. Transvaal and Mozambique Plain, entering E. Zimbabwe, and to Kenya. **Subspecies:** Two races are recognized, with only the typical race occurring on the subcontinent, *D.a. flavida* has a yellow head and is of slender build; it is restricted to S. Malawi.

OOMSLANG
spholidus

This snake is distinguished by its distinct head and very large eyes, keeled body scales in oblique rows, and large back fangs. Males develop bright colours, while the juveniles and females are brown-olive. Its venom prevents blood clotting, causing death. There is one species in the genus.

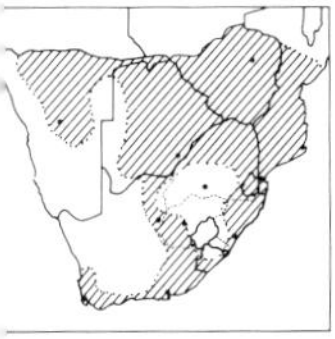

typus

Boomslang *Dispholidus typus* (Pl. 31)

120-150 cm; max. 200 cm
A large snake with a distinct head, and very large eyes with round pupils. The body scales are keeled, in 19 oblique rows at midbody. Coloration is very variable. Juveniles are twig-coloured, often with blue spots on the forebody; the head is dark brown above and white below. They have emerald-green eyes and a yellow throat. Adult females are light olive or brown, with white to brown bellies. Males are more brightly coloured, and occur in various colour phases: they may be leaf-green (and sometimes powder-blue), with a light green belly and occasionally black skin between the scales; or bright green or yellow with black-edged scales, giving a cross-barred appearance, with the head vermiculated in black; or black with dark grey, black-edged belly scales; or brick-red to bright red, with an orange-pink belly.

Biology and breeding: These snakes are dangerous but shy. They hunt during the day. Their excellent vision aids them in catching prey, which consists of small vertebrates, particularly chamaeleons and birds. Prey is actively pursued, seized and chewed to introduce the toxic venom, which quickly kills the victim. When cornered, a boomslang will inflate its neck to expose the brightly coloured skin, and may strike. The amount of venom injected is minute (1 mg), but it is a potent haemotoxin that prevents blood clotting, causing death from haemorrhage. Symptoms of poisoning may not develop for 24-48 hours. A specific antivenom is required for the treatment of this snakebite, and is available only at major city hospitals. Blood or plasma transfusions may be given to replace lost clotting factors if antivenom is unavailable. Envenomation is rare, and it is usually snake handlers who are bitten. Mating occurs in spring, and 10-14 (occasionally up to 25) eggs (40 × 20 mm) are laid in tree hollows or leaf litter in summer. Incubation takes 3-4 months. The young measure 290-330 mm. **Habitat:** Open bush and savannah; also found in sparsely wooded grassland in Albany district of E. Cape. **Range:** Found throughout subSaharan Africa, occurring in the northern parts of the region and extending along the east and south coast to Cape Town. **Subspecies:** Two races are recognized. *D.t. typus* occurs in southern Africa and throughout subSaharan Africa, excluding deserts. It is replaced by *D.t. punctatus* in S. Zaire, Angola and Zambia.

RD SNAKES
elotornis

These arboreal snakes are distinguished by their lance-shaped head, large eyes with keyhole-shaped pupils, extremely long tail and large back fangs.

Diurnal, they rely on their cryptic coloration and slow movement to ambush lizards and birds.

The genus contains two species; *T. kirtlandii* is restricted to the rain forests of central and W. Africa, while the other, savannah, species enters the northern regions of the subcontinent.

Bird or Twig Snake *Thelotornis capensis* (Pls. 18 and 31)

60-100 cm; max. 168 cm
This extremely thin snake has a lance-shaped head and large eyes with keyhole-shaped pupils. The body scales are feebly keeled, in 19 oblique rows

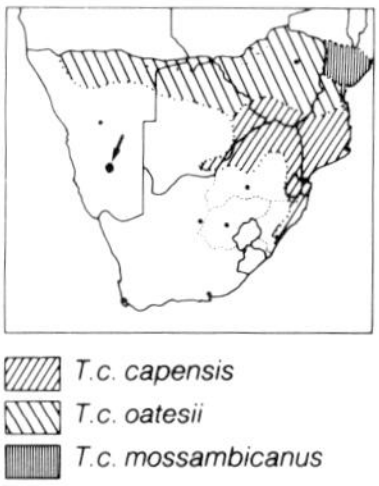

at midbody. The tail is very long. The body is twig-coloured, being grey-brov with black and pink flecks, with a series of diagonal pale blotches. The top o the head is green or blue-green, with varied markings (see Subspecies). Th belly is pink-grey, with many dark grey blotches. Males have longer tails anc fewer ventrals than females. **Biology and breeding:** These beautifully camouflaged snakes can move swiftly if disturbed, but hunt by ambush. The pupil shape and loreal groove allow stereoscopic vision. Prey (mainly lizards but also small birds) is killed by envenomation, and is swallowed while the snake hangs downwards. It inflates its neck when threatened. The venom is potently haemotoxic, causing similar symptoms to that of the boomslang's, and, as with envenomation by the boomslang, a victim can experience a 24- to 48-hour 'latent' period before symptoms become evident, although blood tests will reveal clotting abnormalities. This venom is not neutralized b the boomslang-specific antivenom; treatment must rely on blood or plasma transfusions to replace clotting factors. Fatalities are known, but are rare.

Males engage in combat in mating season. The female lays 4-18 small, elongate eggs (36 x 16 mm) in summer (December-January); these hatch ir March, after 60-90 days. The young measure 230-250 mm. A female may la two clutches per season. **Habitat:** Savannah and coastal forests. **Range:** E. and S. Africa, replaced in the rain forests of central and W. Africa *T. kirtlandii.* **Subspecies:** Three races are recognized, and all occur on the subcontinent. *T.c. capensis* is a small race (80-136 cm), with fewer than 162 ventrals, and a blue-green head that is heavily speckled with black; it occurs in Natal, Swaziland, Transvaal and northern adjacent areas. *T.c. oatesii* is th largest race (90-168 cm), with more than 162 ventrals, and a blue-green hea with a black Y-shaped mark; it is found in northern areas of Namibia and Botswana, and most of Zimbabwe. *T.c. mossambicanus* is 80-140 cm long, and its head is uniform green; it occurs in Mozambique, intergrading with other races along the southern border.

COBRAS, MAMBAS AND THEIR RELATIVES
Family Elapidae

These snakes are usually medium to large in size, with well-developed, fixec fangs at the front of the mouth, and venom glands. The head is covered with large shields, but lacks a loreal. The eyes usually have round pupils. The bo is covered with shiny, pitless scales, which are keeled in the rinkhals. They a often brightly banded, particularly the juveniles.

Elapids are mostly terrestrial, and there are many burrowing forms; some species are arboreal and others are aquatic. They usually lay eggs althougl few species are viviparous. The family includes many of the most poisonous snakes, although these are rarely common enough to cause a large inciden of snakebite. However, the bigger species are 'confident' of their abilities ar will stand their ground. Many have defensive displays.

Elapids are distributed throughout the tropical regions, subSaharan Afric and Australia. Four to five subfamilies are recognized, including the completely aquatic sea snakes (which are sometimes treated as a separate family). There are 236 species in 61 genera; six genera containing 14 speci (four of which are endemic) occur on the subcontinent.

Key to the southern African genera in the Elapidae

1. Tail flattened and oar-like; body scales bead-like, non-overlapping; ventrals much narrower than body; marine .... *Pelamis* (Yellow-bellied sea snake page 96)

   Tail cylindrical; body scales overlapping; ventrals as broad as body 2

2. 3 preoculars, widely separated from nasal; prefrontals in contact with labials; more than 90 subcaudals .... *Dendroaspis* (Mambas, page 95)

1-2 preoculars, in contact with nasals and separating prefrontals from labials; fewer than 75 subcaudals 3

3. Internasals not bordering nostril; dorsal scales in 13 rows at midbody .... *Elapsoidea* (Garter snakes, page 90)

Internasals bordering nostril; dorsal scales in 17 or more (rarely 15) rows at midbody 4

4. Rostral very large and shield-like .... *Aspidelaps* (Coral and Shield-nose snakes, page 89)

Rostral not enlarged 5

5. Dorsal scales strongly keeled; body and tail short (116-150 ventrals, 33-47 subcaudals) .... *Hemachatus* (Rinkhals, page 94)

Dorsal scales smooth; body and tail long (175-288 ventrals; 50-78 subcaudals) .... *Naja* (Cobras, page 92)

RICAN ELAPIDS
bfamily Najinae

RAL AND
IELD-NOSE
AKES
pidelaps

These small elapids are distinguished by their enlarged rostral shields.

They live underground in burrows or beneath stones, emerging at night to feed on small vertebrates. When threatened, they rear up and spread a narrow hood. They huff and puff a lot, and may strike with the mouth closed – but not always. This is a comic performance, but one to heed: the venom is neurotoxic, although few deaths have been recorded.

There are two species in the genus, one of which is endemic to the region.

A.l. lubricus
A.l. infuscatus
A.l. cowlesi

Coral Snake *Aspidelaps lubricus* (Pls. 20 and 26)

30-50 cm; max. 80 cm
A short, solid snake with a large rostral shield on the nose, smooth scales in 19 rows at midbody, and usually a characteristic banded pattern. The head is reddish, with a black crossbar between the eyes, an arrow shape on the top, and a broad nuchal collar. The body is orange to coral-red, with 20-47 black crossbands that decrease in width towards the tail. The belly is yellowish, with the crossbars completely encircling the body in the young, but fading in adults, leaving only the first 2-3 intact. In the northern races, the colour pattern fades (see Subspecies). **Biology and breeding:** A bad-tempered clown, it does well in captivity but rarely forgives its captor. It is fond of rocky outcrops. The diet consists mainly of small vertebrates, particularly other reptiles. Its venom is poorly known, but is apparently neurotoxic. Venom yield for *A.l. infuscatus* has been recorded at 50 mg, with an LD_{50} 300 µg/kg. The effectiveness of antivenom is unknown. Bites from the southern race have not resulted in serious symptoms, but a bite from *A.l. infuscatus* was reported to have killed two children.

The female lays 3-11 eggs (50-54 x 15 mm) in December; these hatch in 59-71 days. The young measure 170-180 mm. In captivity, the female may lay clutches every two months during summer. **Habitat:** Karroid and sandveld regions, entering dry valley plains in S. and E. Cape. **Range:** Western regions of the subcontinent; occurs throughout karroid regions, but reaches Port Elizabeth and Cape Town, and extends through Namibia to S. Angola. **Subspecies:** There are three races. *A.l. lubricus* rarely grows longer than

65 cm, is conspicuously banded, and has 20-28 subcaudals; it is found through Karoo to S. Namibia. *A.l. infuscatus* grows larger, and has a black head, and a grey-brown body with vague, dark crossbands; it occurs in central Namibia. *A.l. cowlesi* has a pale head and a uniform grey-brown bo(it occurs in Kaokoveld and SW. Angola.

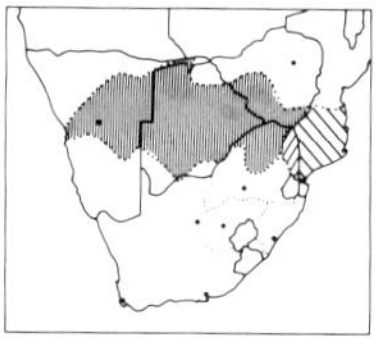

A.s. scutatus
A.s. intermedius
A.s. fulafula

Shield-nose Snake *Aspidelaps scutatus* (Pls. 18 and 26)

(Endemic) 45-55 cm; max. 74 cm
A short, thickset snake with a short, broad head and a very large rostral, 21-25 scale rows at midbody, and keeled posterior body scales. The back i pale grey-brown, yellowish, or buff to orange, with a series of dark blotches that are well-developed in the eastern race, and fade towards the tail. The head and neck are mostly black, with a white eye stripe and throat band. Th belly is white.

Biology and breeding: This snake burrows in sandy soil, using its nose a 'bulldozer'. It feeds at night, and has a varied diet, taking small mammals, amphibians, lizards, and even other snakes. In defence, it may sham death The effects of its venom are unknown; the few case histories recorded exhibited pain and mild neurological symptoms (ptosis, slurred speech anc partial paralysis) that persisted for 2-3 days. There are no recorded deaths from a bite by this species. The effectiveness of antivenom in the treatment the bite is unknown. The female lays eggs. **Habitat:** Savannah and sandvel **Range:** In a wide band through the northern regions of the subcontinent.

Subspecies: Three races are recognized, and all occur on the subcontinent. *A.s. scutatus* has poorly defined markings, a short tail (25-30 subcaudals in males, 20-24 in females), and averages 45 cm in length, rare growing to longer than 60 cm; it occurs through N. Namibia, Botswana, W. Zimbabwe and NW. Transvaal. *A.s. intermedius* is the same colour and size as the typical race but has a medium-length tail (32-35 subcaudals in males, 27-31 in females); it is found in E. Transvaal. *A.s. fulafula* has well-defined markings, a long tail (33-39 subcaudals in males, 30-33 in females) and is a large snake, often growing to longer than 60 cm; it is found in SE. Zimbabwe and S. Mozambique.

GARTER SNAKES
Elapsoidea

These small to medium-sized burrowing elapids have a small head, a short tail, and 13 scale rows at midbody. They are often brightly banded, particul when young. The different species are easily confused due to the uniform scale counts and the different juvenile and adult colour patterns.

Garter snakes burrow in sandy or humic soils, coming to the surface at night. Most eat other reptiles, although small mammals and amphibians are also taken. They are oviparous.

There are seven species widely distributed through subSaharan Africa, w three occurring on the southern African subcontinent, one of which, Sundevall's Garter Snake, is endemic.

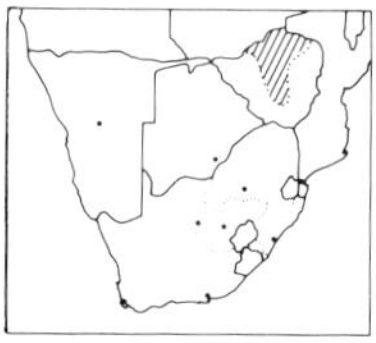

Günther's Garter Snake *Elapsoidea guentheri* (Pl. 21)

40-50 cm; max. 60 cm
This solid, medium-sized snake has a rounded snout, 131-156 ventrals in b(sexes, and four lower labials touching the anterior chin shields. It is banded with pale and dark crossbands of the same width. In juveniles, the body is black, with 16-20 crossbands, and 2-4 on the tail; in adults, the bands fade, and the back becomes a uniform grey-black, with the belly steel-grey.
Biology and breeding: These snakes are known to eat other snakes, skink occasionally amphibians and even termite alates. No details are known abc their venom, and no case histories have been recorded. Up to 10 eggs are in late summer. **Habitat:** Miombo woodland. **Range:** Central plateau of Zimbabwe, extending into adjacent Zaire, Angola and Zambia.

E.s. semiannulata

E.s. boulengeri

Angolan and Boulenger's Garter Snakes

Elapsoidea semiannulata (Pls. 21 and 37)

50-55 cm; max. 65 cm

Medium-sized snakes with a moderately rounded snout, 131-161 ventrals in both sexes, and three lower labials touching the anterior chin shields. They are banded, the pale crossbands being much narrower than the dark ones. In juveniles, the head is white and the back black, with 12-17 narrow white or yellow bands; in adults (longer than 20 cm), the pale bands darken in the centre to leave paired, narrow rings which may disappear completely in large adults (longer than 60 cm). The belly is white or grey-black (see Subspecies). **Biology and breeding:** Not common, this slow-moving snake is inoffensive. Its diet consists of skinks, geckos, amphibians, and especially small snakes; it is prone to cannibalism. Its venom has not been studied; a bite from *E.s. boulengeri* caused pain and swelling, and transient nasal congestion. The female lays up to 10 small eggs (20 x 8-10 mm). **Habitat:** Varied; including arid and mesic savannah, but absent from very arid areas. **Range:** Widely distributed, from Senegal in the west, through central and E. Africa, reaching as far south as Zululand and N. Cape. **Subspecies:** Three races are recognized, with two occurring on the subcontinent (*E.s. moebiusi* occurs extralimitally, in the north). The Angolan garter snake, *E.s. semiannulata*, has a uniform white belly; it is found in N. Namibia and Caprivi Strip. Boulenger's garter snake, *E.s. boulengeri*, has a grey to black belly; it is found in SE. Africa, through Zimbabwe to Zululand and N. Cape.

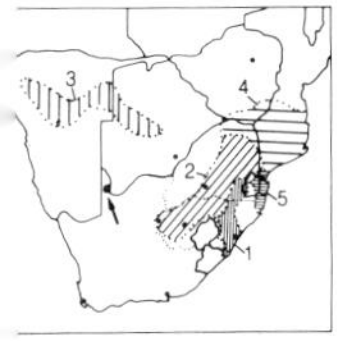

s. sundevallii

s. media

s. fitzsimonsi

s. longicauda

s. decosteri

Sundevall's Garter Snake

Elapsoidea sundevallii (Pls. 21, 32 and 37)

(Endemic) 50-75 cm; max. 99 cm

This snake is medium to large in size, depending on the race. It has a slightly pointed snout. Males have 152-180 ventrals, while females have 138-161. The juveniles are strongly marked with almost equal-sized bands of cream or pink and dark chocolate, and the head and belly are pale, with a dark extension on to the crown of the head. In adults, the pale bands darken from the centre to form paired rings that usually disappear, but persist in the typical race; the body becomes slate grey to black, and the belly and adjacent body scales pink-buff, and sometimes mottled.

Biology and breeding: These are slow-moving snakes, and are reluctant to bite. They settle well in captivity. The diet is varied, and includes snakes, lizards and their eggs, rain frogs, rodents and moles. Rarely seen, they are usually collected in old termitaria, under stones or on roads at night. The venom has not been studied. There have been few recorded bites; symptoms for a bite from *E.s. longicauda* included nausea, vomiting, loss of consciousness and blurred vision; these symptoms resolved in two or three days without the use of antivenom. A bite from *E.s. fitzsimonsi* caused only pain and swelling. The female lays up to 10 small eggs.

Habitat: Varied; coastal forest, highveld grassland and arid and mesic savannah. **Range:** Southern Africa, from central Namibia to S. Mozambique and Natal. **Subspecies:** There are five races. *E.s. sundevallii* is the largest, with juveniles and adults both having 19-34 pale bands; it occurs in Natal to SE. Transvaal. The other races are distinguished by scalation. *E.s. media* has 13-23 subcaudals, and 157-168 ventrals in males and 140-154 ventrals in females; it is found in Transvaal, OFS and N. Cape. *E.s. fitzsimonsi* also has 13-23 subcaudals, and 167-180 ventrals in males and 156-161 ventrals in females; it is found in Botswana and Namibia. *E.s. longicauda* has 22-33 subcaudals, and 152-159 ventrals in males and 138-144 ventrals in females; it is found in N. Transvaal, SE. Zimbabwe and S. Mozambique. *E.s. decosteri* also has 22-33 subcaudals, and 164-179 ventrals in males and 148-156 ventrals in females; it occurs in Zululand and S. Mozambique.

COBRAS
Naja

The cobras are large, stockily built, terrestrial snakes with smooth scales.
They are alert, active foragers, pursuing and capturing small vertebrates. When threatened, they lift the forebody and spread a characteristic hood. F species have modified fangs and can 'spit' venom up to three metres. Bites from spitters and non-spitters present with different symptoms, but all are potentially dangerous. All cobras are oviparous.
They are distributed through Asia and Africa. Upper Miocene fossils are known from France. There are seven African species, and all but two of thes occur in southern Africa; one is endemic.

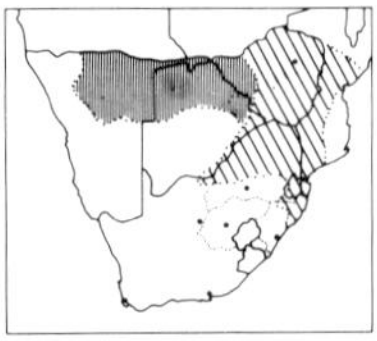

N.h. annulifera
N.h. anchietae

Egyptian Cobra *Naja haje* (Pls. 20 and 27)

150-200 cm; max. 240 cm
This thick-set snake has a large head and 17-19 smooth scale rows at midbody. Suboculars are present. Coloration is variable, but is usually yello grey to brown or blue-black; old specimens are darker. The belly is yellowis with dark blotches. It has a dark throat band which is more conspicuous in juveniles. The banded phase of this species, which is found throughout the range, has 7-9 yellowish bands on the body and two on the tail. This bandin develops in snakes that are longer than 60 cm, and is more frequent in male Males grow slightly larger than females. **Biology and breeding:** Nocturnal, the Egyptian cobra emerges at dusk to forage for small vertebrates. It is ofte seen basking at its retreat (old termite nests, hollow logs, etc.) in the mornin sun. The diet is varied, often including toads and other snakes, and occasionally birds' eggs; it may become a pest in poultry runs. It spreads a broad hood. The eastern race is less aggressive, whereas the western *N.h. anchietae* is more willing to continue an argument. When cornered, it m sham death. This cobra is a non-spitter. Bites are common in some districts and are usually inflicted on the lower leg, at night. Initial symptoms include a burning pain and slight swelling, followed rapidly by neurological symptom and often death from respiratory failure. The venom yield is 175-300 mg; 25-35 mg is fatal in humans. A victim of its bite requires large doses of antivenom. The female lays 8-33 large, oval eggs (47-55 x 25-30 mm) in loo soil or disused termitaria in early summer. Hatchlings measure 240-340 mm
Habitat: Savannah, particularly common in bushveld and lowveld. **Range:** Widely distributed in Africa. **Subspecies:** There are four races, two which occur on the subcontinent. *N.h. annulifera* has 19 scale rows; it occu in the eastern regions, from N. Natal, through lowveld of Transvaal to W. Botswana. *N.h. anchietae* has 17 scale rows and does not grow longer than 210 cm; it is found in N. and central Namibia, through N. Botswana to NW. Zimbabwe. Extralimitally, the typical *N.h. haje* occurs in NE. Africa, up Nile River valley, and north and south of Sahara Desert to Atlantic Ocean. *N.h. arabica* is found in the semi-desert of W. and S. Arabia.

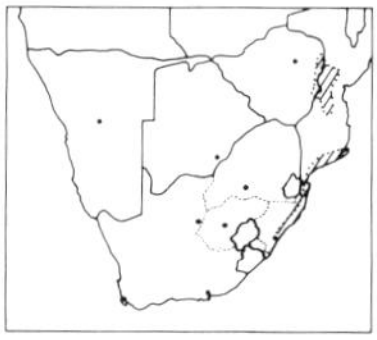

Forest Cobra *Naja melanoleuca* (Pl. 27)

180-210 cm; max. 270 cm
This is the largest of the cobras, but is a slender snake, with highly polished scales in 19 rows at midbody. It has one preocular and a broad rostral. The head and foreparts are yellow-brown, heavily mottled in black, becoming shiny black towards the tail. The lower labials are often white with black edg The belly is pale white-cream, often with dark blotches. **Biology and breeding:** This cobra is fond of water. It may climb into low bush. It is nocturnal, but may forage on overcast days. The diet is varied, and include most small vertebrates, even fish. This alert snake is rarely cornered, but it w spread a narrow hood and bite readily if necessary. It is long-lived; 28 year known in captivity. (SA RDB, Peripheral.) It is a non-spitter. Its venom is neurotoxic, but it bites very rarely and there have been no deaths recorded locally. Males have been observed in combat in breeding season. The fema

lays 15-26 large eggs (60 x 30 mm) in leaf litter, hollow logs, etc. Incubation takes 75-91 days. The hatchlings measure approximately 380 mm. **Habitat:** Tropical and subtropical rain forest. **Range:** Forests of Africa, from Senegal to Ethiopia, south to Angola and Mozambique Plain, just entering E. Zimbabwe and Zululand (to Durban).

Cape Cobra *Naja nivea* (Pls. 20 and 27)

(Endemic) 120-150 cm; max. 170 cm
This relatively small, slender cobra has a broad head, and smooth but dull scales in 19-21 rows. There is one preocular and a narrow rostral. Coloration is varied, with some phases common in certain regions. The yellow cobra is butter-coloured to dirty yellow, sometimes speckled with brown (particularly in Botswana); the brown or speckled cobra is bright reddish-brown to mahogany, with darker and paler flecks (common in SW. Cape); and the black cobra is purplish-black (found in Great and Little Namaqualand). Juveniles are dirty yellow, often finely speckled in dark brown, and have a black throat band. **Biology and breeding:** Active during the day and early evening, this snake feeds on a wide spectrum of prey, including other snakes. It will climb low trees and raid sociable weaver colonies. It is common around Karoo farms, to which it is attracted by rodents. It is a memorable sight when foraging in the veld. Unfortunately, it is both nervous and deadly. It spreads a broad hood and confidently disputes its right of way. It is a non-spitter. The venom is syrupy and as toxic as the black mamba's. The average venom yield is 120 mg (max. 250 mg); 15-20 mg is fatal in humans. The venom is neurotoxic, and death usually occurs from the rapid onset of paralysis. Large volumes of antivenom are urgently required in the treatment of the bite. Bites are not uncommon; this species is responsible for the majority of fatal snakebites in the Cape. The female lays 8-20 large eggs (60 x 30 mm) in a burrow. **Habitat:** Arid karroid regions, particularly along river courses, entering well-drained open areas along the southern coast. **Range:** Through the Cape, in the east to East London, and north to S. Namibia and adjacent Botswana and OFS.

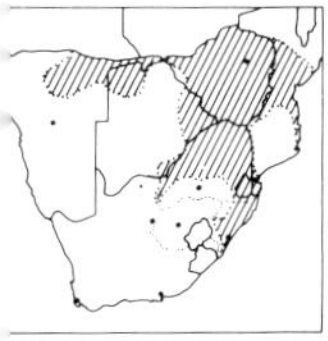

Mozambique Spitting Cobra or M'fezi
Naja mossambica (Pl. 27)

90-120 cm; max. 150 cm
A small cobra with a blunt head and 23-25 scale rows. It has two preoculars and 11-14 scales bordering the parietals. The back is pale grey to dark olive above, with each scale edged in black. The belly is salmon-pink to yellowish. There are irregular black crossbands or blotches on the throat. **Biology and breeding:** A nocturnal species, although juveniles may forage during the day. The varied diet includes rodents, lizards, toads, and grasshoppers. It is very common in lowveld regions and often forages around houses at night. It spreads a broad hood and 'spits' readily. Although it feeds well in captivity, it rarely tames. It produces copious amounts (200-300 mg) of dilute venom which it sprays at the intruder's eyes, causing agonizing and instant pain; 40-50 mg of this venom is fatal in humans. Bites are frequent in Zululand and the Transvaal lowveld. The venom causes skin necrosis; neurotoxic symptoms are minor, and fatalities are rare. It lays 10-22 small eggs (35 x 20 mm) in summer. Hatchlings measure 230-250 mm. **Habitat:** Savannah regions, and cleared areas in former forest. **Range:** Eastern regions, from S. Natal through Transvaal and N. Botswana to NE. Namibia. Elsewhere, to Tanzania.

Black-necked Spitting Cobra *Naja nigricollis* (Pls. 20 and 35)

120-220 cm; max. 280 cm
This cobra has a broad head with a rounded snout, 17-21 scale rows, two preoculars and 7-10 scales bordering the parietals. Coloration is varied (see

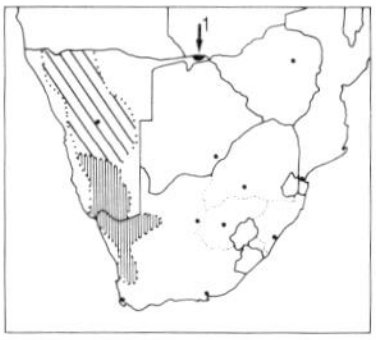

1. *N.n. nigricollis*
N.n. nigricincta
N.n. woodi

Subspecies). **Biology and breeding:** The typical race is common in moist savannah, and is more aquatic than the other subspecies. It is mainly nocturnal, although the juveniles and *N.n. woodi* are active during the day. The diet is varied, and is similar to that of the Mozambique spitting cobra, *N. mossambica* (page 93), except that the western and southern races take more lizards and rodents. It is a nervous species that readily spreads a hoo and 'spits'. The venom is relatively toxic, causing minor neurological symptoms and extensive necrosis, bleeding and haematological abnormalities. Venom yields are large (200-350 mg); 40-50 mg is fatal in humans. Bites are common in N. Namibia, but have been poorly studied. Th typical race causes many severe snakebites in W. Africa. The female lays 10-22 eggs (35 x 20 mm). **Habitat:** The typical race favours moist savannah the two southern races are found in rocky arid regions. **Range:** SubSaharan Africa, from Senegal to Kenya, through W. Zambia to Caprivi Strip, and sou through Namibia to W. Cape.

Subspecies: Three races are recognized, two of which are endemic to th subcontinent, with the typical race just entering the region. *N.n. nigricollis* (t black-necked spitting cobra) is large, with 176-219 ventrals, a uniform olive brown to black back, a yellow to red belly, and a broad, dark throat band; it occurs throughout most of the range, just entering Caprivi Strip. *N.n. nigricincta* (the western barred spitting cobra) is small (max. 150 cm), with 192-226 ventrals, a light grey to pink-brown back and belly, ringed with 51-86 black bands on the body and 13-32 black bands on the tail, and a wic dark throat band that is prominent in juveniles; it occurs in central and N. Namibia. *N.n. woodi* (the black spitting cobra) is of medium size (max. 180 cm), with 223-228 ventrals, a uniform black back, and a dark grey belly streaked with black, while juveniles are grey with a black head and neck; it occurs in S. Namibia, through Namaqualand to Citrusdal in W. Cape.

RINKHALS
Hemachatus

The rinkhals is a close relative of the true cobras, but has keeled scales, no solid teeth on the maxilla, and is viviparous.

There is a single species in the genus, and it is endemic to southern Afric

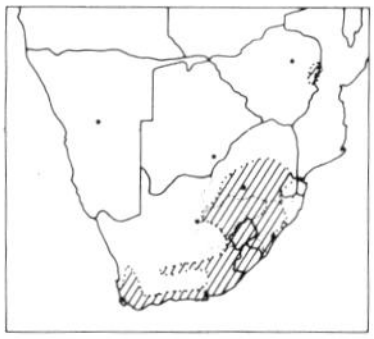

Rinkhals *Hemachatus haemachatus* (Pls. 20 and 35)

(Endemic) 90-110 cm; max. 150 cm

This stocky snake has a broad head and keeled scales in 17-19 rows. Coloration is varied, with specimens in E. Cape to Pondoland and the Inyan Highlands of Zimbabwe being conspicuously banded in dark-brown to blac alternating with pale grey, yellowish or orange; inland, particularly on the highveld, the back is uniform dark brown to black, sometimes speckled with lighter greys and browns. The belly is dark, with 1-2 pale crossbands on the throat (hence its common name) that may fade in old specimens. Juveniles are conspicuously banded. **Biology and breeding:** A nocturnal species, it sometimes active on overcast days. It has catholic tastes, eating most smal vertebrates, particularly toads. In defence, it rears and spreads a broad hoc It can spray venom up to 2,5 m and usually aims it at the intruder's face. If th fails, it will sham death, rolling on to its back with the mouth agape. It settles well and has lived for more than 11 years in captivity, often being a gluttono feeder. Its venom is neurotoxic, with deaths resulting from respiratory paralysis. It is less dangerous than the true cobras, as its venom is more dil for spitting. If the venom enters the eyes, it causes great pain and sometime blindness; the eyes must be washed immediately with a bland liquid. The lo antivenom is effective. Fatalities are rare. Twenty to 30 (max. 63) live babies are born in late summer (January-March). Hatchlings measure approximate 180 mm. **Habitat:** Grassland, from the coast up to 2 500 m. **Range:** Commc on highveld and in Natal and Transkei grasslands; relict populations in montane grasslands of old escarpment in the Cape and Zimbabwe (Inyang and along S. Cape coast.

MBAS
ndroaspis

These large, agile, diurnal elapids have a long, flat-sided head.

All except the black mamba are arboreal. They actively pursue their prey, striking rapidly and often until it succumbs to the toxic venom. These are probably the most feared of all the African snakes, but only the black mamba commonly bites. They are oviparous.

There are four species, which are found throughout most of tropical Africa, with two entering the subcontinent.

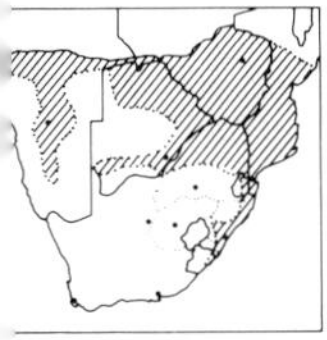

Black Mamba ***Dendroaspis polylepis*** (Pl. 35)

200-250 cm; max. 430 cm
A large, streamlined snake with a narrow, coffin-shaped head and smooth scales in 23-25 oblique rows. The back is uniform gunmetal to olive-brown, but never really black; the belly is pale grey-green, sometimes with dark blotches, and the mouth lining is black. **Biology and breeding:** These active, terrestrial snakes eat birds and small mammals (eg. rats and dassies). Prey is pursued and stabbed with the fangs until it collapses from the venom. Digestion is rapid. The black mamba is territorial, having a favoured home in a termite nest, a hollow log or a rock crevice. If disturbed, it will retreat unless cornered. It is confident in defence; it rears the front third of its body, spreads a narrow hood and gapes the mouth, revealing the black lining. It will bite readily and often. Its hollow 'hiss' is best heeded – if you step back, the snake will also retreat. The venom is neurotoxic and cardiotoxic. The venom yield is 100-400 mg; 10-15 mg is fatal in humans. A bite from this snake is extremely serious, and requires large volumes of antivenom (up to 10 ampoules) to counteract the venom. The victim may be fully conscious, but all the muscles are paralyzed; death from respiratory failure usually occurs in 7-15 hours. In spring, males fight by raising and intertwining their bodies; this combat is often mistaken for mating. The female lays 12-14 large eggs (70 x 35 mm) in termite nests, etc. These hatch in 80-90 days. The young measure up to 600 mm; growth is rapid, and a black mamba may reach 200 cm in length in its first year. **Habitat:** Savannah and open coastal bush, usually below 1 500 m. **Range:** Northern parts of the subcontinent (absent from desert), extending along Natal coast to Port St Johns. Elsewhere, to Senegal and Somalia.

Green Mamba ***Dendroaspis angusticeps*** (Pl. 30)

180-200 cm; max. 250 cm
This large, slender snake has a narrow, coffin-shaped head and smooth scales in 19 oblique rows. The back is uniform bright green (bluish-green in hatchlings). The belly is yellowish-green, and the mouth lining is white. **Biology and breeding:** This active, arboreal species feeds almost exclusively on birds and small mammals. It is shy and rarely seen. When cornered, it is less belligerent than the black mamba, threatening and biting only as a last resort. It has lived for more than 14 years in captivity. The venom is neurotoxic; it is less potent than the black mamba's, and the venom yield is 60-100 mg. Bites are uncommon, and cause mild paralysis and, rarely, death. Males engage in combat in breeding season. Up to 10 eggs (58 x 26 mm) are laid in a hollow log or leaf litter in summer. They hatch after 70-80 days. The young measure 350-400 mm. **Habitat:** Coastal bush, and dune and montane forest. **Range:** Restricted to Natal coastal regions and forests along E. Zimbabwe escarpment. Elsewhere, to E. Africa.

A SNAKES
family
rophiinae

As their name implies, sea snakes are completely aquatic. They have numerous adaptations to marine life, including a nasal salt gland to purge excess salt, a flattened, oar-like body, valved nostrils, and a large lung.

They feed on marine life, particularly eels and fish eggs. All are venomous and many have very toxic venoms. Fortunately, they are docile, but are very common in some waters (eg. Vietnam and Malaya), and bites and fatalities are regionally common. A specific antivenom is required in the treatment of th bite. Viviparous, they give birth in the surface waves. They never normally come ashore.

Sea snakes are closely related to Australian terrestrial elapids. There ar 45-50 species in 13-14 genera, restricted mainly to the coastal waters of Australasia. A single, vagrant species occasionally washes ashore on our southern coast.

YELLOW-BELLIED SEA SNAKE
Pelamis

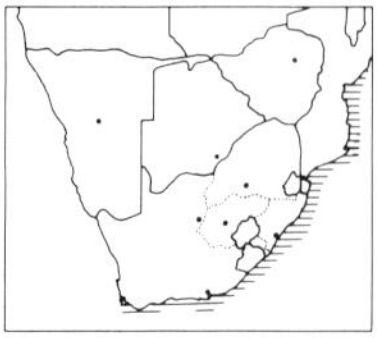

An unusual sea snake, adapted to feeding on small fish in surface waters. There is a single species in the genus. It occurs throughout the Indo-Pacif

Yellow-bellied Sea Snake *Pelamis platurus* (Pl. 25)

60-70 cm; max. 100 cm
The body of this snake is laterally flattened, with an oar-like tail. The head i narrow, with an elongate, flat-topped snout which has valved nostrils. The are no enlarged belly scales. Its striped coloration is very characteristic, although uniform yellow specimens do occur. The dark brown or black ba contrasts with the cream, yellow or light-brown belly. The tail is yellow, wit black spots.

Biology and breeding: Sea snakes gather in slicks where surface curr meet, sheltering in floating seaweed and debris, and ambushing small fis They shed their skin with a 'knotting' behaviour, rubbing one part of the bo against another. They are excellent swimmers, being able to move either backwards or forwards with ease, but are helpless when washed ashore. have few predators as their venom is very toxic to fish; it is possible that th bright coloration is aposematic. The venom of sea snakes occurring in loc waters has not been studied. Elsewhere (in Panama), it is mildly neurotoxi (LD_{50} 130 µg/kg), watery, and yields are small (1-2 mg, max. 4,4 mg). Myotoxic symptoms do not present. No bites or deaths have been record southern Africa. It is unlikely that the local antivenom is effective in the treatment of a bite from this species. They give birth to 3-8 live young (250 mm) in March-October.

Habitat: Drifts in ocean surface waters. **Range:** Throughout warm Indo-Pacific, from Kenya to Baja California. Vagrants are washed south in Agul Current and strand on Natal and Cape beaches. Yellow-bellied sea snake are excluded from Atlantic Ocean by cold Benguela Current along W. Ca coast.

ADDERS AND VIPERS
Family Viperidae

These snakes have large, erectile fangs at the front of the mouth. The hea distinct and usually covered with small, irregular, overlapping scales (nig adders have large head shields like those of the colubrids). The eyes usu have vertical pupils. The body scales are often strongly keeled (but smoot night adders). The body is stocky, with a short tail and usually with a blotc colour pattern.

These short to medium-sized snakes are mainly terrestrial, although so are arboreal, one is a burrowing species, and several are semi-aquatic. T are usually nocturnal or crepuscular, and feed on small vertebrates. Prey i ambushed and killed by the venom. Most species are viviparous, but som primitive groups lay eggs.

Adders and vipers are found throughout the world, except on most oce islands, in Australia and on Madagascar. They are frequently found in temperate climates (the European adder, *Vipera berus*, extends almost to Arctic Circle). There are four subfamilies: the Azemiophinae (consisting of single rare viper, *Azemiops feae*, from SE. Asia); the Causinae (the night adders, from Africa); the Viperinae (the true vipers, from Eurasia and Afric and the Crotalinae (the pit-vipers, from Asia and the Americas). There are species in 17 genera; 12 species (five of which are endemic) in three gen occur on the subcontinent.

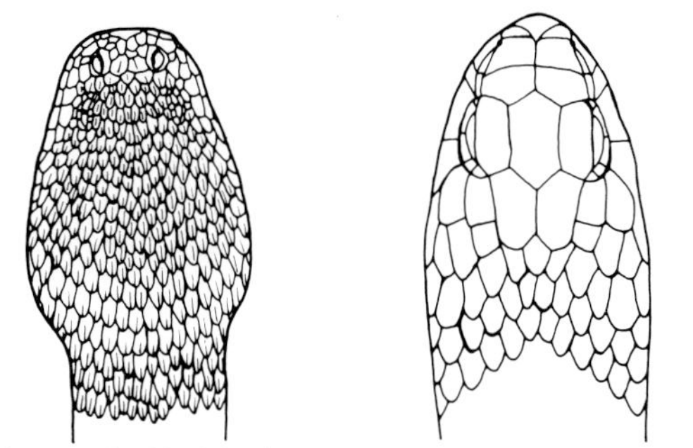

fragmented and keeled scales (typical vipers) | smooth, 'colubrid'-type scales (night adders)

The head scales of southern African vipers

y to the southern
'can genera in the
eridae

1. Head large, much broader than neck, and covered with small, keeled scales; pupils vertical; viviparous	2
Head small, only slightly larger than neck, and covered with large, symmetrical shields; pupils round; oviparous	*Causus* (Night adders, page 97)
2. Nasal separated from rostral by small scales; no enlarged supraorbital shield	*Bitis* (African adders, page 98)
Nasal in contact with rostral or separated by nasorostral shield; large supraorbital shield present	*Atheris* (Bush vipers, page 102)

HT ADDERS
family Causinae

HT ADDERS
usus

These small, primitive vipers are distributed throughout subSaharan Africa. As their common name suggests, they are nocturnal, but they have round pupils. They rarely grow to longer than a metre, and have a short, blunt head and large head shields. The scales are smooth or feebly keeled; the tail is short.

Night adders live in forests or moist savannah, feeding mainly on frogs and toads. Their venom glands are large, extending 5-7 cm into the neck in some species. Venom yields are large, but the venom is weak and causes mainly localized pain and swelling. Fatalities are exceptionally rare. The venom can be neutralized with the use of polyvalent antivenom, but symptoms rarely merit its use. All night adders are oviparous.

There are six species, two of which occur in our region. Neither is endemic.

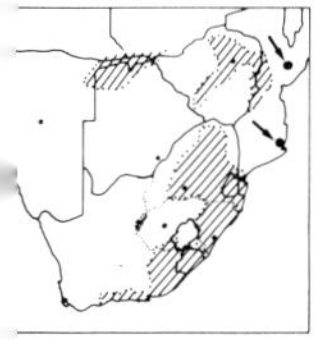

Common or Rhombic Night Adder

Causus rhombeatus (Pl. 15)

40-60 cm; max. 100 cm

A medium-sized adder with a rounded snout, and soft, feebly keeled scales in 17-21 rows at midbody. Ventrals number 134-155. The back is various shades of grey to olive or light brown to pinkish-brown, with 20-30 dark, pale-edged rhombic blotches over the body and tail (these are sometimes absent, particularly in northern populations). The back of the head has a characteristic, dark V-shape. The belly is cream to pinkish-grey. **Biology and breeding:** It rests during the day in undergrowth under stones or logs, in termitaria, etc., and forages at night. Its eyesight is poor, and prey is detect mainly by smell. Sometimes aggressive, it huffs and strikes readily, but tam quickly. The venom glands are large, extending into the neck; they yield 20-30 mg of mild, dilute venom that causes pain and swelling. There are no recorded fatalities for a bite by this snake. Antivenom is effective, but rarely necessary. It lays 15-26 eggs (26 x 16 mm) in summer; these take 70-85 da to hatch. The young measure 130-160 mm. In captivity, clutches may be lai 2-3 times a year. **Habitat:** Mesic savannah. **Range:** Eastern regions, from Riversdale in the Cape, through Natal and Transvaal to Zimbabwe, N. Botswana and Caprivi Strip. Elsewhere, to Nigeria and Sudan.

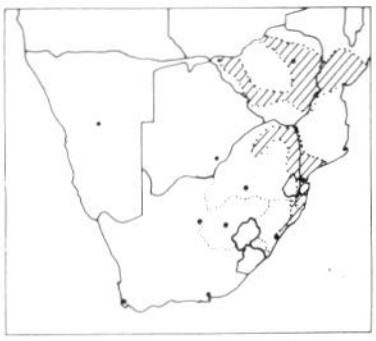

Snouted Night Adder *Causus defilippii* (Pl. 15)

30-40 cm; max. 44 cm
This small adder has an upturned snout, a stout build and a more distinct h than that of the common night adder, *C. rhombeatus* (page 97). It has 17 sc rows at midbody, and 108-126 ventrals. The body is pale brown to pinkish-brown on the flanks, with a darker brown back, and a series of 20-30 black triangular blotches on the back and tail. The back of the head bears the characteristic V-shape. The belly is pearl-white, and grey in juveniles. **Biol and breeding:** It feeds almost exclusively on small amphibians. It does not burrow, despite its snout shape. The symptoms and toxicity of its venom ar similar to those of the common night adder, but the yields are smaller. The venom glands are short, not extending into the neck. There are no known fatalities for a bite by this species. Antivenom is effective, but rarely necess Males engage in combat in breeding season. It lays 6-8 eggs (23 x 15 mm) summer; these take 90-100 days to hatch. The young measure about 100 n **Habitat:** Common in lowveld, tolerating arid savannah. **Range:** Low-lying Natal, Transvaal lowveld, Mozambique and Zimbabwe, to Tanzania.

TRUE VIPERS
Subfamily Viperinae

AFRICAN ADDERS
Bitis

This is the largest group of African vipers, containing both the largest and t smallest viperines. All are terrestrial, stocky species, and some attain gross proportions. The head shields are fragmented. The tail is very short, especi in the females. There is a well-developed sac above the nostril that may be similar to the heat-sensitive organ of pit-vipers.

Prey is captured by ambush; if it is small, it is held in the mouth, but it is usually released until it succumbs to the venom, and then eaten at leisure. African adders are viviparous.

African adders are present throughout subSaharan Africa, with one spec (*Bitis arietans*) also occurring in Arabia. There are 13 species, with nine on subcontinent, five of which are endemic.

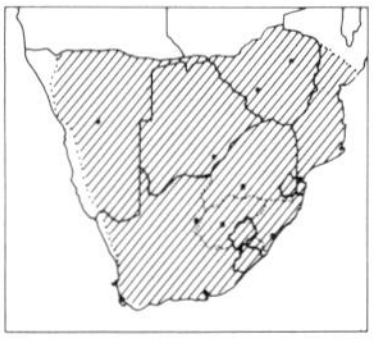

B.a. arietans

Puff Adder *Bitis arietans* (Pl. 12)

70-90 cm; max. 120 cm
This thick, heavily built snake has a large, flattened, triangular head and lar nostrils which point vertically upwards. The scales are heavily keeled, in 29 rows at midbody. The body is yellow-brown to light brown, with black, pale edged chevrons on the back and bars on the tail; sometimes the chevrons lost in the general speckled colouring. There are dark blotches on the crow the head and between the eyes, and two oblique bars from the eye to the li The belly is white or yellow, with a few scattered blotches. Snakes from E. Cape and Natal are more brightly coloured. Unusual patterns are known to occur. Males are smaller and more brightly coloured than females.

Biology and breeding: This common adder is a sluggish snake. It emer at dusk, lying in cover and ambushing prey, which includes rodents and bi and even other snakes and, once, a tortoise. It has effective camouflage. If disturbed, it adopts a striking posture and usually warns by giving a deep, hollow hiss; once heard, it is not easily forgotten. It strikes readily. The puff adder normally moves in 'caterpillar' fashion, leaving a straight, deep track in sand. It may climb into low scrub to bask, particularly when gravid. It often swims, and lies on warm roads at night (the latter usually with dire consequences). It lives for up to 14 years in captivity if it is provided with plenty of warmth and sunshine. The venom is cytotoxic, often causing extensive swelling, pain and necrosis. Yields are large (100-350 mg); 100 mg is fatal in humans. The long fangs (12-18 mm) inject the venom deeply, and bites are usually inflicted on the lower leg. Bites are common, but only a small proportion proves fatal; nonetheless, this snake causes over 60% of serious bites in the region, and is responsible for most of the fatalities. Death usually results from kidney failure and other complications caused by the extensive swelling. It is essential to treat a victim of its bite for fluid loss, and antivenom should be used in serious cases. Males engage in combat and trail females in spring. Large litters, usually consisting of 20-40 young (150-200 mm) are born in late summer. Very large females from E. Africa may have up to 156 young, the most for any snake.

Habitat: Absent only from desert and mountain tops. **Range:** Throughout the subcontinent, north through the whole of Africa. **Subspecies:** Two races are recognized. The typical race occurs throughout Africa, and is replaced in Somalia by *B.a. somalica*, which has keeled subcaudals.

. gabonica

Gaboon Adder *Bitis gabonica* (Pl. 12)

80-120 cm; max. 180 cm

This large, fat adder is redeemed by its beautiful colours. It has a big, flat, triangular head, with a pair of enlarged, horn-like scales on the snout, and 33-46 scale rows at midbody. The back has an attractive geometric pattern of rich purple and brown, interspersed with pastel colours. The head is buff, with a dark mid-dorsal line and a spot above the jaw angle, and dark and light stripes that radiate from the eye to the lip. The belly is buff, with dark grey blotches. **Biology and breeding:** In suitable habitat, and even in captivity, this snake is a memorable sight. Its complex coloration is ideal camouflage among leaf litter. It is a surprisingly tolerant snake; it does a lot of huffing and puffing but rarely bites, and adapts well to captivity, enjoying the good life and growing fat. In nature, it ambushes large rodents, ground birds, and even toads. Its strike is lightning fast, and its fangs are massive (up to 4 cm). The Gaboon adder is rare in the region and is deservedly protected (SA RDB, Vulnerable). The venom is cytotoxic and is produced in large amounts (450-600 mg); 90-100 mg is fatal in humans. Fortunately, bites are very rare. Symptoms include massive swelling and pain, and extensive necrosis. The use of antivenom is essential, and the victim should be treated for fluid loss.

Males engage in combat, and only dominant males mate with females. Gestation may take 12 months; 16-43 young (and up to 60 in the W. African race) are born in late summer; these measure 240-370 mm and weigh 25-45 g. It breeds every 2-3 years. It has been known to hybridize with the puff adder. **Habitat:** Coastal dune and remnant montane forest. **Range:** Locally restricted to coastal forests of N. Zululand and eastern escarpment forests of E. Zimbabwe. Elsewhere, through central and E. Africa to Nigeria. **Subspecies:** Two races are recognized. The typical race occurs in the region, and is replaced in W. Africa by *B.g. rhinoceros*, which has larger nasal horns.

Berg Adder *Bitis atropos* (Pl. 12)

(Endemic) 35-45 cm; max. 60 cm

A small adder with an elongate head that lacks raised ridges or horns above the eyes. The scales of the head and body are keeled and in 29-33 rows at

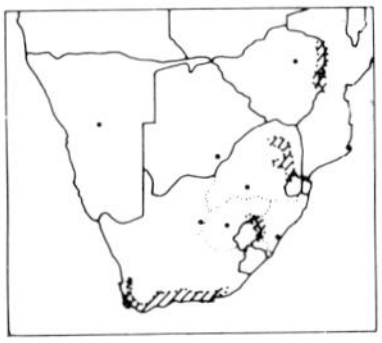

midbody. The subcaudals are smooth. Coloration varies according to the region. The back is dark brown to greyish-olive, with a paired series of triangular, black, pale-edged blotches on the sides, separated by a narrow, yellow-white streak. There is a dark arrowhead on the crown, and two pale streaks on either side of the head. The belly is off-white with grey smears, anc occasionally grey-black overall. Snakes from SE. Transvaal are reddish and poorly marked. **Biology and breeding:** This snake is common in suitable habitat. Irascible when first caught, it rarely tames. It is fond of basking in grass tussocks on rocky ledges. Juveniles eat amphibians, while adults also take lizards and rodents. It yields 22-28 mg of venom. Unusual minor neurological symptoms occur in victims of its bite, including drooping of the eyelids and loss of smell and taste; these symptoms resolve in 1-2 days, and the bite rarely results in swelling or necrosis. No fatalities are known. Antivenom is ineffective in treatment. Mating may occur in autumn, prior to hibernation; 4-9 (max. 15) young are born in late summer. The young measur 120-145 mm. **Habitat:** Montane grassland (up to 3 000 m), and coastal and montane fynbos. **Range:** S. Cape fold mountains from Cedarberg to Port Elizabeth, then in isolated populations on Natal and Transvaal Drakensberg, and Chimanimani and Inyanga mountains in Zimbabwe.

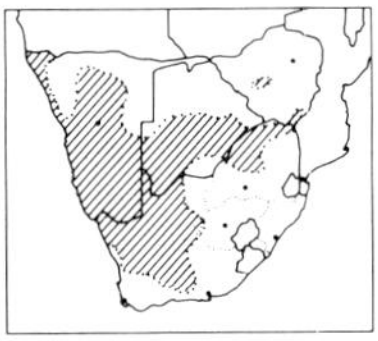

Horned Adder *Bitis caudalis* (Pl. 13)

30-35 cm; max. 51 cm
This small, squat adder has a single horn on the ridge above each eye (although this may rarely be absent). The head is triangular. The scales are strongly keeled, in 23-31 rows at midbody. Coloration is blotched, and regional variations occur. The back varies in colour from light grey (Etosha Pan) through buff to reddish (N. Cape and Kalahari) or greyish-olive to light brown (Karoo), with a series of dark (sometimes pale-edged and/or pale-centred) dorsal and lateral blotches. The top of the head has a broad, dark V-shape or hourglass-shape, and dark bars may radiate from the eye to the jaw angle. The tail tip is often black. The belly is uniform cream-white. Femal have short tails, keeled subcaudals and duller colours.

Biology and breeding: This is a common species. It prefers to lie in the shade of small scrub and shuffles down into the sand, which aids concealment and protects the snake from heat. Active at dusk, it may sidewind in loose sand. The diet is mainly small lizards (rodents and amphibians are also eaten) that are captured by ambush, and may be attracted into the snake's range by its waving the black tail tip. The snake usually holds on to the prey. The horned adder hisses and strikes at first, bu tames well (except those from N. Transvaal). The venom is mild, causing swelling, pain and some necrosis. No fatalities have been recorded, and antivenom is rarely necessary in the treatment of its bite. Males engage in combat in breeding season. Mating occurs in October-November, and gestation takes 90-110 days; 4-15 (max. 27) young, measuring 100-150 mr are born in December-February, coinciding with the hatching of lizards' egg **Habitat:** Sandy regions, but absent from mobile dunes. **Range:** Widespreac in the arid western region, throughout Karoo and Kalahari, reaching S. Zimbabwe, W. and N. Transvaal and S. Angola.

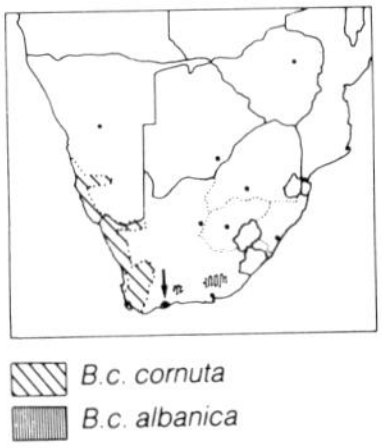

Many-horned Adder *Bitis cornuta* (Pls. 13 and 14)

(Endemic) 30-40 cm; max. 60 cm
A small, stocky adder with a tuft of 2-4 horns above each eye (these are sm or absent in the southern race). The scales are in 25-29 rows at midbody, w the subcaudals keeled at the tail tip in males and keeled throughout in females. The back is grey, and sometimes brown, with four rows (the centra two may fuse) of dark, angular, pale-edged blotches. The top of the head h symmetrical dark markings that may form an arrowhead. The belly is dirty

white, and sometimes speckled. **Biology and breeding:** This snake shelters in rock cracks or rodent burrows in rocky areas. It rarely sidewinds or shuffles into loose sand. It is active at dusk and in the early morning. The diet consists of lizards (mainly lacertids and skinks), supplemented by rodents and amphibians. Irascible, it rarely tames in captivity but may feed well. Experimentally, its venom is as toxic as that of the puff adder, *B. arietans* (page 98), but yields are small. Symptoms from the few known bites are similar to those that present after a bite by the horned adder, *B. caudalis* (page 100); no fatalities are known. Mating takes place in October-November; 7-12 (max. 20) young, measuring 130-150 mm, are born in late summer.

Habitat: Mountains or sandy plains. **Range:** S. Namibia, through W. Cape to E. Cape. **Subspecies:** Two races are recognized. *B.c. cornuta* (the western many-horned adder) has prominent horns, a well-defined pattern, and a higher ventral count (128-150); it occurs in Namibia to W. Cape. In *B.c. albanica* (the eastern many-horned adder), the horns are reduced or absent, and it has a duller pattern, and a lower ventral count (120-130); it occurs in E. and S. Cape. Snakes from scattered localities in the intervening region may lack horns.

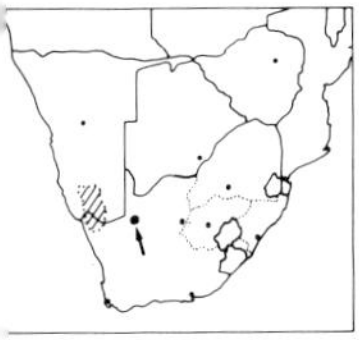

Desert Mountain Adder *Bitis xeropaga* (Pl. 13)

(Endemic) 40-50 cm; max. 61 cm
This small, relatively slender adder has raised ridges over the eyes, but no horns. The number of scale rows around the neck may be the same as or greater than that at midbody (25-27). The subcaudals are usually smooth. The back is dirty buff or ash to dark grey, with 16-34 bars, each consisting of a median dark brown rectangle flanked on each side by a white spot and a light brown region. There is no dark mark on the top of the head. Three to four white bars radiate from the eye to the mouth. The belly is light dusky-grey with dark speckles. **Biology and breeding:** This adder is similar in habits and behaviour to the many-horned adder, *B. cornuta* (page 100); they are not found together. Neither sidewinds or buries in sand. The desert mountain adder accepts skinks and mice in captivity. Nothing is known about its venom, and there have been no recorded bites. (SA RDB, Restricted.) Little is known about this snake's breeding; 4-5 young are born in late summer. **Habitat:** Sparsely vegetated rocky hillsides and mountain slopes. **Range:** Lower Orange River, from Augrabies Falls to Richtersveld and Aus in Namibia.

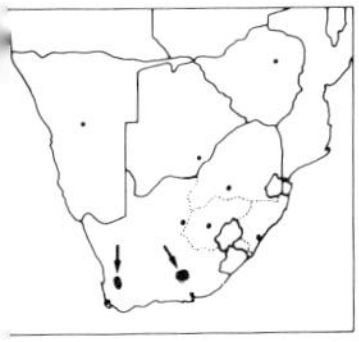

Plain Mountain Adder *Bitis inornata* (Pl. 14)

(Endemic) 25-32 cm; max. 34 cm
A small, squat adder that lacks horns above the eyes. The scales are in 27-31 rows at midbody. The posterior subcaudals are faintly keeled. The back is dull brown to red-brown, with faint darker blotches. The belly is light brown, with the blotches restricted to the sides. Males are smaller than females. **Biology and breeding:** This snake shelters in grass tussocks and beneath rock slabs on mountain plateaus. It is active during the early morning, feeding on lizards, (skinks and lacertids) and rodents. It settles well in captivity, but is sensitive to high temperatures. It hibernates during the winter snows. Nothing is known about its venom; there have been no recorded bites. (SA RDB, Restricted.) Six to eight young (125-152 mm) are born in late summer. **Habitat:** Grassveld of rocky mountain plateaus. **Range:** Inland old escarpment, with relict populations on Cedarberg, and Sneeuberg near Graaff-Reinet.

Namaqua Dwarf Adder *Bitis schneideri* (Pl. 14)

(Endemic) 18-24 cm; max. 28 cm
This is the smallest adder. Its eyes are situated on either side of its rounded head. There is a small ridge over each eye bearing a small, hornlike scale.

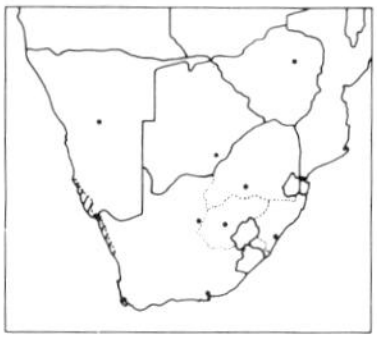

Scales are in 23-27 rows at midbody. Ventrals number 104-129 and subcaudals are keeled (but are smooth proximally in males). The back is grey to brownish-grey, with three series of rounded, dark, pale-centred blotches, and is speckled with black. The head has a pale arrowhead. The tail tip is sometimes black. The belly is dirty yellow, speckled with black. **Biology and breeding:** This snake shuffles into loose sand at the base of grass tussocks, leaving only its head exposed. Mainly nocturnal, it eats small geckos and frogs. It sidewinds readily. Its habitat is threatened by alluvial diamond mining (SA RDB, Vulnerable). Its venom is very mildly cytotoxic; symptoms include local swelling and pain. Antivenom is unnecessary in the treatment of its bite, and there have been no recorded fatalities. Three to four young, measuring 110-130 mm, are born in late summer. **Habitat:** Semi-stable, vegetated coastal sand dunes. **Range:** Southern regions of Namib Desert, from Lüderitz Bay in Namibia to Little Namaqualand.

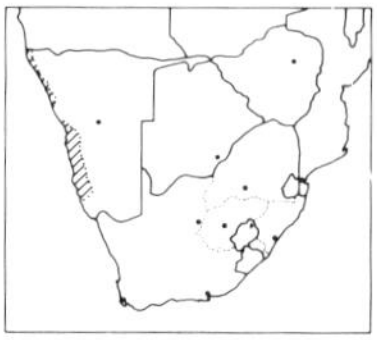

Péringuey's Adder *Bitis peringueyi* (Pl. 14)

20-25 cm; max. 29 cm
This very small adder has its eyes situated on top of its rounded, flat head. There is no supraorbital ridge. Scales are in 23-31 rows at midbody. Ventrals number 117-144, and the subcaudals are smooth except towards the tail tip. The back is pale buff to orange-brown, or pale greyish-yellow, with three series of faint, dark spots (those on the flanks are pale-centered), and irregularly stippled with pale and dark spots. The back of the head sometimes has faint dark marks. The belly is whitish, occasionally with dark reddish-brown spots on the sides; 25% of these snakes have a black tail tip. **Biology and breeding:** This snake is famous for its ability to sidewind; this unusual locomotion involves undulating the body in smooth, lateral curves, lifting most of the body from the sand. Sidewinding enables the snake to move swiftly over hot, loose dunes. It shuffles completely into sand, leaving only its eyes exposed, and sometimes its black tail tip, which it waves to attract deser and sand lizards. It drinks fog that condenses on its flattened body. It is endangered by over-collecting for the pet trade. Its venom is very mild, causing pain and local swelling; antivenom is unnecessary in the treatment of its bite. There have been no recorded fatalities. Three to 10 minute young (80-110 mm) are born in March-April. **Habitat:** Fine, wind-blown sand of true Namib Desert. **Range:** West coast, from Rotkuppe in S. Namibia to S. Angola.

BUSH VIPERS
Atheris

These are small vipers with fragmented head shields, strongly keeled scales and prehensile tails.

They are usually arboreal (but two species are terrestrial), and feed on sma tree frogs and lizards. They are viviparous.

There are 10 species distributed through the tropical rain forests of W. and central Africa. A single, terrestrial species just enters the subcontinent.

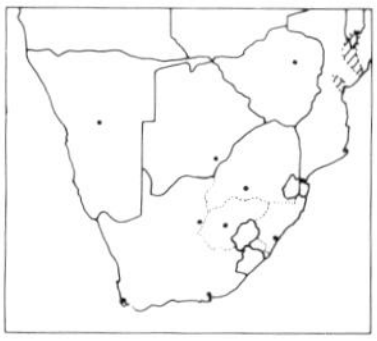

Lowland Swamp Viper *Atheris superciliaris* (Pl. 14)

40-60 cm; max. 61 cm
A robust snake with an elongate head and a large supraorbital scale. The pupils are vertical. The scales are strongly keeled, in 27-29 rows at midbody. The tail is longish (32-45 paired subcaudals). The back is grey-brown, with three rows of blackish spots, separated by a series of yellowish bars that form an interrupted lateral line. The top of the head has three black chevrons. The belly is off-white, with black markings. The undersurface of the tail is straw-yellow to orange. Males rarely grow longer than 40 cm. **Biology and breeding:** The habits and behaviour of this snake are poorly known. It inhabits rodent burrows, emerging at dusk to feed on small frogs (particularly reed frogs). It is aggressive when first captured, but tames readily. Its venom has not been studied; there have been no recorded bites. Three to eight young

(135-155 mm) are born in November-December. **Habitat:** Low-lying marshes and floodplains. **Range:** Mozambique Plain from Beira north to Quissanga, up Zambezi River to Lake Malawi and Tanzania.

Worm Lizards

Suborder Amphisbaenia

These very unusual reptiles were for a long time classed as lizards but have now been placed in a separate suborder within the Squamata. Fossils date back to the N. American Palaeocene epoch (65 million years ago). Worm lizards are distinguished by an enlarged medial tooth on the premaxillary bone, a reduced right lung (in snakes and legless lizards the left lung is reduced), and a unique middle ear, among many other features. They are the most specialized burrowing reptiles, capable of driving tunnels through hard soils, and are rarely seen above ground.

All are limbless (except the Mexican *Bipes*, which has front legs) and most retain only vestiges of the limb girdles. They lack external ears and have backward-facing nostrils, both being adaptations to prevent the entry of sand. Their eyelids are fused and the very reduced eyes lie deep below the translucent skin. The body is usually cylindrical, and is covered with smooth, squarish scales which are arranged in distinctive rings (annuli), causing these lizards to look like worms – hence the common name. These annuli are divided into dorsal and ventral sets by a faint longitudinal groove. The skin is separated from the underlying body, being attached only by three muscle sets per vertebra. This unique specialization allows the head and body to move within the skin so that a ramming motion can be generated for burrowing. To withstand these pressures, the skull is short and robust, with the braincase completely ringed by frontal bones for extra strength. Primitive species, and those inhabiting loose soils, are round-snouted, and the head is covered with close-fitting, enlarged head shields. Some species have evolved a hardened cutting edge to the snout that may form a horizontal spade or a vertical keel. This is used to scrape soil from the front of the burrow and compact it into the walls.

These squamates are small to medium-sized, the largest, at 75 cm, being from S. America. Most feed on invertebrates, especially ants and termites. Large prey is located by scent and vibration, gripped between the few, large teeth, and torn to bits by being dragged into the burrow. Some S. American species are large enough to eat small vertebrates. Like other squamates, males have hemipenes and fertilization is internal. The majority lay eggs, although some species retain these within the body and give birth to live young. Most are flesh-coloured, and some have darker dorsal pigmentation. The tail in larger species is usually short and rounded, sometimes pigmented and often heavily scarred. This is because in defence it may be waved in the air, mimicking the head and deflecting attacks away from the vital organs; in fact, the name 'amphisbaena' means 'two-headed'. In many smaller species the tail is longer and can be shed, but unlike that of lizards cannot be regenerated.

There are about 140 species in 21 genera and four families. These are distributed throughout most of subSaharan Africa and tropical S. America, with scattered populations in Arabia, Spain and N. America. Two families occur in Africa, with only the Amphisbaenidae reaching southern Africa.

OPICAL WORM ARDS
nily
phisbaenidae

This is by far the largest family of worm lizards, with 15 genera containing about 130 species, most of which are distributed in Africa and S. America. The family contains both round-headed and spade-snouted forms. There are nine African genera, containing 58 species, with 12 species (three of which are endemic) in four genera occurring on the subcontinent.

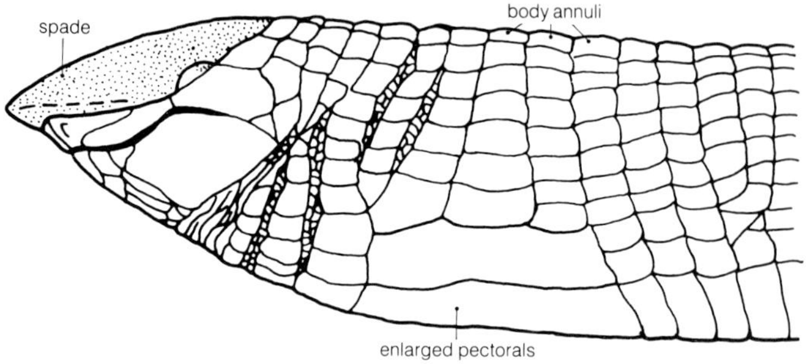

The head of a spade-snouted amphisbaenid

Key to the southern African genera in the Amphisbaenidae

1. Segments of the pectoral region not elongate; snout rounded, without a sharp cutting edge 2
   Segments of pectoral region elongate; snout with sharp horizontal cutting edge 3

2. Body slender; head shields fused into 1 or 2 shields .... *Chirindia* (Pink round-headed worm lizards, page 104)
   Body not slender; head shields all distinct ..... *Zygaspis* (Purple round-headed wo lizards, page 105)

3. Nasals well separated by rostral; tail short, bluntly rounded .... *Monopeltis* (Spade-snouted worm lizards, page 106)
   Nasals usually touching above rostral; tail long, terminating abruptly in a callous pad ..... *Dalophia* (Blunt-tailed worm lizards page 108)

PINK ROUND-HEADED WORM LIZARDS *Chirindia*

These small, elongate worm lizards have a rounded head and extensive fus of the head shields. The nasal, first upper labial and prefrontal, and sometin other shields, are all fused behind the rostral into a large shield.

Restricted to E. and S. Africa, there are four Tanzanian species and two ir southern Africa; one of these is endemic.

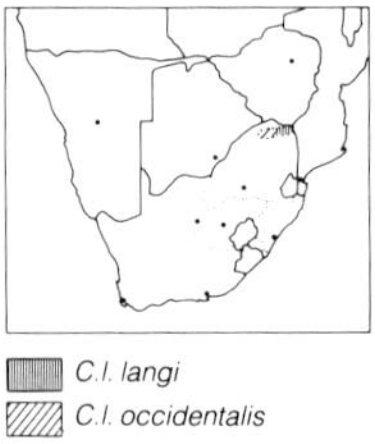

Lang's Round-headed Worm Lizard *Chirindia langi* (Pl. 4

(Endemic) 12-14 cm; max. 16,5 cm
A small, elongate worm lizard with a rounded head. There are three upper labials, and the ocular is fused with the second upper labial. There are 242-309 body annuli, with 27-35 segments per annulus (see Subspecies). Males have 4-6 preanal pores (see Subspecies). The tapering tail has a rounded tip and 24-28 caudal annuli. The body is uniform unpigmented flesh-pink throughout. **Biology:** This worm lizard burrows in loose soil, and is usually discovered beneath stones or rotting logs. It eats mainly termites. When caught, it wiggles wildly to escape and will shed its tail if grasped (17-19% of specimens have truncated tails). Predators include snake (particularly purple-glossed and quill-snouted snakes), jackals and ratels. **Habitat:** Sandy Kalahari soils, entering mopane woodland on clay soils.

Range: NE. Transvaal, SE. Zimbabwe and adjacent Mozambique.
Subspecies: Two races are recognized. *C.l. langi* is longer (av. 13,3 cm), with more body annuli (275-309), but fewer segments (usually 30) per annulus, and only four preanal pores in males; it occurs in the eastern part of the range, from N. Kruger National Park. *C.l. occidentalis* is smaller (av. 12,5 cm), with fewer body annuli (242-262), but more segments (usually 33-34) per annulus, and six preanal pores in males; it is found in N. Transvaal, west of Kruger National Park.

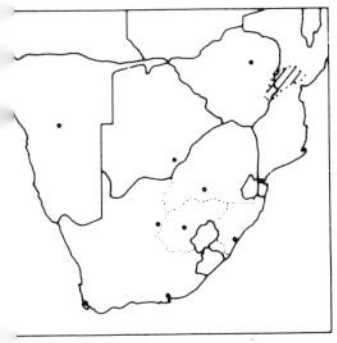

Swynnerton's Round-headed Worm Lizard
Chirindia swynnertoni **(Pl. 42)**

12-14 cm; max. 17 cm
This short, elongate worm lizard has a rounded head and two upper labials (the nasal, first and second upper labials, prefrontal and ocular are all fused). There are 235-265 body annuli, with 24-28 segments per annulus. Males have six preanal pores. The tail tapers to a rounded tip, and has 19-26 caudal annuli. The body is uniform unpigmented flesh-pink, with a purple sheen. **Biology and breeding:** This species is similar in habits and behaviour to Lang's round-headed worm lizard, *C. langi* (page 104). Predators include the dwarf wolf snake, which is a specialist feeder on this worm lizard. The first specimen was found in the stomach of a kingfisher. Its breeding is poorly known; a single egg (22 x 3 mm) was found in a female in December. **Habitat:** Thicket and grassland on alluvial soils. **Range:** Mozambique Plain and adjacent Zimbabwe.

RPLE ROUND-
EADED WORM
ZARDS
gaspis

These small, stoutish worm lizards have a rounded head, and distinct nasal, prefrontal and ocular scales.
Three species are distributed from S. Zaire to N. Cape, and all occur on the subcontinent. One is endemic.

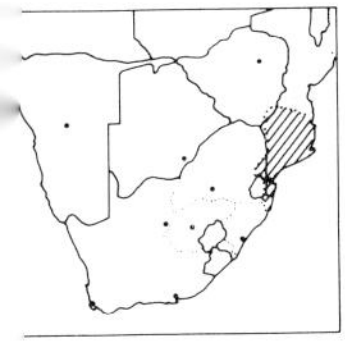

Violet Round-headed Worm Lizard
Zygaspis violacea **(Pl. 42)**

(Endemic) 14-16 cm; max. 20 cm
A small, stout worm lizard with a rounded snout. The preocular is fused with the prefrontals. It has three upper labials. There are 179-211 body annuli, with 12-22 (usually 14-16) dorsals and 12-20 (usually 14-16) ventrals per midbody annulus. The tail is longish, tapering slightly, and has 39-52 caudal annuli. The back is uniform dark purple-brown, the belly is light purple, sometimes with the pigment restricted to the scale edges, and the chin and cloacal region are white. **Biology:** This worm lizard lives under stones or rotting logs, in aeolian and alluvial sands. It feeds on termites. Predators include Bibron's burrowing asp. **Habitat:** Varied; savannah and coastal thicket. **Range:** Mozambique Plain south of Save River, into N. Zululand, SE. Zimbabwe and SE. Transvaal.

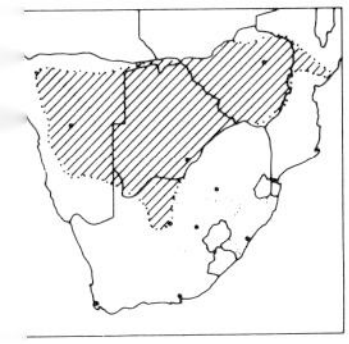

Kalahari Round-headed Worm Lizard
Zygaspis quadrifrons **(Pl. 42)**

16-20 cm; max. 23 cm
This small worm lizard has a rounded snout and elongate preoculars. The third upper labial is bordered by a temporal shield above a smaller post-upper labial. There are 198-242 body annuli, with 13-22 dorsals and 13-19 ventrals per midbody annulus. Males have four preanal pores. The tail is longish, with 32-49 caudal annuli. The body is uniform purple-brown above, with a lighter belly; it is often darker towards the tail. **Biology:** This worm lizard is common under stones in suitable habitat. It feeds on small insects and their larvae, particularly termites. **Habitat:** Sandy scrub and bushveld. **Range:** Through Kalahari, from N. Cape to Zaire, and along Limpopo River to Mozambique.

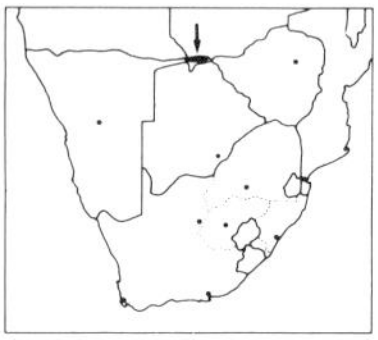

Black Round-headed Worm Lizard *Zygaspis niger* (Pl. 42)

20-24 cm; max. 30 cm
A large worm lizard with a rounded snout and elongate preoculars. The third upper labial is bordered by a large temporal shield. There are 185-205 body annuli, with 16-22 dorsals and 13-16 ventrals per midbody annulus. Males have four preanal pores. The tail is longish, with 43-54 caudal annuli. The ba is ivory, with a black base to each scale (90% on dorsals, 50% on ventrals), giving a speckled effect. The chin and scattered areas are ivory.
Biology: This species is similar in habits and behaviour to the Kalahari round-headed worm lizard, *Z. quadrifrons* (page 105). **Habitat:** Savannah o sandy soils. **Range:** SW. Zambia and adjacent Angola, just entering Caprivi Strip.

SPADE-SNOUTED WORM LIZARDS
Monopeltis

These are large worm lizards with a broad, horizontal, spade-shaped snout that is covered with 1-2 large, horny shields. The nasal shields are always separated by the rostral. The pectoral region usually has very enlarged, lon smooth shields. The body is cylindrical, with fewer than 300 (very rarely mor than 300) annuli (counted along the belly, as extra half-annuli are often inserted on the dorsal surface). The tail is usually short (4-24 caudal annuli, extralimitally 61 in *M. adercae* from Zaire), with a rounded tip. Preanal pores may number up to 13 (but there are usually only two in southern African species), and sometimes there are none.

Capable of burrowing in hard soils, these worm lizards are found in a wide variety of habitats, although they rarely enter forest. Most are exposed durin earth moving, either in road construction or ploughing. Floods and attacks b carnivorous ants also force them to the surface. In suitable habitat they may common. Reproduction is poorly known, but at least one species gives birth live babies. The tail may be shed in long-tailed species. They feed almost exclusively on termites, although adult and larval insects are also taken. Predators include snakes, particularly quill-snouted snakes, but when force to the surface they are also eaten by small carnivores and even birds of prey.

There are 16 species in central and southern Africa, with five species occurring on the subcontinent, one of which is endemic.

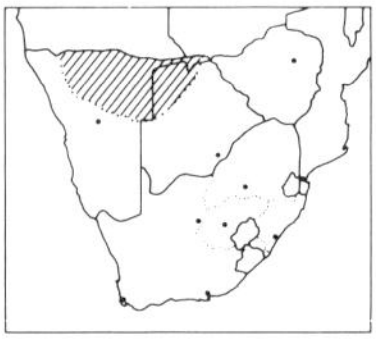

Angolan Spade-snouted Worm Lizard
Monopeltis anchietae (Pl. 41)

20-30 cm; max. 35 cm
This robust species has two shields that form a 'spade'. There are four pectorals. It has 170-198 body annuli, with 6-28 extra dorsal half-annuli. The tail is short (5-9 caudal annuli), with a rounded tip that cannot be shed. There are two preanal pores. The back, from the snout to the tail tip, is reddish-brown, being darkest behind the head and over the tail and lighter c the flanks. The belly is unpigmented. **Habitat:** Moist savannah, entering drie regions. **Range:** N. Namibia and Caprivi Strip, and into adjacent Angola an Botswana.

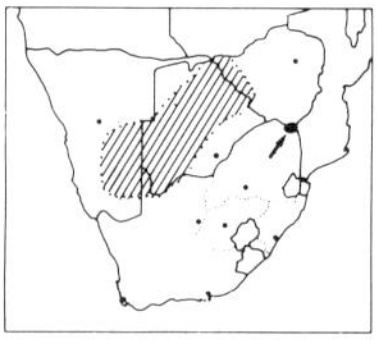

Leonhard's Spade-snouted Worm Lizard
Monopeltis leonhardi (Pl. 41)

(Endemic) 20-25 cm; max. 30 cm
A medium-sized species with two shields that form a 'spade'. It has six pectorals. There are 170-213 body annuli, with 2-22 extra dorsal half-annuli. The tail is short (5-9 caudal annuli), with a rounded tip that cannot be shed. There are no preanal pores. Coloration is similar to that of the Angolan spade-snouted worm lizard, *M. anchietae* (above); juveniles are less strong pigmented. **Biology:** Predators on this species include jackals and yellow-

billed kites. **Habitat:** Kalahari sands. **Range:** Central Botswana, into adjacent Namibia, N. Cape and SW. Zimbabwe, extending along Limpopo River to Kruger National Park.

Zambezi Spade-snouted Worm Lizard
Monopeltis zambezensis

9-15 cm; max. 18 cm
This very small, slender species resembles the Cape spade-snouted worm lizard, *M. capensis* (below). It has one shield forming a 'spade'. It has six pectorals. There are 226-263 body annuli, with 18-20 dorsals and 13-16 ventrals, and few extra dorsal half-annuli. The tail is short (6-8 caudal annuli), with a rounded tip that cannot be shed. There are two preanal pores. The back is unpigmented anteriorly, but is speckled with pigment over the rear, particularly on the tail. **Habitat:** Mopane on red soils. **Range:** Zambezi River valley around Kariba Dam.

M.c. capensis
M.c. rhodesianus

Cape Spade-snouted Worm Lizard
Monopeltis capensis (Pl. 41)

25-30 cm; max. 35 cm
This robust, medium-sized species has one shield forming a 'spade'. There are 4-6 pectorals. It has 172-221 body annuli, with 13-33 dorsals and 10-32 ventrals (see Subspecies), and numerous extra dorsal half-annuli. The tail is short (7-11 caudal annuli), with a rounded tip that cannot be shed. There are two preanal pores. The body is uniform pink-white, in some localities having a darker tail (see Subspecies).

Biology and breeding: This worm lizard is very common in suitable habitat. It burrows in red soils and feeds on termites, beetle larvae, etc. For many predators it must be as tasty as the worms it resembles; enemies include birds (the yellow-billed kite, bateleur and hornbill), jackals, and particularly certain snakes (burrowing asps and quill-snouted snakes, among others). Floods drown many, as well as pushing them to the surface where predators wait. One to three babies (90-100 mm) are born in summer. **Habitat:** Kalahari sandveld and alluvial plains. **Range:** Kalahari region, south to Orange River and along Zambezi and Limpopo river valleys, extending into S. Mozambique Plain.

Subspecies: Two races are recognized. *M.c. capensis* has more than 40 dorsals and ventrals per body annulus, and a plain tail; it occurs through Transvaal, N. Cape and S. Botswana, to Namibia, Angola and Zambia. *M.c. rhodesiana* has fewer than 40 dorsals and ventrals per body annulus, and the tail is pigmented on top; it is found in central Zimbabwe and adjacent regions.

M.s. sphenorhynchus
M.s. mauricei

Slender Spade-snouted Worm Lizard
Monopeltis sphenorhynchus (Pl. 41)

22-27 cm; max. 28 cm
A smallish, slender species with one shield forming a 'spade'. It has six pectorals. There are 228-316 body annuli (see Subspecies), with 20-45 dorsals and 16-29 ventrals, and few or no extra dorsal half-annuli. The tail is short (7-11 caudal annuli), with a rounded tip that cannot be shed. There are two preanal pores. The body is pale pink to whitish all over. **Habitat:** Deep Kalahari sand or coastal alluvium. **Range:** Northern part of the subcontinent. **Subspecies:** Two races are recognized. *M.s. sphenorhynchus* has its ocular separated from the nasal, and 228-284 body annuli; it is found on S. Mozambique Plain, into N. Zululand and along Limpopo River to N. Transvaal. *M.s. mauricei* has its ocular touching the nasal, and 275-316 body annuli; it occurs in Botswana, into adjacent regions.

BLUNT-TAILED WORM LIZARDS
Dalophia

These are large worm lizards, very similar in appearance to the spade-snouted worm lizards, *Monopeltis* (page 106), having a broad, horizontal, spade-shaped snout, usually with a single large, horny shield. The nasal shields are usually in contact with each other or are only narrow separated. The pectoral region has 4-6 enlarged, smooth, long shields. Th body is slender, usually with more than 300 body annuli (counted along th belly). There are no preanal pores. The tail is slender (17-46 caudal annuli and always truncated, with a characteristic flattened terminal pad.

Although they are similar in most respects of their behaviour and habits the spade-snouted worm lizards, they feed more heavily on beetles and th larvae, rather than termites. Reproduction is poorly known, but some spec are known to lay eggs. Three species (none of which occurs on the subcontinent) can shed their tail in defence.

Seven species occur in Africa south of Zaire, with only two entering the subcontinent.

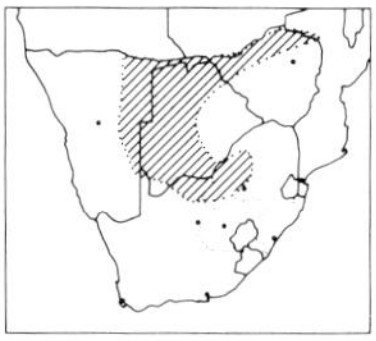

Blunt-tailed Worm Lizard *Dalophia pistillum* (Pl. 41)

35-45 cm; max. 56 cm
A large worm lizard with a broad, horizontal 'spade' that is covered with a single horny shield, partially divided by lateral sutures. The nasals are in contact. There are six pectorals, and 280-352 body annuli. The tail is blunt-ended, with 19-33 caudal annuli but without 'herringbone' segmenta on the dorsal surface. There are no preanal pores. The body is uniform flesh-coloured, with light grey speckles on the back and tail, that may rare extend on to the lower flanks. **Biology and breeding:** This species is prey on by quill-snouted snakes, ratels and polecats. (SA RDB, Peripheral.) Th female lays four eggs (32-35 x 8-10 mm) in September. **Habitat:** Varied; includes Kalahari sand and coastal alluvium. **Range:** Northern border of t region, south to N. Cape and OFS. Elsewhere, to Zambia and along Zamb River to mid-Mozambique Plain.

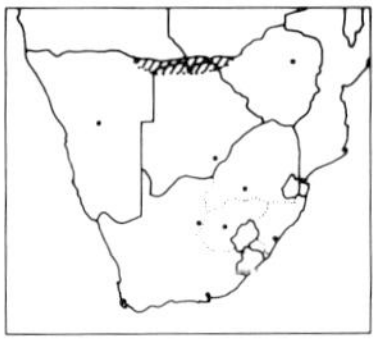

Long-tailed Worm Lizard *Dalophia longicauda*

38-45 cm; max. 52 cm
This large worm lizard resembles the blunt-tailed worm lizard, *D. pistillum* (above). It has a broad, roundly-pointed, horizontal 'spade' that is coverec with a single horny shield, partially divided by lateral sutures. The nasals a contact. There are six pectorals, and 307-338 body annuli. The tail is long blunt-ended, with 33-42 caudal annuli, and with 'herringbone' segmentati on the dorsal surface. There are no preanal pores. The body is uniform fle pink anteriorly, with light grey speckles on rest of the body and tail, that m rarely extend on to the lower flanks; the belly is unpigmented. **Habitat:** Mc alluvial soils along permanent river courses. **Range:** Caprivi Strip and adjacent regions.

Lizards

Suborder Sauria (Lacertilia)

Lizards are the most familiar reptiles, although many are highly specialize and the legless species are often confused with snakes. Typically, they ha well-developed limbs, but even in the most highly specialized legless forn internal vestiges of limb girdles are always present. The body is covered v scales that may be granular and non-overlapping. They usually have exte ears and movable eyelids, although in most geckos and some skinks the are transparent and fused, forming a 'spectacle' over the eye. The halves the lower jaw are fused. The tongue cannot be withdrawn into a sheath, although in some families (e.g. monitors) it is retractile. Many lizards (e.g. geckos and skinks) can shed their tail and regenerate a new one.

Lizards have radiated into many environments, and are particularly common and diverse in deserts. There are not many aquatic species, with only the marine iguanas of the Galápagos Islands and a few freshwater monitors being at home in water. Most feed on insects and other invertebrates, but a number of species are herbivorous, while some large species regularly eat small mammals. The Komodo dragon may even kill and eat humans, albeit rarely. The only venomous lizards are the two species of Gila monster restricted to the south of North America. Most lizards are egg-laying, and a number of American species brood their eggs after laying. Viviparity is common in some families, particularly the skinks.

Lizards occur throughout the world; some species have even reached the small oceanic islands of the Indo-Pacific. They form the largest group of living reptiles, with over 3 700 species in nearly 400 genera. At least 16 families are recognized, and these are sometimes grouped into four infraorders, all of which are represented in southern Africa. The seven local families are placed in the Anguimorpha (Varanidae), Gekkota (Gekkonidae), Iguania (Agamidae and Chamaeleonidae) and Scincomorpha (Scincidae, Lacertidae and Cordylidae). There is still some debate as to the relationships of these groups.

to the Lizard
ilies in southern
ca

1. Top of head covered with granules or small, irregularly arranged scales 2

   Top of head covered with large, symmetrical shields 5

2. Toes bound in opposed bundles for grasping; tongue extensile, with club-shaped tip; eyes set in independently movable turrets .... Chamaeleonidae (Chamaeleons, page 179)

   Toes separate; tongue not extensile; eyes normal 3

3. Dorsal scales overlapping and strongly keeled .... Agamidae (Agamas, page 174)

   Dorsal scales juxtaposed, smooth or granular, sometimes with scattered tubercles 4

4. Eyelids usually not or hardly movable, cannot close over eye; tongue short and broad, covered with soft papillae; tail easily shed .... Gekkonidae (Geckos, page 187)

   Eyelids movable, can close over eye; tongue forked at tip; tail cannot be shed ...... Varanidae (Monitors, page 172)

5. Dorsal scales overlapping, cycloid, smooth or with 3 or more keels; femoral pores absent .... Scincidae (Skinks, page 111)

   Dorsal scales small and granular or juxtaposed and with a strong keel; with femoral pores or a row of glandular scales on posterior face of thigh 6

6. Lateral granular fold present and/or limbs vestigial .... Cordylidae (Plated lizards, girdled liz and their relatives, page 1

Lateral granular fold absent; limbs well developed .... Lacertidae (Old World lizards, page 1

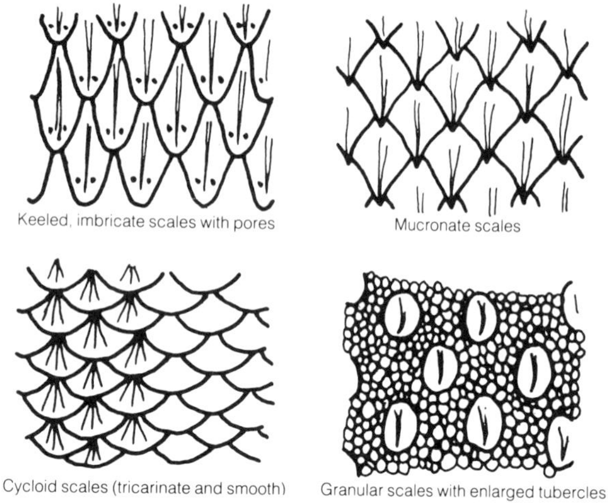

Lizard scales

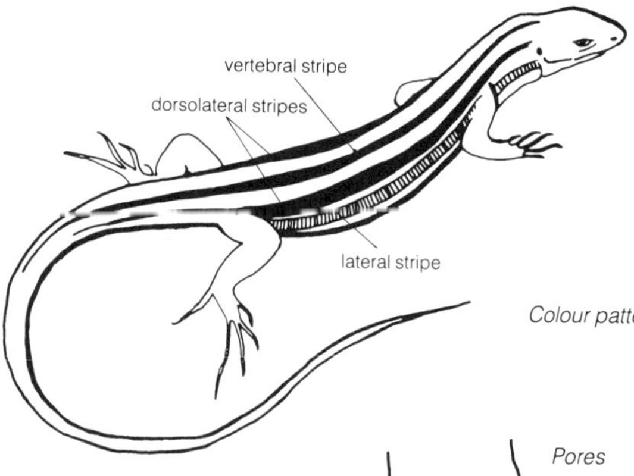

Colour patterns

Pores

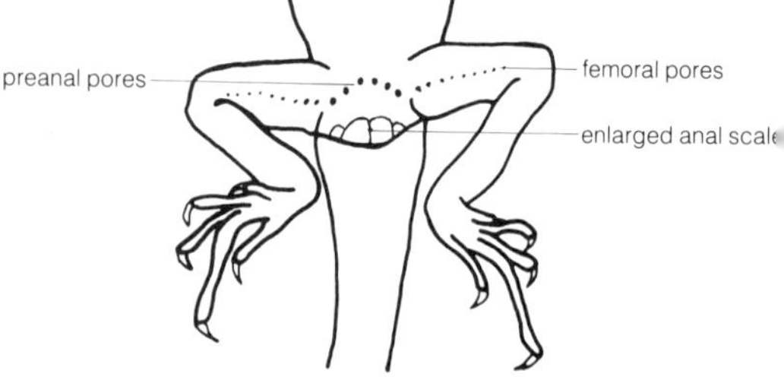

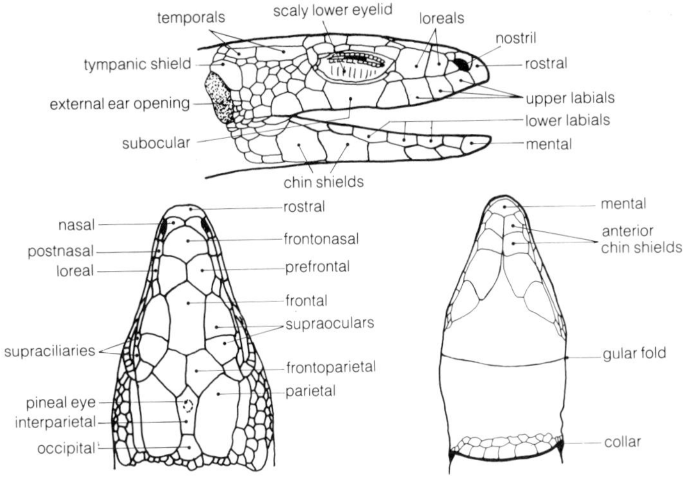

The head scales of a scincomorph lizard

KINKS
amily Scincidae

Although most of these diverse lizards have well-developed limbs, some have either no limbs or only vestiges of limbs. The dorsal scales are usually smooth, flat and highly polished, often appearing iridescent. They overlap strongly and are underlaid by osteoderms. This gives skinks a highly flexible but rigid and strong coat, ideal for burrowing or living in rock cracks. The head is relatively small and usually lacks an obvious neck. It is covered with large, symmetrical head shields. Ear holes are usually present, and often protected by scales. The eyes are often small (and absent in some burrowing forms) and the pupils are round. Functional eyelids are usually present, but in some species the lower eyelids have transparent windows and are immovable (like those of snakes). Femoral pores are absent. The tail is easily shed and quickly regenerated.

Most skinks may be called 'typical' lizards. They are usually terrestrial, although many climb trees and rocks, and are active during the day. Most are of medium size, with very few growing large. They feed almost exclusively on small insects, which they seize after a short rush from cover. Most maintain a high body temperature by shuttling between sunny spots and shade. Some territorial species, which are mainly rock-living, are brightly coloured and develop dominance hierarchies, with brilliant, tyrant males. However, most species are solitary and drably patterned in browns and greys, and forage among loose leaves and wooded litter. Many of these have elongate bodies and reduced limbs, and in some burrowing genera the limbs are completely lost. In southern Africa the dwarf burrowing skinks, *Scelotes* (page 119) show a clear evolutionary progression towards limb loss, and demonstrate how snakes must have evolved. Most skinks lay small clutches of soft-shelled eggs underground. Viviparity has evolved on many occasions. Individuals in some local species may lay eggs or give birth to live young.

This family is found throughout the tropical and temperate regions of the world. Despite this diversity, it has a very poor fossil history, and only two genera are known prior to the Pleistocene epoch (two million years ago). One of these dates from the late Cretaceous period (about 80 million years ago), and the other from the Oligocene epoch onwards (38 million years ago). The family is divided into three subfamilies, all of which occur in southern Africa. Approximately 75 genera with over 600 species occur worldwide. This is the second most diverse group of lizards in southern Africa (exceeded only by the geckos), with 59 species in 11 genera; 31 species and one genus are endemic.

Key to the southern African genera in the Scincidae

1.	Each nostril pierced between the rostral and a very small nasal shield; prefrontals and frontoparietals (if present) very small; scales smooth; limbs reduced or absent	2
	Each nostril pierced in a small nasal shield and well-separated from the rostral; prefrontals and frontoparietal(s) present	4
	Each nostril pierced in the front of a very large rostral and with a groove running to rear border of rostral; scales smooth; limbs absent	7
2.	Interparietal scale not touching supraocular scales	3
	Interparietal touching supraocular scales	*Scelotes* (Dwarf burrowing skinks, page 119)
3.	Five toes on each forelimb and hind limb	*Proscelotes* (Slender skinks, page 11
	Three or four toes on each forelimb and hind limb	*Sepsina* (Savannah burrowing skinks, page 123)
4.	Eyelids movable	6
	Eyelids immovable; each eye covered with a transparent spectacle; scales smooth; limbs present but small	5
5.	Three supraoculars; 24-26 scale rows at midbody; terrestrial	*Panaspis* (Snake-eyed skinks, page 134)
	Five supraoculars; 26-29 scale rows at midbody; inhabiting intertidal zone	*Cryptoblepharus* (Coastal skinks, page 12
6.	Transparent window in each lower eyelid; limbs well-developed; scales with fine keels	*Mabuya* (Typical skinks, page 127
	Lower eyelids scaly, but may be translucent; limbs short; scales smooth or weakly striated	*Lygosoma* (Writhing skinks, page 12
7.	Eyes completely exposed, without eyelids; no enlarged preanal plate; three enlarged, transverse scales on front of head	*Typhlacontias* (Western legless skinks, page 124)
	Eyes with eyelids, or situated under head shields; an enlarged preanal plate; two enlarged, transverse scales on front of head	8

8 No eyelids present; eyes appearing as dark spots below head shields .... *Typhlosaurus* (Blind legless skinks, page 116)

Lower eyelids movable and elongate; 3-4 supraciliaries .... *Acontias* (Greater legless skinks, page 113)

Lower eyelids immovable, oval and transparent; two supraciliaries .... *Acontophiops* (Woodbush legless skink, page 115)

FRICAN LEGLESS KINKS
bfamily Acontiinae

These specialized, legless, burrowing skinks are characterized by a divided frontal bone in the skull. They lack all traces of external limbs and have a short, stubby tail (less than 22% of the total length). The body scales are smooth, tightly fitting and not enlarged on the belly. There is a single enlarged anal plate. The front scales on the head (rostral and mental) are enlarged, and each nostril is placed in the front of the rostral and is connected to its rear border by a long groove. There are no external ear openings. Coloration is varied, and many have a striped pattern.

All African legless skinks are burrowing reptiles, most being restricted to loose sandy soil or leaf litter. A few of the larger, more robust species are able to burrow in harder, clayey soils. They feed on earthworms, beetle larvae and termites. Details of reproduction are known for few species; these give birth to 1-14 babies.

This subfamily is restricted mainly to southern Africa, with a few species extending into adjacent countries. There are three genera with 17 species, and all but two are endemic to the subcontinent.

REATER LEGLESS
INKS
ontias

This is a group of medium-sized to large legless skinks that have elongate, movable lower eyelids and three to four supraciliary scales above each eye. Greater legless skinks do not have external ear openings. The tail is short and stubby.

All are burrowing and are normally found under stones or dead logs on loose soil. They feed on small invertebrates, but the bigger species may take snakes and other burrowing lizards. They rarely drink from standing water, and appear to obtain most of their moisture from the surrounding soil and their food. It is possible that the long groove on the rostral may funnel soil water to the nostril. They are viviparous, giving birth to a single brood of young in late summer. They are often eaten by burrowing snakes (eg. the burrowing asp and harlequin snake) and small carnivores.

There are seven species in the genus, and all occur in the region; six are endemic, the remaining species having an isolated, relict population in SE. Kenya.

Short-headed Legless Skink *Acontias breviceps*

(Endemic) 20-23 cm; max. 24 cm
A medium-sized legless skink, similar in appearance to the thin-tailed legless skink, *A. gracilicauda* (page 114). It has a broad head, a rounded snout, and a slender body. The lower eyelids are opaque, and each usually has three suboculars. The tail is cylindrical, and has a middle row of enlarged subcaudals. The body is olive to olive-brown, with the scales spotted with dark brown to black. The belly is light olive-yellow, with spotted scales. Juveniles have a creamy, plain belly.

Breeding: Two females from the Transvaal each contained two embryos (measuring up to 64 mm) in January. **Habitat:** Montane grassland. **Range:** E. Cape and E. Transvaal.

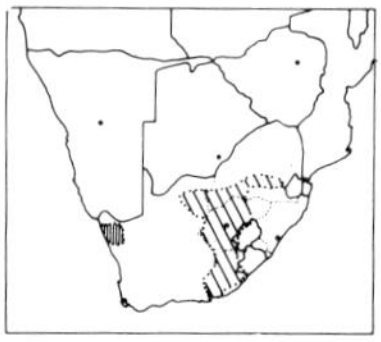

A.g. gracilicauda
A.g. namaquensis

Thin-tailed Legless Skink *Acontias gracilicauda* (Pl. 44)

(Endemic) 23-28 cm; max. 31 cm
A medium-sized legless skink with a broad head, a rounded snout, a slende body and a thin, tapering tail. The lower eyelids are opaque, and each has three suboculars. The second upper labial usually enters the eye. There are 18 scale rows at midbody. The body ranges in colour from pale golden-olive olive to grey-brown, but the body scales are always dark-edged, giving a speckled appearance. The belly is plain pale golden-yellow. **Biology:** Thes skinks show a preference for compact, hard soils. **Habitat:** Valley bushveld grassland, and entering sandy regions. **Range:** Two isolated populations in Little Namaqualand, and E. Cape, OFS and S. Transvaal. **Subspecies:** Two races are recognized. *A.g. gracilicauda* has 149-171 ventrals, and its body olive to olive-grey; it occurs in E. Cape and highveld of OFS and S. Transva *A.g. namaquensis* has 158-179 ventrals, and its body is pale olive-brown, w dark scale margins; it is found in Little Namaqualand.

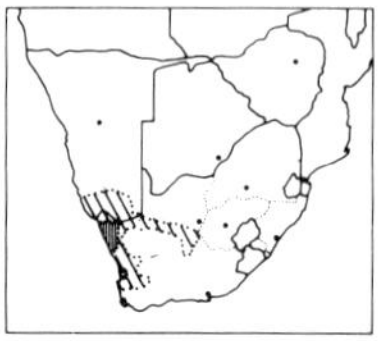

A.l. lineatus
A.l. tristis
A.l. grayi

Striped Legless Skink *Acontias lineatus* (Pl. 45)

(Endemic) 15-17 cm; max. 18 cm
A small legless skink with a flattened snout that has a horizontal, spade-like edge. The lower eyelids are transparent. The tail is flattened below. There ar a number of colour variations (see Subspecies); in addition, some specimer are uniform black above and below, while others are black-backed but have pale belly or a few pale spots around the head or vent. **Biology and breeding:** These skinks are common in sandy flats, burrowing in loose soil a the base of vegetation. They possibly have only a single baby.
Habitat: Sandy, arid soils. **Range:** W. and N. Cape, and into Little and Grea Namaqualand. **Subspecies:** Three races are recognized. *A.l. lineatus* usua has five upper labials, and a yellow back with 4-10 dark stripes or rows of spots, and a plain, flesh-coloured belly; it occurs in Great Namaqualand, N. Cape and inland regions of W. Cape. *A.l. tristis* has similar coloration to th typical race, but has only four upper labials; it is found in Little Namaqualanc In *A.l. grayi*, there are usually five upper labials, and each scale on the back has a dark, transverse mark; it occurs in the vicinity of Graafwater in SW. Cape.

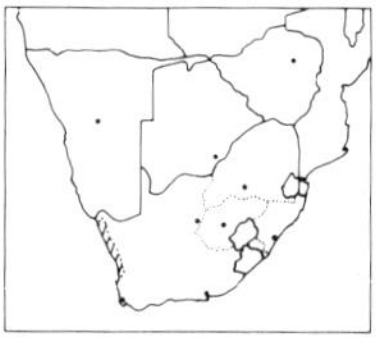

Coastal Legless Skink *Acontias litoralis* (Pls. 43 and 44)

(Endemic) 11-13 cm; max. 15 cm
A small legless skink with a flattened snout that has a horizontal, spade-like edge. The lower eyelids are transparent. There are four upper labials. Ventra number 145-160. The tail is flattened below. The head and body are yellow, with a broad, dark purple-brown stripe down the back. Some specimens are all yellow-orange except for a dark eye stripe. **Habitat:** Sparsely vegetated coastal sands. **Range:** W. Cape coastal strip from Daberas, south to Elandsbaai.

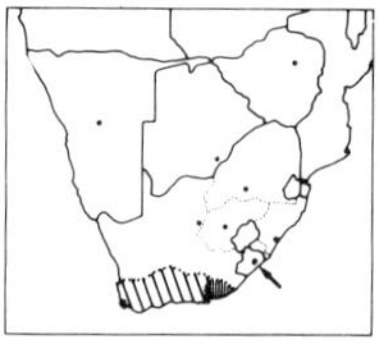

A.m. meleagris
A.m. orientalis

Cape Legless Skink *Acontias meleagris* (Pl. 45)

(Endemic) 20-26 cm; max. 30 cm
A medium-sized legless skink with a slender head and body, a rounded sno and a blunt tail that tapers only slightly. The lower eyelids are opaque. There are three suboculars. The second upper labial does not enter the eye. There are 14-16 scale rows at midbody. **Biology and breeding:** These skinks are found in dry sandy soils, often below stones or dead trees. They usually give birth to four babies (up to 80 mm) in late summer. **Habitat:** Coastal and fynb vegetation, and richer soils associated with dry river courses and inland escarpment. **Range:** W. and E. Cape, with isolated populations extending

inland along Karoo escarpment. **Subspecies:** Two races are recognized. In *A.m. meleagris*, the tail is not tapered, and the back is plain or has a series of dark spots that may form ragged stripes; it occurs in W. Cape, through Cape fold mountains and escarpment mountains of Karoo, to Cradock and Tsitsikamma. *A.m. orientalis* has a slightly tapered tail, and the back has six distinct black stripes; it is found in E. Cape.

Giant Legless Skink *Acontias plumbeus* (Pl. 44)

(Endemic) 35-45 cm; max. 55 cm
The world's largest legless skink, with a stout body, elongate snout and broad head, and a cylindrical tail. The lower eyelids are opaque. There are two suboculars. The body is blue-black to black, often with a steel-grey snout. The belly is paler. **Biology and breeding:** These skinks prefer accumulated leaf litter and humic soils in damp situations. They usually eat large invertebrates (worms, centipedes and larvae) but will also take frogs and other small vertebrates. The female gives birth to 2-14 babies (106-127 mm) in March-April. They grow slowly, and settle well in captivity, eating minced meat and pet food. **Habitat:** Forested areas. **Range:** N. and E. Transvaal, extending through Swaziland to coastal N. Natal and S. Mozambique. Isolated relict populations occur on eastern escarpment of Zimbabwe and at East London.

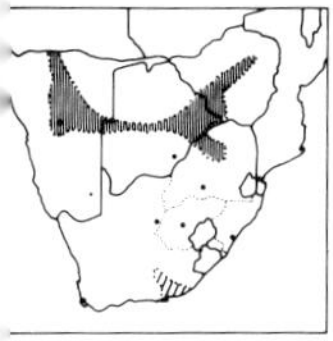

A.p. tasmani
A.p. occidentalis

Percival's Legless Skink *Acontias percivali* (Pl. 44)

22-26 cm; max. 30 cm
A medium-sized legless skink with a slender head and body and a rounded snout. The lower eyelids are opaque. There are three suboculars. The second upper labial is well separated from the eye. There are 14-16 scale rows at midbody. The tail is cylindrical and moderately tapered. Coloration is varied (see Subspecies). **Biology and breeding:** This skink prefers to burrow in loose leaf litter and soil around the base of trees. It is often found under dead logs, where it hunts for grubs and earthworms. A single brood of 1-3 babies is born in summer. **Habitat:** Coastal valley bushveld and mesic savannah. **Range:** Three widely separated populations occur (see Subspecies). **Subspecies:** Three races are recognized, two of which occur in the region. In *A.p. tasmani*, the interparietal is an equilateral triangle, and the body is diffuse red-brown above and yellowish below; it is found in E. Cape. *A.p. occidentalis* has a long, narrow, triangular interparietal, and the body is diffuse olive to grey-brown (and occasionally all-black); it occurs in N. Transvaal and Zimbabwe, west to Namibia and into S. Angola. The typical race, *A.p. percivali*, is uniform blackish-brown, it is restricted to Voi in SE. Kenya.

OODBUSH
GLESS SKINK
ontophiops

This is an unusual legless burrowing skink that seems to be an intermediate stage between the greater legless skinks (page 113), and the blind legless skinks (page 116). The only species in the genus is endemic to NE. Transvaal.

Woodbush Legless Skink *Acontophiops lineatus* (Pl. 46)

(Endemic) 16-18 cm; max. 19 cm
A small, stout-bodied legless skink with immovable lower eyelids that each have an oval, semi-transparent window. Two supraciliaries are present. The external ear openings are hidden. There is a single enlarged anal plate. The tail is short and blunt. The body is yellow-white to creamy white, with thin, dark brown to black stripes on the back and belly. The head and tail are usually darker than the body. **Biology and breeding:** They are usually found singly under stones on rocky hillsides. Much of their habitat has been destroyed by pine plantations at Woodbush, but there are healthy populations on the Wolkberg (SA RDB, Restricted). They give birth to two young. **Habitat:** Montane grassland. **Range:** N. Transvaal, from Woodbush to Wolkberg.

BLIND LEGLESS SKINKS
Typhlosaurus

These small legless skinks show many adaptations to life underground. The eyes are almost lost, and remain only as dark spots under the head shields. The rostral is very large and oval-shaped, and is pierced at the front by the nostril, from which a long groove extends to the back of the rostral. The nost can be sealed by a plug that protrudes from its rear wall. The external ear openings are hidden. The body is thin, and covered with large, smooth, close-fitting scales. There is a single, enlarged preanal plate. Like the greate legless skinks, *Acontias* (page 113), they have varied colour patterns, both within and between species. They show an interesting evolutionary progression, the more specialized species that live in the western deserts having developed thinner bodies and fewer head shields.

All are burrowing, most living in sandy regions. They are usually found un stones or among dead bark and fallen branches beneath trees. Others may forced to the surface when pans flood or during ploughing and road construction. They feed mainly on small beetle larvae and termites. Their reproduction is poorly known; 1-2 large babies have been reported for a few species. Their major predators include burrowing snakes (burrowing asps, quill-snouted snakes and garter snakes).

This genus contains nine species, eight of which are endemic (a single species reaches Zambia). Most species are found along the deserts and coastal sands of the west coast.

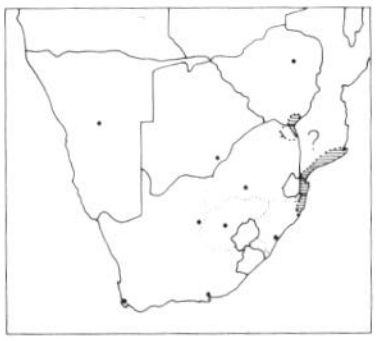

T.a. aurantiacus
T.a. fitzsimonsi

Golden Blind Legless Skink *Typhlosaurus aurantiacus* (Pl. 46

(Endemic) 16-21 cm; max. 25 cm
A small blind legless skink with a rounded (southern populations) to slightly flattened (northern populations) snout. The head shields are reduced in number, and there are only two unpaired head shields behind the rostral. Th frontal is pentagonal in shape, and the prefrontal is separated from the lorea by the frontonasals. There are 12 scale rows at midbody. The tail is of mediu length (26-34 subcaudals). Typically, the body is uniform pale orange or yellow, and paler below, sometimes with scattered spots below the tail. Som populations develop dorsal stripes on the back (see Subspecies), and all-black specimens also occur.

Biology: This skink has been found under logs and in the roots of a Cassava plant. **Habitat:** Coastal sands and sandveld. **Range:** Eastern coas plain of Mozambique and N. Zululand, extreme SE. Zimbabwe and adjacen Transvaal. **Subspecies:** Two races are recognized, both of which occur on the subcontinent. *T.a. aurantiacus* is usually plain golden in colour and has two temporals; it occurs on the coastal plain of N. Zululand and Mozambiqu extending into extreme SE. Zimbabwe. FitzSimon's blind legless skink, *T.a. fitzsimonsi*, is usually striped, has a single temporal, and the first supraciliary is fused with the prefrontal; it is restricted to NE. Kruger Nationa Park and adjacent NE. Transvaal.

Brain's Blind Legless Skink *Typhlosaurus braini* (Pl. 43)

(Endemic) 20-23 cm; max. 25 cm
The thinnest of all the blind legless skinks. All the head shields between the rostral and parietals have fused into a single large shield. The first 8-10 sca behind the head are wider than the other body scales. There are 12-14 scal rows at midbody, and more than 250 ventrals. The tail is relatively long (51-5 subcaudals). The whole body is light pink, with only a few faint brown marks the forehead. **Biology:** This skink is active in the early evening during winte but forages later in the evening during summer. Its tracks are very conspicuous, but it quickly disappears deep into the sand when disturbed. It feeds on termites and insect larvae. The Namib golden mole is a major predator. **Habitat:** Semi-stable sand dunes. **Range:** Central Namib Desert from Kuiseb River to Koichab River.

Cuvier's Blind Legless Skink *Typhlosaurus caecus* (Pl. 44)

(Endemic) 18-22 cm; max. 24 cm
A small blind legless skink with a slightly flattened but rounded snout. There are few head shields, with only two undivided shields behind the large rostral, which is more than twice as long as the other head shields together. The frontal is reduced to a transverse band and is smaller than the prefrontal. The loreal is in contact with the prefrontal, and there is a small preocular. There are 12-14 scale rows at midbody, and 207-230 ventrals. The body lacks pigmentation; it is usually flesh-pink, often with a yellow-orange tinge on the back. **Habitat:** Sparsely vegetated coastal dunes. **Range:** W. Cape coastal area, from Port Nolloth to near Cape Town.

Cregoi's Blind Legless Skink *Typhlosaurus cregoi* (Pl. 46)

T.c. cregoi
T.c. bicolor

(Endemic) 17-21 cm; max. 24 cm
This primitive blind legless skink has a rounded snout, and numerous head shields that include three unpaired head shields behind the rostral, and separate frontonasals and loreals. There are 16-20 scale rows at midbody, and 170-192 ventrals. The tail is relatively long (31-43 subcaudals). The body is golden, with 6-10 thin black stripes extending to the tail tip. The belly is usually plain or faintly spotted in northern populations, but striped in those from Transvaal (where all-black specimens are common). **Biology:** These skinks are found under rocks in montane grassland, and occasionally in evergreen forest. **Habitat:** Montane rocky hillsides. **Range:** Two isolated populations, one in N. Transvaal and the other in Zimbabwe highlands. **Subspecies:** Two races are recognized. *T.c. cregoi* usually has two supraoculars, the frontal as wide as the prefrontal, and 35-43 subcaudals; it is found on Soutpansberg and in NW. corner of Kruger National Park. *T.c. bicolor* has three supraoculars, the frontal wider than the prefrontal, and 31-37 subcaudals; it is restricted to highlands of Zimbabwe.

Gariep Blind Legless Skink *Typhlosaurus gariepensis*

(Endemic) 12-14 cm; max. 15 cm
A small blind legless skink that resembles the striped blind legless skink, *T. lineatus* (below). It has two unpaired head shields behind the rostral. The frontal is pentagonal in shape, and is much larger than the prefrontal. The rostral is bordered by five shields, while the prefrontal touches the loreals. There are three upper labials, a single supraciliary, and no subocular. There are 12 scale rows at midbody. The back is yellow, with four rows of dark streaks that form stripes on the tail. The belly is plain light yellow. **Biology and breeding:** These skinks live mainly among the roots of bunch grass, feeding almost exclusively on termites. Their tracks (thin, wavy, humped lines) can often be seen early in the morning on the dune surface. Mating probably occurs in August-September. In late summer, after a five-month gestation period, the female gives birth to a single baby (58-60 mm). The young grow rapidly during their first year, and may reach sexual maturity in less than two years. **Habitat:** Vegetated sand ridges in Kalahari sand. **Range:** Kalahari Gemsbok National Park and adjacent Namibia, Botswana and N. Cape.

Striped Blind Legless Skink *Typhlosaurus lineatus* (Pl. 46)

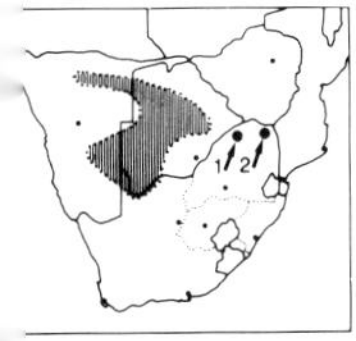

T.l. lineatus
T.l. subtaeniatus
T.l. richardi

14-18 cm; max. 20 cm
A small lizard that is very similar to the Gariep blind legless skink, *T. gariepensis* (above), but differs in having four upper labials, a subocular, and usually two supraciliaries. There are 14 scale rows at midbody. Coloration is varied (see Subspecies), and all-black specimens occur frequently. **Biology and breeding:** These skinks are common in heaps of wind-blown

sand at the base of grass tufts and bushes, and desert dune streets and san ridges. The diet is composed mainly of termites. One or two large babies (56-74 mm) are born in mid-January through early March, after a five-month gestation period. The young grow rapidly during the first year, and reach sexual maturity in 2-3 years (exceptionally in less than two years). **Habitat:** Kalahari sands.

Range: Kalahari region, extending through most of Botswana and adjacen Namibia, with isolated races in W. Zambia and N. Transvaal. **Subspecies:** Four races are recognized, three of which occur on the subcontinent. *T.l. lineatus* has a yellow-golden back with 4-8 thin dark stripes and a pure white belly; this race is found throughout the Kalahari region. *T.l. subtaeniatus* is striped above and below; this race is restricted to the Gre Saltpan in Waterpoort in N. Transvaal. *T.l. richardi* has a white belly, a single supraciliary, and few (160-168) ventrals; this race is restricted to sandveld in N. Venda. *T.l. jappi* has only two broad dorsal stripes and a plain belly; this race is restricted to the western border of Zambia.

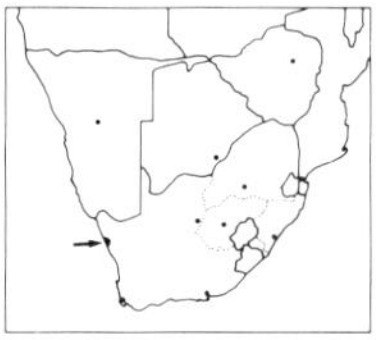

Lomi's Blind Legless Skink *Typhlosaurus lomii* (Pl. 43)

(Endemic) 13-14 cm; max 15 cm
A small, slender blind legless skink with an elongate and flattened snout. The head shields are reduced in number, with only two unpaired shields behind the large rostral, and three upper labials and two lower labials. There are 12 scale rows at midbody, and 160-167 ventrals.

The body is bright pink above, with a golden infusion on the front quarter that on the rear of the body is confined to a vertebral line. The belly is unpigmented and almost translucent. **Biology:** This species is poorly known it was discovered in 1986. **Habitat:** Sandy soils in succulent veld. **Range:** Little Namaqualand.

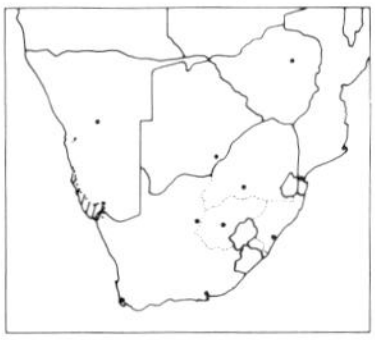

Meyer's Blind Legless Skink *Typhlosaurus meyeri* (Pl. 43)

(Endemic) 18-21 cm; max. 22 cm
A small blind legless skink with a snout that has a sharp, horizontal edge and rostral that is flattened below. The head shields are reduced in number, with only two undivided shields behind the large rostral, which is longer than the other head shields together. The posterior supraciliary is fused with the postocular, and five chin shields border the mental. There are 12-14 scale rows at midbody, and 207-225 ventrals.

The body has a broad yellow vertebral band, and a dark stripe on each flank. There is a dark band from the eye to the nostril, the tail has a black tip, and the lateral stripes often fuse. In some specimens the lateral stripes brea into spots, while others lack all coloration except the dark streak on the head **Habitat:** Sparsely vegetated coastal dunes. **Range:** Spencer Bay in Namibia south to Orange River and into Richtersveld.

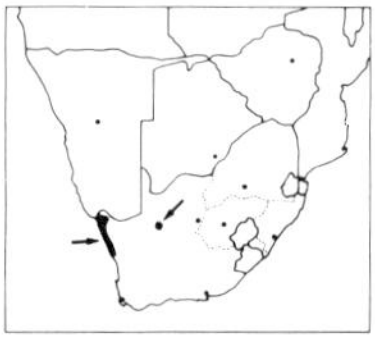

Boulenger's Blind Legless Skink *Typhlosaurus vermis* (Pl. 4

(Endemic) 25-30 cm; max. 32 cm
A long, thin blind legless skink with a slightly flattened but rounded snout. There are few head shields, with only two undivided shields behind the large rostral, which is more than twice as long as the other head shields together. The frontal is reduced to a transverse band and is smaller than the prefronta The loreal is separated from the prefrontal and there is no preocular. There a 12-14 scale rows at midbody, and 215-244 ventrals. The body is unpigmented, flesh-pink in colour. **Breeding:** Three elongate eggs (17 x 5 mm) were present in a female from Port Nolloth. **Habitat:** Sparsely vegetated coastal dunes. **Range:** Orange River to Spoeg River in Little Namaqualand. An unconfirmed report exists from near Prieska in N. Cape.

LD WORLD KINKS
ubfamily Scincinae

These are primitive skinks that usually have a smooth, cylindrical body and small, often vestigial legs. The frontal bones and nasal bones are separate. The body scales are generally smooth and overlapping, and there are two or more preanal scales. The tail is often long (always more than 30% of the total length), and in the species with reduced legs is sometimes used for food storage. It is readily shed and quickly regenerated. Coloration is usually drab, although a few species develop bright blue tails, and there is little polymorphism.

Many are either burrowing or secretive species, hunting among loose leaves and dead branches. Limb reduction is common in the subfamily, although only about 28 species totally lack external limbs. Reproduction is varied, with approximately half of the species being viviparous. A number of American species brood their eggs, protecting them from predators.

This subfamily is restricted to the Old World, except for a few species which enter North America, and with a relict distribution in south-central and E. Asia. They form an important part of the skink fauna in subSaharan Africa and on the islands of the Indian Ocean. There are approximately 180 species in 25 genera, with only four genera and 20 species occurring in southern Africa, of which 15 species are endemic.

ENDER SKINKS
oscelotes

This is a group of small skinks with reduced limbs that each bear five toes. External ear openings are present. The interparietal is usually small and does not touch the supraoculars. The supranasals touch behind the rostral.

There are three species in the genus, distributed in isolated populations in SE. Africa, with only one reaching the subcontinent.

arnoldi

Arnold's Skink *Proscelotes arnoldi* (Pl. 51)

16-19 cm; max. 22 cm
A small skink with small, five-toed feet and a tail that is much longer than the body. Each nostril is bordered by the rostral and a small, ring-like nasal. The body scales are smooth and in 22-24 rows at midbody. The body is brown, often with a dark spot on each scale, forming long lines. There is a distinct, grey dorsolateral stripe, and the belly is salmon-pink. Juveniles have a bright blue tail. The northern race lacks the sharply defined dorsolateral stripe.

Biology and breeding: This skink lives under stones and logs, foraging during the day among vegetation and feeding on small insects. It gives birth to up to five young in late summer. Predators include the wolf snake. **Habitat:** Montane grassland and evergreen forest. **Range:** Relict populations on eastern escarpment of Zimbabwe and Malawi.

Subspecies: Two races are recognized, one of which occurs in the region. In *P.a. arnoldi*, the limbs are relatively large and have only 6-8 lamellae under the fourth toe; it occurs in eastern highlands of Zimbabwe.
P.a. mlanjensis has smaller limbs, with 11-12 lamellae under the fourth toe; it is found on Mulanje Plateau in Malawi.

VARF RROWING INKS
elotes

This is a diverse genus of small burrowing skinks that show a clear evolutionary progression towards limb loss. It includes species with front and hind limbs, each with five toes, as well as those without any vestiges of external limbs. The body is covered with small, smooth scales. The tail is longer than the body in those species with fully developed feet, and often slightly shorter in the legless species. The interparietal is large and touches the supraoculars, and the paired supranasals are in contact behind the rostral. External ear openings may be present or absent.

All dwarf burrowing skinks either burrow in sandy soil or forage in loose leaf litter or grassland. The mode of reproduction is known for a number of species; these are viviparous.

There are 15 species in the genus, and all but one are endemic to southern Africa. *S. uluguruensis* is restricted to the Uluguru Mountains in Tanzania.

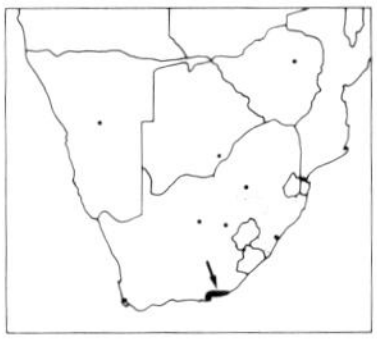

Algoa Dwarf Burrowing Skink *Scelotes anguina* (Pl. 47)

(Endemic) 12-14 cm; max. 16 cm
A small burrowing skink that lacks all traces of external limbs and has a tail t is slightly shorter than the body. The lower eyelids are scaly, and the ear openings are hidden. The frontal is short. Each nostril is pierced between th rostral and a small, ring-like nasal. There are four supraoculars. The back is silvery, sometimes with a bluish tinge, and with a dark band along the backbone. The flanks are dark brown. The belly is much paler and is often spotted. **Biology and breeding:** Common in the extensive vegetated sand dunes of Algoa Bay, this skink burrows in the upper layers of dry, sandy soil beneath the cover of stones or rotting spekboom bushes, etc. It feeds on sm insect larvae and termites. Two to four babies are born in February-March. Predators include the harlequin snake and small carnivores. **Habitat:** Coast dunes and valley bushveld. **Range:** Restricted to Algoa Basin in E. Cape.

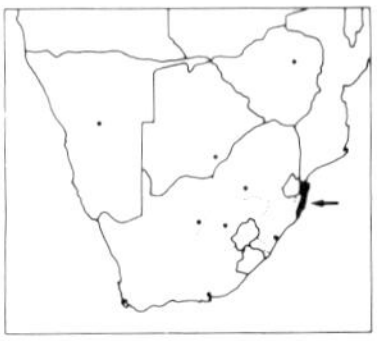

Zululand Dwarf Burrowing Skink *Scelotes arenicola* (Pl. 47

(Endemic) 10-12 cm; max. 14 cm
This small skink is very similar to the Algoa dwarf burrowing skink, *S. anguin* (above). It lacks all traces of external limbs and has a tail that is slightly shor than the body. The lower eyelids are scaly, and the ear openings are hidder The frontal is short. Each nostril is pierced between the rostral and a small, elongate, oval nasal. There are three supraoculars. The body is pale brown above, with a dark stripe along the backbone that becomes a double row of dark spots towards the rear. There are numerous similar stripes on the flank The belly is yellowish-white, and the tail is tinged with blue-violet. Juveniles darker, and have a bluer tail. **Biology and breeding:** This skink lives in sand soils in exposed and shaded grassland, and in humus and under logs in coastal dune forest. Two to four babies (43-52 mm) are born in February-March. **Habitat:** Vegetated coastal dunes. **Range:** Angoche in S. Mozambique, to Zululand.

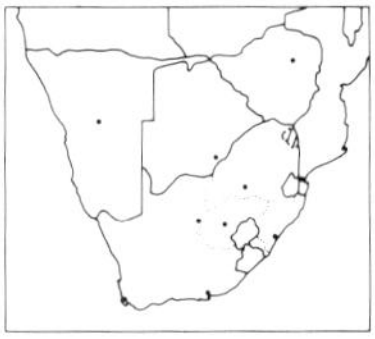

Lowveld Dwarf Burrowing Skink *Scelotes bidigittatus* (Pl. 4

(Endemic) 12-15 cm; max. 16 cm
A small burrowing skink that lacks forelimbs. The small hind limbs each hav two toes. The head is rounded, with scaly lower eyelids and minute ear openings. There are four supraoculars, all touching the frontal, no postnasa and five supraciliaries. There are 20 scale rows at midbody. The tail is sligh longer than the body. It is brown to greyish-brown on the back, each scale having a dark spot that together form almost continuous stripes. There is a well-defined, pale dorsolateral stripe that is straw-yellow on the head and body. The flanks are darker, and the belly is greyish. The tail is bluish, and sometimes bright metallic blue. **Biology and breeding:** Very common in suitable habitat, this skink is found among dead leaves and rotting logs. Or or two young are born in late summer or early autumn. **Habitat:** Lowveld bu **Range:** NE. Transvaal lowveld, between Soutpansberg and Olifants River.

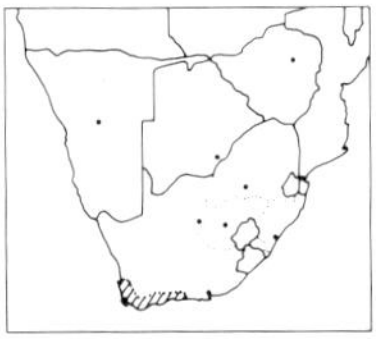

Silvery Dwarf Burrowing Skink *Scelotes bipes* (Pl. 48)

(Endemic) 12-14 cm; max. 15 cm
A small burrowing skink that lacks forelimbs. Each minute hind limb has twc toes. The head is flattened, with scaly lower eyelids and minute ear openinc There are three supraoculars, two of which touch the frontal, and four supraciliaries. There are 18 scale rows at midbody. The tail is shorter than t body, which is silvery-grey above, tinged with pale buff, the scales having dark centres that give a faint stippled or striped pattern. The belly is paler, v faint spots. **Biology and breeding:** Usually found under stones or burrowir

in sandy soils, these skinks feed on small invertebrates. Two young (60-65 mm) are born in March. **Habitat:** Coastal strandveld. **Range:** SW. Cape, from Mossel Bay to near Saldanha Bay.

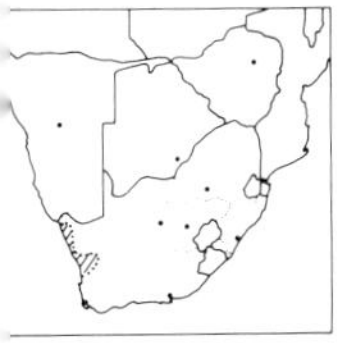

Striped Dwarf Burrowing Skink *Scelotes sexlineatus* (Pl. 48)

(Endemic) 12-14 cm; max. 15 cm
This lizard is very similar to the silvery dwarf burrowing skink, *S. bipes* (page 120), but has slightly longer hind limbs and toes, a shorter frontal, the tail longer than the body, and a different colour pattern. The body is pale silvery-grey, often tinged with rich buff. All the scales are spotted with brown, giving a finely striped appearance; the dorsolateral stripes are particularly prominent. The tail is bright pinkish-blue, and the belly is greyish-white. **Biology and breeding:** This species is similar in habits and behaviour to the silvery dwarf burrowing skink. **Habitat:** Succulent veld. **Range:** Little Namaqualand, from Port Nolloth to Clanwilliam.

Hewitt's Dwarf Burrowing Skink *Scelotes brevipes* (Pl. 47)

(Endemic) 10-12 cm; max. 13 cm
A small burrowing skink that lacks forelimbs. Each of the minute hind limbs has a single toe. The snout is rounded, the lower eyelids are scaly, and the ear openings are minute. The postnasal is absent, and there are four supraoculars, three of which touch the frontal, and four supraciliaries. There are 18 scale rows at midbody. The tail is as long as or slightly longer than the body. The body is pale bronze above, with each scale dark-centred, fading to slate-grey on the flanks and gunmetal-blue on the tail. The belly is grey-white. **Biology and breeding:** This skink is found under stones on mountain slopes, or logs on alluvial sand. Two or three young are born in late summer. **Habitat:** Rocky grassland and alluvial sand. **Range:** SE. Transvaal, into Zululand and S. Mozambique.

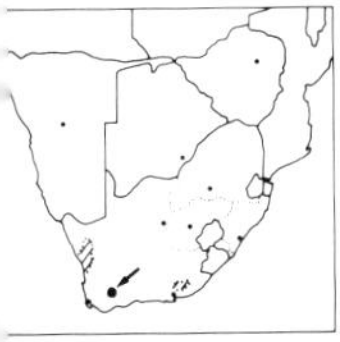

Cape Dwarf Burrowing Skink *Scelotes caffer* (Pl. 48)

(Endemic) 9-11 cm; max. 12 cm
A small skink with well-developed forelimbs and hind limbs, each limb having three toes. There are 20 scale rows at midbody. The lower eyelids are scaly and the ear openings are hidden. The tail is cylindrical and slightly longer than the body. The back is drab silver with a tinge of olive, and each scale is dark-centred. The flanks are darker. A pale, buff-coloured lateral stripe may be present. The tail is bluish-grey at its base and blue at the tip. The belly is bluish-grey. **Biology:** In the east, this skink is usually found under stones and among dead plants, particularly the rotting, hollow stems of 'noors' (*Euphorbia*). On the west coast it is found under litter on flat, sandy spots. When exposed, it wriggles rapidly into cover. The tail is readily shed. **Habitat:** Dry Fish River bush in E. Cape; karroid, stony veld in W. Cape; strandveld on W. Cape coast; succulent karroid veld in Namaqualand. **Range:** Scattered populations in E. Cape, at Matjiesfontein, on the west coast (at Elands Bay) and in Little Namaqualand (at Garies).

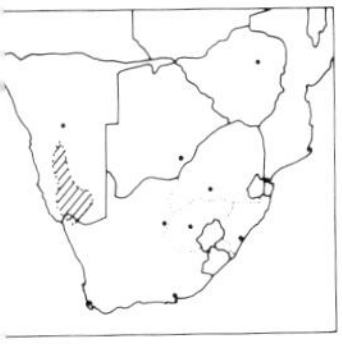

Western Dwarf Burrowing Skink *Scelotes capensis* (Pl. 49)

(Endemic) 9-10 cm; max. 12 cm
A small burrowing skink with small but well-developed front and hind limbs that each have five toes. The head is distinctly flattened, and the lower eyelids each have a transparent disc. There is a small postnasal. The ear openings are round and minute. There are 22 scale rows at midbody. The tail is a little longer than the body. The body is light olive to olive-brown, with a distinct coppery sheen, merging to dark brown to black on the tail. A pale olive-yellow

to yellowish-brown dorsolateral stripe extends from the snout to the tail, whe it is often blue to bluish-grey (white and more distinct in juveniles). The belly greenish-yellow to pale yellowish-brown, with a paler chin and throat. **Biology:** This species is usually found under stones on sandy soil, or amon the leaf mould around succulent bushes. **Habitat:** Rocky, succulent veld. **Range:** Little Namaqualand to central Namibia.

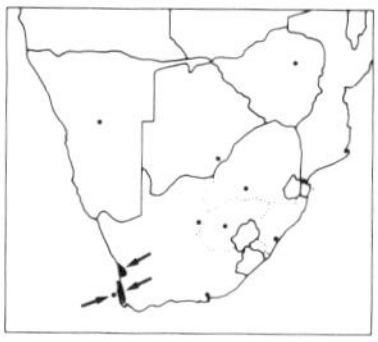

Gronovi's Dwarf Burrowing Skink *Scelotes gronovii* (Pl. 47

(Endemic) 10-13 cm; max. 15 cm
A small burrowing skink that lacks forelimbs, and has a single toe on each h limb. The snout is flattened. The lower eyelids are scaly, and the ear openin are minute. There are three supraoculars, two of which touch the frontal, an four supraciliaries. There are 18 scale rows at midbody The tail is slightly shorter than the body, which is silvery-grey above, with the four middle scal rows spotted in brown and appearing as thin stripes. The flanks also appear faintly striped due to spots on the scales. The belly is yellowish-white to greyish-white, and is often heavily speckled. **Biology and breeding:** It is usually found under litter among the sand dunes, often close to the high-wat mark. (SA RDB, Restricted.) One or two young (45-50 mm) are born in Marc April. **Habitat:** Sparsely vegetated coastal dunes. **Range:** W. Cape coast, from Doringbaai to Graafwater; also on Dassen and Robben islands.

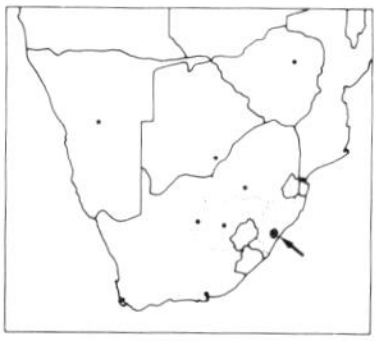

Günther's Dwarf Burrowing Skink *Scelotes guentheri*

(Endemic) 16-19 cm; max. 21 cm
A medium-sized burrowing skink that is similar to Hewitt's dwarf burrowing skink, *S. brevipes* (page 121). The head is rounded, with scaly lower eyelids four supraoculars, and hidden ear openings. There are 20 scale rows at midbody. The tail is shorter than the body. The back is pale brown to greyish brown, with a darker spot on each scale. The belly is plain white, with brown spots beneath the tail. **Biology and breeding:** A very rare species (SA RDB Rare), it lives beneath rocks and logs in damp soil near water. A female had five small, developing embryos in June. **Habitat:** Primary and secondary grassland. **Range:** Natal midlands between Howick and Nottingham Road.

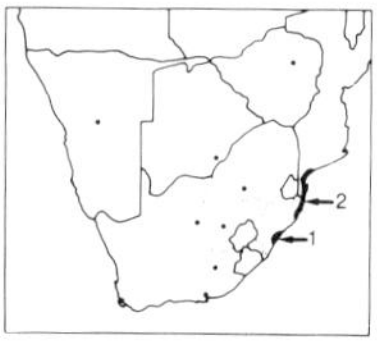

1. *S.i. inornatus*
2. *S.i. mossambicus*

Smith's Dwarf Burrowing Skink
Scelotes inornatus (Pls. 47 and 48)

(Endemic) 11-13 cm; max. 15 cm
This small burrowing skink lacks all traces of external limbs. The tail is slightl shorter than the body. The lower eyelids are scaly, and the ear openings are almost hidden. Each nostril is pierced between the rostral and a small, ring-like nasal. There are four supraoculars. The anterior chin shields are in contact. The back has a pale buff stripe that is four scales wide, flanked by rows of dark purplish-brown spots. The belly is paler than the back, and eac scale has a small, dark spot, except on the chin and throat, which are plain. **Biology and breeding:** It is found in humus and sandy soils. Females had 1 embryos in October-December. **Habitat:** Coastal sand with dune thicket. **Range:** Coastal plain of N. Natal and adjacent S. Mozambique. **Subspecies** Two races are recognized. *S.i. inornatus* has 20 scale rows at midbody; it occurs in the coastal districts of Natal to S. Maputaland. *S.i. mossambicus* h 18 scale rows at midbody; it is found in N. Maputaland, into S. Mozambique.

Kasner's Dwarf Burrowing Skink *Scelotes kasneri* (Pl. 48)

(Endemic) 15-18 cm; max. 21 cm
This medium-sized burrowing skink lacks forelimbs. Its small hind limbs eac have two clawed toes. The head is flattened, with well-developed eyes and

opaque, but not scaly, lower eyelids. The ear openings are very small. There are three supraoculars, two of which touch the frontal, and four supraciliaries. There are 22 scale rows at midbody. The back has a pale buff to straw-coloured dorsal stripe, bordered by three rows of scales that bear dark purple-brown spots, giving the impression of three thin, dark stripes. The belly is yellowish-white to greyish white, and dark-spotted beneath the tail. **Biology:** This skink is common under stones and litter on coastal dunes. (SA RDB, Restricted.) **Habitat:** Sparsely vegetated coastal dunes. **Range:** W. Cape coast, from Lambert's Bay to Vredenburg.

S.l. limpopoensis
S.l. albiventris

Limpopo Dwarf Burrowing Skink
Scelotes limpopoensis (Pl. 47)

(Endemic) 10-13 cm; max. 14 cm
A small burrowing skink with a slightly flattened head. There are 2-3 toes on each forelimb (see Subspecies), and four toes on each hind limb. The eyes are well developed, and there is a large, semi-transparent scale in each lower eyelid. The ear openings are small but distinct. There are 20-22 scale rows at midbody. The tail is as long as or slightly longer than the body. The back has a broad, dark brown stripe, bordered by a buff dorsolateral streak that in turn is flanked by a broad, dark brown stripe. The tail is tinged with blue, and the belly may be plain or spotted, particularly under the tail (see Subspecies). **Biology and breeding:** This skink forages among rocks and dead logs on sandy soil. Two young are born in November-December. **Habitat:** Alluvial sand with mesic savannah. **Range:** Limpopo River valley and adjacent regions. **Subspecies:** Two races are recognized, both of which occur on the subcontinent. *S.l. limpopoensis* has spotted belly scales, five supraciliaries, and three toes on each forelimb; this race is found in N. Transvaal and adjacent Zimbabwe and Botswana. *S.l. albiventris* has an unspotted belly, six supraciliaries, and two toes on each forelimb; this race is restricted to the vicinity of Langjan Nature Reserve in the Soutpansberg District of N. Transvaal.

Montane Dwarf Burrowing Skink *Scelotes mira* (Pl. 49)

(Endemic) 16-18 cm; max. 19 cm
A small burrowing skink with well-developed but small front and hind limbs, each with five toes. The head is slightly flattened, with well-developed eyes that each have a scaly lower eyelid. The ear openings are small, vertical and oval. There is no postnasal, and the supranasal touches the first upper labial. The tail is longer than the body. The back varies from light brown to greyish-brown, and all the scales are dark-centred, particularly on the tail, which has a steel-blue tinge. The belly is dirty white to yellowish-brown, usually with feeble spotting. **Biology and breeding:** This species lives in grass among rocks on upper mountain slopes and summits. Usually four young (50-60 mm) are born in late summer. **Habitat:** Rocky montane grassland. **Range:** E. Transvaal Drakensberg, with isolated populations on Wolkberg and Steenkampsberg, south to N. Natal midlands.

VANNAH
RROWING
INKS
osina

These small skinks have reduced limbs. The presence of 3-4 toes on each limb is the only external feature that distinguishes this genus from Arnold's Skink, *Proscelotes arnoldi* (page 119). The interparietal is small, and does not touch the supranasal scales, which are in contact behind the rostral. External ear openings are present.

Nothing is known of the reproduction, habitat preferences or biology of these rare and secretive skinks.

The genus contains four or five species which are distributed through central Africa to Tanzania and Namibia. Two species are found in the northern parts of southern Africa; one of these is endemic.

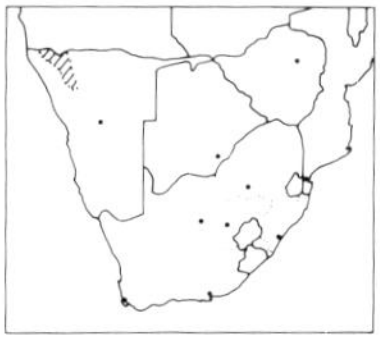

Albert's Burrowing Skink *Sepsina alberti* (Pl. 49)

(Endemic) 7-9 cm; max. 11 cm
A beautiful, small skink with short legs, each with four small, clawed toes. T blunt head has largish eyes with scaly lower eyelids. The preocular is much smaller than the loreal. There are 22 scale rows at midbody. The back is pa greenish with a golden-yellow tinge, and each scale has a dark brown to bl edge. The tail is bright blue at the tip, and the belly is grey-white. **Habitat:** Rocky, arid savannah. **Range:** Restricted to Kaokoveld in Namibi

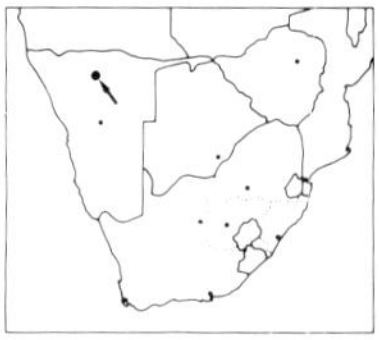

Angola Burrowing Skink *Sepsina angolensis*

12-14 cm; max. 16 cm
A small, stout-bodied skink that resembles Albert's burrowing skink, *S. alb* (above). It has small limbs, each with three toes. The head has a rounded snout, and small but distinct ear openings, and the lower eyelids are transparent and divided into septae. There are 22-24 scale rows at midbod The back is yellow-brown, with dark-centred scales that give a reticulated appearance. The belly is creamy to yellowish-white. **Habitat:** Mesic savann **Range:** From lower Congo, through Angola and W. Zambia to N. Namibia.

WESTERN BURROWING SKINKS
Typhlacontias

This small genus of burrowing skinks lacks external limbs, except for minut hind limbs in one species. The head has a flattened snout, and the large rostral is pierced by the nostril, from which a long groove extends backwar There are three large, unpaired plates on top of the head, between the rost and interparietal. The eyes are small and lack eyelids, and there are no external ear openings. The body scales are smooth and overlapping. There are no femoral or preanal pores.

They burrow in sandy soils under leaf litter and rotting vegetable matter. Little is known of their reproduction. A female contained large embryos, wh suggests that they are viviparous.

The taxonomy of the group is confused. Provisionally, there are 3-4 spec three of which occur in southern Africa; a new species, as yet undescribed known from the Kaokoveld.

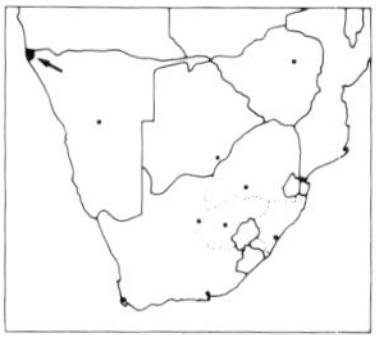

Bogert's Burrowing Skink *Typhlacontias bogerti*

8-10 cm; max. 12 cm
A small, slender burrowing skink that is similar in appearance to the Kalaha burrowing skink, *T. gracilis* (page 125). It lacks external limbs and a postna scale. The second and third upper labials enter the eye, which lacks eyelid and is not covered by the head shields. There are no external ear openings There are 18 scale rows at midbody. The body is pale buff-brown, with two rows of brown-spotted scales along the backbone. A strongly pigmented, dark stripe extends along the flanks on to the tail. The belly is pale cream-b **Biology and breeding:** This species may be collected in sand at the base trees, and under stones and the leaves of Welwitschia plants. A female contained well-developed embryos. **Habitat:** Sparsely vegetated desert. **Range:** N. Namib Desert, from NW. Kaokoveld to SW. Angola.

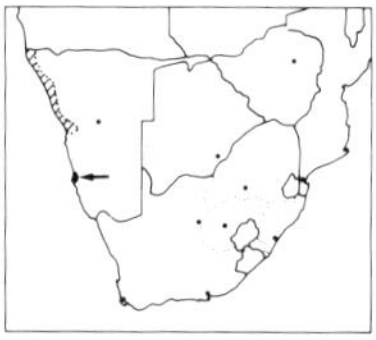

FitzSimons's Burrowing Skink *Typhlacontias brevipes* (Pl. 4

12-14 cm; max. 15 cm
A small, slender burrowing skink that has minute, rudimentary hind limbs a a small postnasal scale. The eyes have no eyelids and are not covered by t head shields, and there are no external ear openings. There are 18 scale r at midbody. The tail is 35-41% of the total length. In colour, the body is vari shades of light buff to sulphur-yellow. The scales along the back and upper flanks are often dark-centred, sometimes forming vague stripes. These

stripes, when present, are usually restricted to the front of the body and particularly the tail, which is often blue-grey between the dark lines. **Biology:** These skinks forage at night in semi-stable sand, usually at the base of the leeward side of a dune, around the roots of grass tufts. They leave characteristic wavy tracks. They are easy to catch by watching for signs of activity, and then making a sudden grab in the sand. Very sensitive to movement, they will disappear quickly into the sand. They eat small insects, including termites, ants, beetles and ant-lions. **Habitat:** Sparsely vegetated desert. **Range:** Isolated populations in Namib Desert, from Lüderitz Bay to S. Angola.

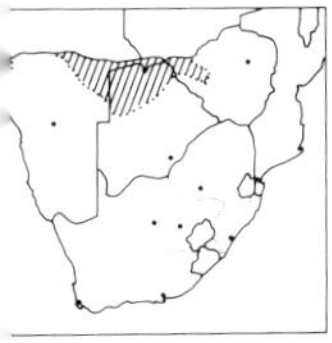

Kalahari Burrowing Skink *Typhlacontias gracilis* (Pl. 45)

9-11 cm; max. 12 cm
A slender burrowing skink that lacks external limbs. The eyes have no eyelids and are not covered by the head shields, and there are no external ear openings. There is no postnasal scale. There are 18 scale rows at midbody. The body is buff-coloured, with a vague, dark stripe along the backbone. A broad, grey-brown lateral stripe extends from the eye along the flanks to the tail. The belly is white at the edges and darker in the centre. **Biology:** These skinks are common under logs and piles of vegetable debris. They are often drowned when flat pans flood after heavy rains. Few specimens are found with their original tail, indicating heavy predation pressures. **Habitat:** Flat, sandy plains. **Range:** Northern parts of the region, from NW. Zimbabwe, through N. Botswana and Caprivi Strip, to adjacent NE. Namibia and into SE. Angola and W. Zambia. **Subspecies:** The status of the forms *T.g. ngamiensis* and *T.g. rhohani*, which have previously been treated as either separate species or subspecies, is confused, and is under investigation.

VANCED SKINKS
bfamily
gosomatiinae

This diverse group of skinks is relatively uniform in form in Africa, but has undergone a tremendous radiation in Australia. There they have developed varied body shapes, with legless and obese species, that may have smooth or rough, almost spiny scales. The head is short, and rarely has a distinct neck. It is always covered with large, symmetrical shields which, in some species, may be fragmented or reduced due to fusion. The eyes are always well developed; some species have developed a transparent window in the lower eyelid, and in others the eyelid may be completely transparent and permanently shut, eg. in some snake-eyed skinks, *Panaspis* (page 134). Ear openings may be absent, or protected by lobes, and each nostril pierces a single, discrete nasal scale. The limbs are almost always present (and are well developed in all southern African species), although limb reduction has evolved many times in the subfamily. The body scales are cycloid and imbricate, and either smooth or finely keeled (carinate). There is always at least one pair of enlarged preanal scales. The tail is more than 30% of the total length.

Although a few species have reduced limbs, and scuttle around in leaf litter, most of our local species are active, diurnal lizards that live in trees, on rock outcrops, and on the ground. The rock-living forms often occur in colonies, and consequently develop territorial behaviour and different colour patterns in the sexes. Most species, however, are drably coloured in browns, greys and blacks. Viviparity has evolved on numerous occasions.

The subfamily is widespread, but is particularly common in subSaharan Africa and the Australasian region. There is much controversy over the number of genera within the subfamily, and numerous new species are still being discovered, particularly in Australia. It contains about 40-60 genera with over 600 species. Of this bewildering array, only four genera containing 22 species occur in southern Africa; only one is endemic (this scarcity of endemics in the region indicates a relatively recent 'colonization' of the subcontinent by the subfamily).

COASTAL SKINKS
Cryptoblepharus

These are small skinks with well-developed limbs, each with five toes, and immovable lower eyelids, each with a transparent window.

They are mainly terrestrial, with some semi-arboreal species. They lay sm clutches of 1-2 soft-shelled eggs.

The genus is found throughout much of the Indo-Pacific region. It was onc considered to contain a single species, with numerous races, but 7-9 of the Australian races, and those from Society and Europa islands, are now thoug to be full species.

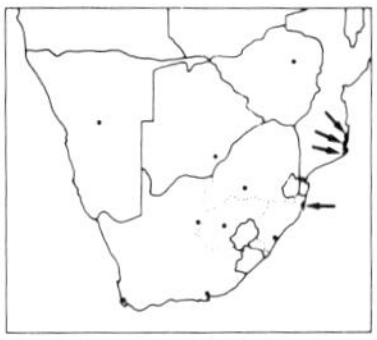

Bouton's Skink *Cryptoblepharus boutonii* (Pl. 51)

8-10 cm; max. 11 cm
A small, slender skink with well-developed limbs, each with five long, clawe toes. The distinct head has large eyes with immovable eyelids, each with a transparent spectacle. The body is covered with smooth, close-fitting scale in 26-29 rows at midbody. The tail is cylindrical, tapers to a fine point and is longer than the body. The back has a blackish-bronze hue, with indistinct lateral bands and numerous pale spots on the flanks and legs. The tail bear large spots on the sides that give a finely banded appearance. **Biology and breeding:** This diurnal skink forages on intertidal rocks among the breaking waves. It feeds on insects, crustaceans and even small fish that it catches o the rocks and in tidal pools. It is often caught by breaking waves, but is an adept swimmer. It may dive into pools to escape predators, which include s birds and crabs. Its ability to withstand salt water has allowed it to disperse throughout the Indo-Pacific by clinging to floating logs, etc., and drifting on ocean currents. (SA RDB, Vulnerable.) One or two soft-shelled eggs are laic moist sand above the high-tide mark.

Habitat: Coastal rock outcrops. **Range:** Widely distributed throughout Indian Ocean, Australia and through Pacific Ocean to Easter Island. On the African mainland it occurs in isolated populations along the east coast, reaching as far north as Mogadiscio in Somalia. In southern Africa it is know from a single colony at Black Rock in N. Zululand, and from three populatior in the vicinity of Inhambane in Mozambique. **Subspecies:** More than 36 subspecies have been proposed for the many isolated populations on oceanic islands, some of which have recently been elevated to full species. The populations in southern Africa have been provisionally referred to *C.b. africanus*, which occurs along the east coast of Africa.

WRITHING SKINKS
Lygosoma

These small to large skinks usually have well-developed, five-toed limbs, ar scaly, movable lower eyelids (in most African species). There are no enlarg preanal scales.

They live in varied habitats, from arid savannah to tropical evergreen fore. They are cryptic inhabitants of leaf litter or burrowers in loose soil, and feed small insects and their larvae. Many lay small clutches of soft-shelled eggs, but a number of specialized species, adapted to the arid regions of Somalia and N. Kenya, are viviparous.

The genus contains about 35 species which are distributed through subSaharan Africa and the Indian subcontinent, with 14 African species. Two of these enter southern Africa; neither is endemic.

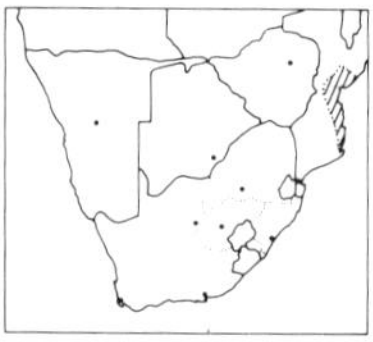

Mozambique Writhing Skink *Lygosoma afer* (Pl. 49)

18-21 cm; max. 23 cm
A medium-sized, stout, burrowing skink with small but well-developed, five-toed limbs, and a thick, smooth tail that is rarely longer than the body. Each eye has a scaly, movable eyelid, and the ear openings are small and deeply sunk. The prefrontals are small and widely separated. Each nostril, which is pierced between 2-3 nasals, is well separated from the rostral. The body is covered with almost smooth scales, in 26-28 rows at midbody. It is

pale to dark brown, heavily marked with irregular, dark brown and white spots and streaks. Some specimens have uniform brown bodies, but usually have spotted tails. The belly is creamy white, sometimes with brown spotting. **Biology and breeding:** Common in sandy soil underneath logs and piles of vegetable debris, these skinks eat small invertebrates, including ants, termites, beetles, cockroaches and millipedes. They lay 4-7 eggs (16-19 x 9-11 mm) in an underground cavity. **Habitat:** Eastern coastal plain. **Range:** E. Africa, from Somalia to Mozambique, reaching southern limit at Imhambane.

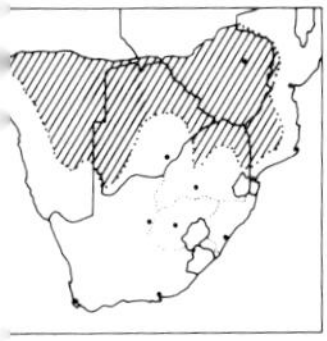

sundevallii

Sundevall's Writhing Skink *Lygosoma sundevallii* (Pl. 49)

15-18 cm; max. 19 cm

This small burrowing skink is very similar to the Mozambique writhing skink, *L. afer* (page 126), but is distinguishable from that species by both its small size and different coloration. The body is light brown to grey, usually with a dark spot at the base of each scale that gives a finely speckled appearance. The belly is uniform creamy white, although it may be speckled beneath the tail. **Biology and breeding:** These skinks burrow in the surface layers of sandy soil, beneath stones and rotting logs, feeding on small invertebrates. In summer they lay 2-6 white, oval, soft-shelled eggs (18-19 x 9-11 mm) underground, particularly in old termitaria. The hatchlings measure 50-55 mm. Many are eaten by small carnivores (genets, Cape foxes and African wild cats) and snakes (Cape wolf snakes and black file snakes). The tail was once prized as a cure for snakebite. **Habitat:** Arid sandy areas and well-drained hillsides. **Range:** Northern part of the subcontinent, reaching southern limit in Kalahari Gemsbok National Park, north to S. Angola and Somalia. **Subspecies:** There are two races, and only the typical race is found in southern Africa. *L.s. somalica*, which is restricted to Somalia, is distinguished by its long fifth toe.

PICAL SKINKS
buya

This is a large genus of small to large skinks that have well-developed limbs, each with five toes, and long, tapering tails. The eyes are large, each with a movable eyelid that may have a transparent window. The ear openings are distinct, but are often deeply sunk. Each nostril pierces a single nasal. Supranasals and prefrontals are present. The body is covered with cycloid, overlapping scales that are usually keeled (they are smooth in only one species from southern Africa). Preanal and femoral pores are absent. The species all look very similar, although some are fatter, some have longer tails, etc. Their colour pattern and habitat are useful field guides; otherwise, details of scalation and foot structure will distinguish them.

These active, diurnal skinks live in varied habitats, and are common in all parts of the subcontinent, foraging on rocks, in trees and on the ground. They feed mainly on insects, which are seized after a short, fast dash from cover. The primitive species lay eggs, but many species are viviparous, and some southern African species may be either live-bearing or egg-laying within different parts of their range.

A cosmopolitan genus containing about 85 species, it is distributed throughout subSaharan Africa and the Indian Ocean (more than 60 species), the lowlands of South America and the Caribbean (eight species), and SE. Asia and the Pacific (13 species). Of the 18 species that occur in southern Africa, only one is endemic.

Wedge-snouted Skink *Mabuya acutilabris* (Pl. 53)

13-15 cm; max. 17 cm

A small skink with a flattened snout and a sharp edge to the upper lip. The ear openings are covered with long, sharp lobes. The subocular is narrowed below and does not reach the lip. The dorsal scales have three keels, and are in 28-32 rows at midbody. The scales on the soles are keeled, and those un the long toes have a single keel. The body is light brown above, with dark brown or black spots and white flecks that form short bands; there is usually pale dorsal stripe and one (sometimes two) well-marked pale stripe on each side. The flanks may be barred in black. The belly is white. **Biology:** These skinks are terrestrial, catching beetles, ant-lions and wasps in sandy areas. They shelter in short burrows dug in accumulated sand at the base of grass tufts or small scrub. **Habitat:** Desert and Karoo-Namib scrubland. **Range:** Little Namaqualand, through Namibia to S. Angola and Zaire.

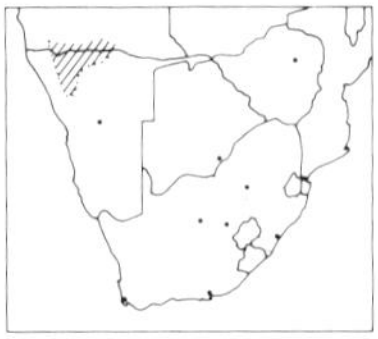

Ovambo Tree Skink *Mabuya binotata* (Pl. 55)

20-25 cm; max. 30 cm
A large, stout skink with a transparent window in each lower eyelid. Each ea opening is partially covered with three lobes. The subocular is narrowed be and reaches the lip. The dorsal scales have three keels, and are in 36-38 ro at midbody. The relatively long forelimbs and hind limbs overlap when pressed against the body. The scales on the soles are tubercular, and those under the long toes are smooth. The back is uniform olive-grey, tinged with buff, with a broad, black band on the side of the neck. The belly is grey-whit spotted with dull brown on the throat. Juveniles have scattered dark spots o the back that may form irregular bands. **Biology:** Usually found in the hollo trunks of mopane trees, these skinks catch beetles and grasshoppers after short rush from cover. They are shy and difficult to approach. **Habitat:** Arid savannah. **Range:** N. Namibia and S. Angola.

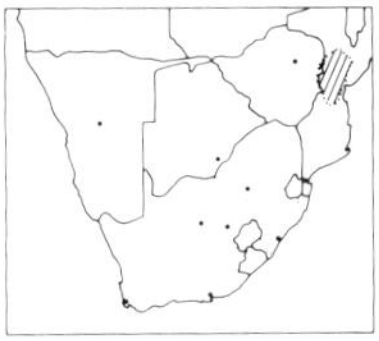

Boulenger's Skink *Mabuya boulengeri* (Pl. 53)

20-25 cm; max. 30 cm
A large, slender skink with a tail that is more than twice the body length. The are four supraciliaries. Adults usually have 7-9 keels on the dorsal scales, b this may range from three to 11, depending on the size of the skink. There a 28-32 scale rows at midbody. The back is grey-brown, sometimes with a fe scattered black flecks. There is a black streak on the side of the head. The is flecked with dark brown and the belly is yellow. **Biology:** Although most a arboreal, southern populations rarely occur on vertical tree trunks, preferrin to forage on logs and in leaf litter; they also climb reeds alongside streams, and long grass (where they may even sleep). The diet includes insects and spiders. **Habitat:** Coastal bush and dry woodland. **Range:** Save River in Mozambique, north to Tanzania and just entering E. Zimbabwe.

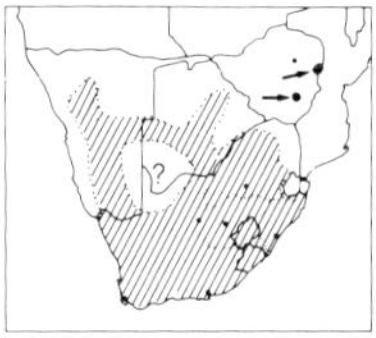

Cape Skink *Mabuya capensis* (Pl. 52)

20-25 cm; max. 27 cm
A large, fat (often obese) skink with a large window in each lower eyelid. Th subocular reaches the upper lip. The ear openings are crescent-shaped ar have small lobes. There are feeble spiny scales on the soles of the feet and keel on the lamellae beneath the toes. The dorsal and lateral scales have th keels (these are strongest on the back), and there are 32-36 scale rows at midbody. There are 15-20 lamellae under the fourth toe. The body is light brown to olive greyish-brown, with three pale stripes. Between the stripes a extending on to the flanks are series of dark brown to black spots or short bars. The belly is uniform yellowish-white to grey. Occasional specimens a uniform grey-brown above, sometimes with vague stripes.

Biology and breeding: This common, gentle skink lives on the ground, hunting large insects in clearings and open, sandy spots. It digs tunnels in loose sand at the base of bushes or boulders, and also favours dead trees aloe stems. Throughout most of its extensive range, it gives birth to 5-18 babies (60-75 mm) in late summer. However, females from some regions (eg. Pretoria and Port Elizabeth) have also laid clutches of eggs (see also the variable skink, *M. varia*, page 133). It tames easily, and would be much more common in gardens if it were not hunted by domestic cats. **Habitat:** Very varied; arid karroid veld, moist coastal bush, montane grassland, etc. **Range:** Throughout the subcontinent (except in Namib Desert, extreme northern regions and Transvaal lowveld). North of Limpopo River, there are relict populations on Inyanga Mountains in Zimbabwe, and Liuwa Plain in Zambia.

Chimba Skink *Mabuya chimbana*

12-14 cm; max. 15 cm
This small, slender skink is similar to the variegated skink, *M. variegata* (page 134). It has a window in each lower eyelid. The subocular reaches the upper lip. The ear openings are lobed. There are spiny scales on the soles of the feet and a keel on the lamellae beneath the toes. The dorsal scales have 5-7 weak keels and are in 34-40 rows at midbody. The body is olive-brown to bronze above, sometimes with black flecks on the scales and a ragged, pale dorsolateral stripe that is more distinct on the tail. The flanks are mottled in black and white, and the belly is white. **Biology:** This rock-living species favours granite outcrops. It is poorly known. **Habitat:** Mixed mesic savannah. **Range:** SW. Angola, extending into N. Namibia.

Hoesch's Skink *Mabuya hoeschi* (Pl. 52)

17-19 cm; max. 20 cm
A medium-sized, slender skink with a window in each lower eyelid and smooth scales on the soles of the feet. The subocular reaches the upper lip. The ear openings have distinct lobes. The dorsal scales have 2-3 weak keels and are in 32 rows at midbody. The tail is approximately twice as long as the body. In colour, it is brown-grey above, with four rows of dark blotches that form transverse bands. There is a pale lateral stripe. The belly is white. **Biology:** This recently described species lives among stones and feeds on wasps, beetles and moths. **Habitat:** Arid savannah. **Range:** N. Namibia to S. Angola.

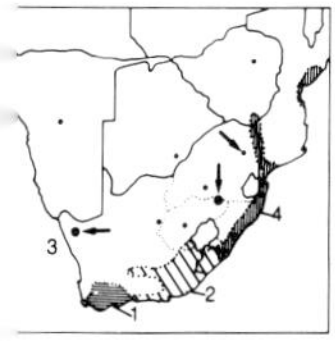

.h. homalocephala
.h. smithii
.h. peringueyi
.h. depressa

Red-sided Skink *Mabuya homalocephala* (Pl. 52)

(Endemic) 15-18 cm; max. 20 cm
A medium-sized, elegant skink with a small, transparent window in each lower eyelid. Each ear opening is partially covered with 2-3 lobes. The subocular reaches the lip, but is not narrowed below. The dorsal scales have three strong keels and are in 28-30 rows at midbody. The relatively short forelimbs and hind limbs hardly overlap when pressed against the body. The scales on the soles are tubercular, and those under the long toes are smooth. Coloration is varied (see Subspecies), but usually the back is olive to olive-brown or pale brown, bordered by a pale dorsolateral stripe and then a dark brown to black lateral band, which itself is bordered by a conspicuous, pale lateral stripe (which in breeding males becomes suffused with bright red). The back may have black spots on the scales that may fuse to form up to seven dark, thin lines. The belly is yellowish to blue-white.

Biology and breeding: These skinks are mainly terrestrial, foraging in leaf litter around the base of scrub thickets. Inland, they are fond of basking on rounded boulders in dry riverbeds. They are difficult to approach and will quickly dive into cover. About six eggs (8-9 x 13-15 mm) are laid in November-December in a small chamber dug in sandy soil under a boulder or dead log. The hatchlings measure 65-75 mm. **Habitat:** Varied; usually in moist situations. In the Cape, in coastal bush, fynbos, and riverine vegetation in montane grassland; enters sandveld in N. Kruger National Park.

Range: Throughout coastal regions of the south and east, from Cape Town t S. Mozambique. Relict populations occur on Cape escarpment mountains, and there is an old record from Little Namaqualand.

Subspecies: There are four poorly defined races. In *M.h. homalocephala* the back scales have three distinct keels, the frontonasals are in contact, an the pale dorsolateral stripes are often absent; it is found in SW. Cape. *M.h. smithii* has a slightly flattened head and body, and the back scales have three faint keels; it occurs in E. Cape, inland to extreme SE. OFS. In *M.h. peringueyi*, the pale dorsolateral and dark lateral stripes are absent, the back scales have 5-7 faint keels, and the frontonasals are separated; it is found in Little Namaqualand. *M.h. depressa* has a flattened head and body, the back scales have 5-7 faint keels, and the frontonasals are in contact; it occurs i coastal Natal to S. Mozambique, extending inland to Kruger National Park.

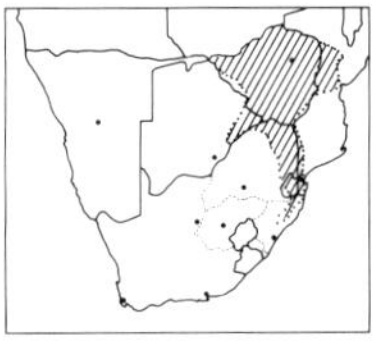

M.q. margaritifer

Five-lined or Rainbow Skink *Mabuya quinquetaeniata* (Pl. 55)

18-24 cm; max. 29 cm

A large, beautifully coloured skink with a small, transparent window in each lower eyelid. The ear openings are oval, and each is bordered by 2-5 lobes. The subocular reaches the lip but is not narrowed below. The dorsal scales have three strong keels, and are in 42-44 rows at midbody. The relatively lor forelimbs and hind limbs overlap when pressed against the body. The scale on the soles are tubercular, and those under the long toes are smooth. Coloration is varied, depending on sex and age. Juveniles and subadult ma are dark olive-brown to black above, with three distinct bluish-white stripes that are brilliant electric-blue on the tail. Adult females retain this coloration, although the stripe may become subdivided and faint. Adult males become buffy-olive to olive-brown above, each scale bearing a pearly white spot. Th pale stripes become indistinct. The tail changes to yellowish to orange-brow and the belly is creamy white.

Biology and breeding: An active, rock-living species, this skink runs around on exposed granite domes and other hard rock faces (paragneiss a some sandstones), feeding on insects. It is territorial; the status of the large breeding males, which become sexually mature in 15-18 months, is display by their different colours. They do not attack young males or females that retain the 'juvenile' colour pattern. The female lays 6-10 eggs (13-15 x 19-21 mm) in summer; the hatchlings measure 75-80 mm. It is possible that two clutches are laid per season, particularly in northern populations. **Habitat:** Mesic and arid savannah. **Range:** Natal, through Transvaal lowvel and Zimbabwe to Algeria and Senegal. **Subspecies:** Three races are recognized, with only *M.q. margaritifer* entering southern Africa. It is replac by the typical race, *M.q. quinquetaeniata*, in N. and E. Africa, and by *M.q. scharica* in W. and central Africa.

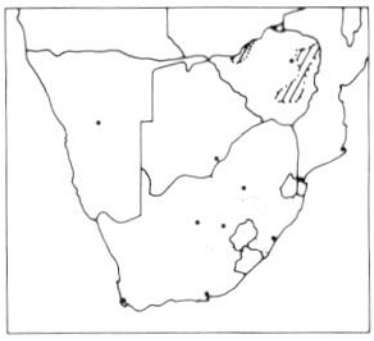

Bronze Rock Skink *Mabuya lacertiformis*

11-13 cm; max. 13,4 cm

This small, slender skink is similar to the variegated skink, *M. variegata* (page 134). It has a window in each lower eyelid. The subocular reaches th upper lip. The ear openings are lobed. There are spiny scales on the soles o the feet and a keel on the lamellae beneath the toes. The dorsal scales have five strong keels, and are in 36-40 rows at midbody. The body is grey-browr bronze above, with a few scattered, dark spots. The tail has a faint pale strip above. The belly is white, with vague, dark marks on the chin and throat. **Biology and breeding:** These skinks are active in the early morning and evening. They are found mainly on hard rock outcrops (granite, paragneiss and some sandstones), living under exfoliating flakes, although they may al shelter in hollow trees. They hunt for insects among the accumulated leaves around rock bases, and prefer boulder-strewn slopes rather than open rock faces. They lay a small clutch of 3-4 eggs (8,5 x 6 mm). **Habitat:** Arid and mesic savannah. **Range:** Lower Zambezi River valley and adjacent regions, extending along eastern escarpment of Zimbabwe, and with an isolated population in SW. Angola.

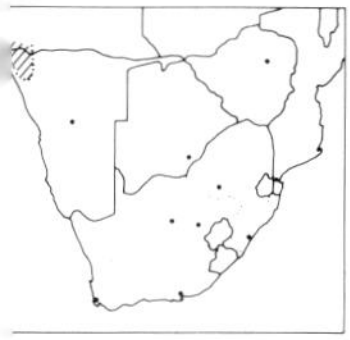

Angolan Blue-tailed Skink *Mabuya laevis* (Pl. 55)

12-15 cm; max. 16 cm
A small, slender, flattened skink with a conspicuous colour pattern and smooth body scales. There is a window in each lower eyelid. The subocular reaches the upper lip. The ear openings are lobed. The scales on the soles of the feet and on the lamellae beneath the toes are smooth. There are 29-33 scale rows at midbody. The body is brilliantly coloured in both sexes, in adults and juveniles. The back is shiny black, with a sky-blue stripe on each side. Scattered sky-blue spots occur on the back between the stripes, and on the flanks. The head is golden carrot-orange, with each head shield black-edged. The tail is sky-blue, and the chest and belly are blue-grey. **Biology:** These shy skinks forage in the early morning in large rock cracks of granite outcrops, feeding on small beetles. **Habitat:** Arid rocky savannah. **Range:** Kaokoveld of N. Namibia and S. Angola.

m. maculilabris

Speckled-lipped Skink *Mabuya maculilabris*

18-21 cm; max. 24 cm
This large skink is very similar to Boulenger's skink, *M. boulengeri* (page 128), but is stouter, and has five supraciliaries and 30-38 scale rows at midbody. It has different coloration, although this may be variable. Usually the back is pale brown, with distinctly darker flanks that are speckled with dark brown and white. The belly is yellow, and sometimes has brown speckles. Males are often more heavily spotted than females. Some specimens are uniform brown above. **Biology and breeding:** These skinks are found on tree trunks and in hollow logs. They climb on to thatched huts and are common in forest clearings. The diet consists of insects and other invertebrates. Over much of its range, the species does not have a definite breeding season, and may lay 5-6 clutches, each of 6-8 eggs (8 x 14 mm), a year. They are laid in moist soil under logs. **Habitat:** Evergreen forest, dry deciduous woodland and coastal bush. **Range:** Liberia in W. Africa, to Angola and Mozambique, just crossing Zambezi River to Inhamitanga.

Subspecies: Five races are recognized, with only the typical race entering the subcontinent. The other four races are restricted to offshore islands: *M.m. infralineata* occurs on Europa Island; *M.m. albotaeniata* is found on Pemba Island; *M.m. comorensis* is restricted to the Comores Islands; and *M.m. casuarinae* occurs on Casuarina Island.

Grass-top Skink *Mabuya megalura*

22-26 cm; max. 30 cm
A large, fat-bodied skink that resembles the Cape skink, *M. capensis* (page 128), but has an exceptionally long tail (well over twice the body length). The subocular reaches the upper lip. The ear openings lack lobes. The scales on the soles of the feet are smooth. The dorsal scales have three weak keels (but may be smooth in old females), and are in 24-26 rows at midbody. The body is light orange-brown to grey-brown, with a pale, dark-edged stripe along the backbone (which may be absent), and a white stripe on the flanks. The belly is white. **Biology and breeding:** This species forages in grass alongside clearings, dashing out to grab large grasshoppers, crickets, mantids, etc. It gives birth to 4-8 babies. **Habitat:** Savannah grassland. **Range:** Ethiopia, through Uganda and Katanga, to Mozambique, just south of Zambezi River.

Western Three-striped Skink *Mabuya occidentalis* (Pl. 52)

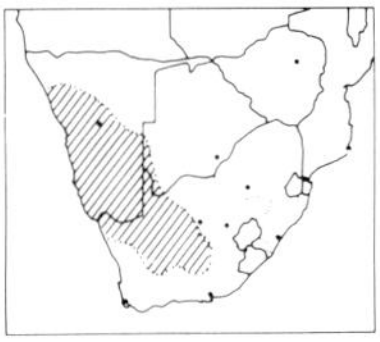

20-25 cm; max. 27 cm
This large skink is very similar in appearance to the Cape skink, *M. capensis* (page 128), but is more slender, has 30-32 scale rows at midbody, smooth scales on the flanks, 2-3 enlarged ear lobes and 21-24 lamellae under the fourth toe. Its coloration is more distinct. The back is rich red-brown to olive-brown, with three pale, dark-edged stripes. The flanks, which are darke are broken by a white stripe that runs from the lips to the groin. The belly is uniform white. **Biology and breeding:** A terrestrial species, it runs in open, sandy veld, sheltering at night in a short burrow that it digs in loose soil at the base of scrub bush. It hibernates during winter. Its reproduction is poorly known; females from the Namib Desert (February) and OFS (November) eac contained seven foetuses, while Kalahari females lay 5-7 eggs. **Habitat:** Aric savannah, karroid veld and desert. **Range:** Karroid areas of the Cape, throu Namibia and SW. Botswana, to S. Angola.

Kalahari Tree Skink *Mabuya spilogaster* (Pl. 54)

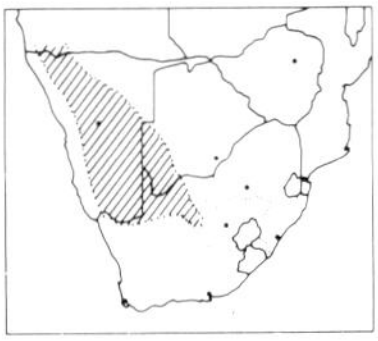

15-17 cm; max. 19 cm
This medium-sized skink has a window in each of its lower eyelids. The subocular is narrowed below and reaches the upper lip. The ear openings each have two short lobes. There are spiny scales on the soles of the feet ar a keel on the lamellae beneath the toes. The dorsal scales have three keels, and are in 32-36 rows at midbody. The body is dark brown, with a pair of broad, pale dorsolateral stripes. There are numerous scattered, small, pale spots between the stripes and on the flanks. The white belly with its irregular dark speckles distinguishes this species from the striped skink, *M. striata* (below). **Biology and breeding:** An aboreal species, it is common on acaci trees along dry river courses. It gives birth to 3-5 babies. **Habitat:** Arid savannah. **Range:** Kimberley in N. Cape, through Botswana and Namibia tc S. Angola.

Striped Skink *Mabuya striata* (Pl. 54)

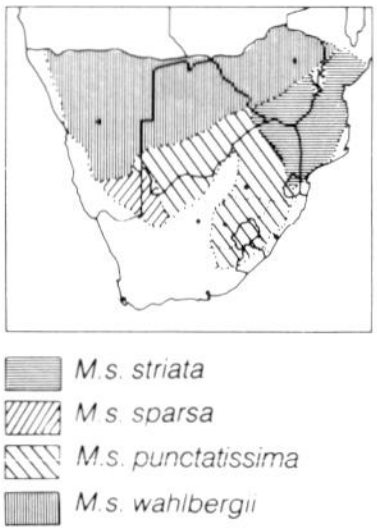

18-22 cm; max. 25 cm
This medium-sized skink has a window in each of its lower eyelids.
The subocular is narrowed below and usually reaches the upper lip (see Subspecies). The ear openings are lobed. There are spiny scales on the so of the feet and a keel on the lamellae beneath the toes. There are 32-43 sca rows at midbody. Coloration differs between the subspecies. In northern populations the head of breeding males is orange-brown, with a yellow-orange throat. **Biology and breeding:** The diet of this species incluc most small insects (beetles, moths, etc.) and other small invertebrates. They are common around towns, and tame readily. Mainly arboreal, they also live rock outcrops and houses. Southern populations give birth to a single litter c 3-9 babies (63-76 mm) in summer. Reproduction may occur throughout the year in northern populations. Growth is relatively fast, and they reach sexua maturity in 15-18 months. **Habitat:** Varied; from mangrove swamp to arid savannah. **Range:** E. Africa, south to Transkei and NE. Cape, and west through Zambia to Angola and S. Namibia.

Subspecies: All four races occur on the subcontinent. *M.s. striata* adults are large (8-11 cm snout-vent length), and the subocular is separated from lip. The back is red-brown, with distinct yellow dorsolateral stripes, and the belly is white, although the chin and throat may be speckled with grey, blac or orange. It occurs in Natal, north through Transvaal lowveld and SE. Zimbabwe to Ethiopia. *M.s. sparsa* adults rarely exceed 8 cm in length from snout to vent, the subocular does not reach the lip, and the prefrontals are i contact. The back is dark brown to black, with numerous small, pale spots, and the belly is white, sometimes with a black or orange throat. It is found in S. Namibia and adjacent Cape and Botswana. *M.s. punctatissima* adults rarely exceed 8 cm in length from snout to vent, and the prefrontals are well separated. The back is dark brown to black, speckled with pale spots and with a pair of pale dorsolateral stripes, and the belly is yellow, usually with black spots and streaks on the chin and throat. It occurs in NE. Cape, through Transkei, W. Natal, OFS, S. Botswana and Transvaal, with relict populations in eastern highlands of Zimbabwe and Malawi. *M.s. wahlbergii* adults are large (8-11 cm snout-vent length), and the subocular usually reaches the lip. The back is pale grey, sometimes with faint, pale dorsolateral stripes; a black band extends from the eye above the shoulder. The belly is white. It is found in N. Namibia and Botswana, into Angola, NW. Zimbabwe and W. Mozambique.

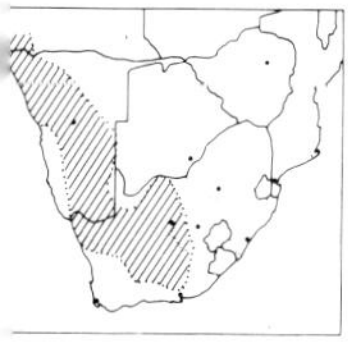

Western Rock Skink *Mabuya sulcata* (Pl. 54)

16-20 cm; max. 23 cm
A medium-sized, slender, flattened skink with a window in each lower eyelid. The subocular does not reach the upper lip. Each ear opening has 3-4 small lobes. There are spiny scales on the soles of the feet, and three keels on the lamellae beneath the toes. The dorsal scales have 3-5 strong keels, and are in 34-40 rows at midbody. Coloration varies between the sexes. In juveniles and adult females the body is pale olive to olive-brown, with six dirty-gold stripes. The chin and throat region may be infused with yellow-orange. The belly is dirty white, and usually heavily spotted with black on the throat. Sexually mature males become suffused with black, and throughout the Karoo, OFS and around Lüderitz Bay, males may become completely jet-black above and below. In Little Namaqualand and Namibia, males are usually dirty bronze on the hind body and black elsewhere, and sometimes all-bronze.

Biology and breeding: Very active, these rock-living skinks can be seen chasing over rock outcrops, feeding on beetles, grasshoppers and other invertebrates. They shelter at night in rock cracks, often in pairs. Three to five babies (58-65 mm) are born in December-February. There are informal reports that they may also occasionally lay eggs. **Habitat:** Karroid veld, desert and arid savannah. **Range:** Karroid areas of the Cape and adjacent OFS, through Namibia to S. Angola. **Suspecies:** In addition to the typical race, two poorly defined races are sometimes recognized. *M.s. nigra* is all-black, even in some juveniles; it is restricted to the vicinity of Lüderitz Bay in Namibia. *M.s. ansorgii* grows larger (up to 10 cm snout-vent length) and has a bigger head; it occurs in Kaokoveld of N. Namibia and S. Angola.

Variable Skink *Mabuya varia* (Pl. 53)

12-16 cm; max. 19 cm
A medium-sized skink with a rounded snout and a window in each lower eyelid. The subocular reaches the upper lip. The ear openings are oval and have short lobes. There are spiny scales on the soles of the feet, and three keels on the lamellae beneath the toes. The dorsal scales have three strong keels, and are in 30-36 rows at midbody. Coloration is variable; the back may be blackish, olive, pale brown or red-brown, with or without black spots. There is always a distinct, white lateral stripe, and sometimes other pale stripes down the backbone and on the upper flanks. The belly is bluish-white.
Biology and breeding: These skinks forage on broken ground, climbing on rocks and tree bases. The diet consists of insects (grasshoppers, caterpillars and termites), spiders and, exceptionally, other lizards. Prey is grabbed after a short dash from cover.

Over most of its range it is viviparous, and embryos have been recorded from the OFS and Tete in the lower Zambezi Valley (where small clutches of 2-4 young are usual). A large female from the Suurberg in E. Cape gave birth to 10 small babies (60-76 mm) in mid-February. However, in N. Transvaal a single clutch of 6-12 eggs is laid in November-December, and they hatch aft two months' incubation. The hatchlings measure 40-50 mm. Growth is rapid, and both sexes reach maturity in eight months. They may live for two years, but males usually die earlier (after 15-16 months) than females (16-17 months). **Habitat:** Varied; grassland to arid and mesic savannah. **Range:** SE. Cape, through E. Africa to Sudan and Somalia, west to Namibia, Angola and Congo.

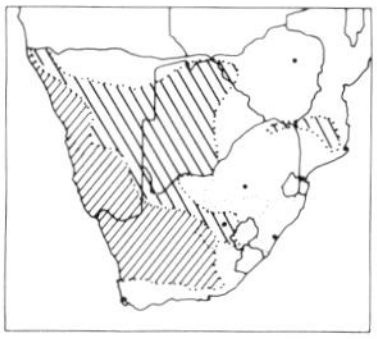

M.v. variegata
M.v. punctulata

Variegated Skink *Mabuya variegata* (Pl. 53)

11-14 cm; max. 15 cm
A small, slender skink with a window in each lower eyelid. It grows larger in th western arid regions. The subocular reaches the upper lip. The ear openings are lobed. There are spiny scales on the soles of the feet, and a keel on the lamellae beneath the toes. The dorsal scales are in 30-36 rows at midbody. Coloration is very varied, particularly in the west. The body is light grey to dar brown above, usually with a pair of pale stripes on the sides. There is sometimes another pale stripe along the backbone, with or without a series o paired black spots (see Subspecies). In Namaqualand and S. Namibia the pale stripes may be obscured or absent, and the back heavily flecked with black. Plain pale grey specimens also occur. The belly is plain white. Breedin males develop a reddish-brown blush below the hind legs and on the tail base.

Biology and breeding: This species is active during the day on small rock outcrops, sheltering in burrows under rocks and logs, and occasionally in soil-filled rock cracks. The diet includes spiders and beetles. It is viviparous, having small litters, each of 2-4 babies (48-57 mm), usually in January-March Births have also occurred in August in the Namib Desert. **Habitat:** Extremely varied; desert, karroid veld, montane grassland, savannah, coastal bush anc valley bushveld. **Range:** Through most of the subcontinent, but absent from S. Cape, Natal and E. Transvaal, and extending into S. Angola.

Subspecies: There are two races, and both occur in the region.
M.v. variegata has three keels on the dorsal scales; it is found in E. Cape, through karroid areas to Great Namaqualand and Namib Desert.
M.v. punctulata has 5-7 (but three in juveniles) keels on the dorsal scales, an a thin dorsal stripe bordered by a series of black spots or streaks that continu on to the tail; it occurs in S. Mozambique, through N. Transvaal, Botswana, N. Cape and Namibia, to S. Angola.

SNAKE-EYED SKINKS *Panaspis*

These small to medium-sized skinks usually have small but well-developed, five-toed limbs. The lower eyelids may be movable or fixed and snake-like.

They are terrestrial or semi-burrowing, and occur in arid savannah to evergreen tropical forest. All lay small clutches of soft-shelled eggs.

Restricted to subSaharan Africa, the genus contains approximately 33 species, only one of which reaches southern Africa.

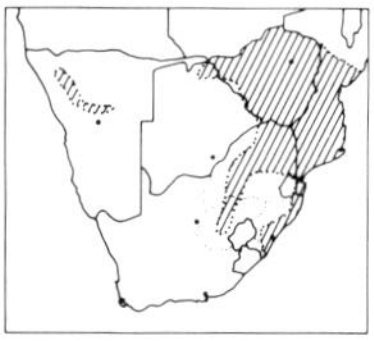

Wahlberg's Snake-eyed Skink *Panaspis wahlbergii* (Pl. 51)

8-10 cm; max. 11 cm
A small, burrowing skink with small but well-developed, five-toed limbs and a unblinking, snake-like eye. The nasals and prefrontals are well separated. Th ear openings are visible. The body scales are smooth, in 24-26 rows at midbody. The tail is cylindrical and slightly longer than the body. Coloration i varied; the back may be light grey, brown or gold, and is sometimes plain, bu usually bears six dark lines. A pale dorsolateral stripe may be present, and a dark brown to blackish lateral band may be conspicuous or faint. The belly is white to greyish-blue, except in breeding males, which are pinkish-orange below. **Biology and breeding:** These skinks are diurnal, scuttling among grass roots and rotting logs, and around stones and old termitaria on broken

ground. They eat termites and other small insects. They are short-lived, males surviving 10-12 months and females a few months longer. Both reach sexual maturity in 8-9 months. Mating takes places in August-October, and females lay 2-6 oval, white eggs (7-9 x 4-5 mm) in November-December, under stones or in logs. In good seasons they may produce another clutch in late summer. The eggs hatch in 40-50 days; hatchlings measure approximately 30 mm. **Habitat:** Arid and mesic savannah. **Range:** Through Natal and Transvaal to central and E. Africa, and entering N. Botswana and Namibia.

LD WORLD IZARDS OR ACERTIDS
amily Lacertidae

These small to medium-sized lizards have a slender body, a long tail and well-developed legs. The dorsal scales are usually small, smooth and granular (although some genera have large, rough, overlapping scales), but large and quadrangular on the belly. The head is covered with large, symmetrical scales that have osteoderms. Femoral pores are present. The tail has whorls of keeled scales, which may be spiny, and it can be shed and regenerated.

Active and diurnal, they live mainly on the ground, but some are rock-living or arboreal. There are no burrowing species; limb reduction is rare and is always associated with grass-living species. Many are brightly coloured, particularly the breeding males. All except the European viviparous lizard (*Lacerta vivipara*) lay eggs, which is surprising as many live in temperate and montane environments where viviparity may be expected to evolve.

The family is restricted to the Old World, mainly Europe and Africa, but one genus (*Takydromus*) extends into the Far East. All look very similar. Although there are 25 genera and 160 species, no subfamilies are recognized. There are nine genera in southern Africa, only one of which is endemic, although 20 of the 30 species are endemic.

ey to the southern frican genera in the acertidae

1. Tail flattened, with a lateral fringe of large, flat scales; arboreal .... *Holaspis* (Blue-tailed tree lizard, page 137)
   Tail cylindrical, without a lateral fringe; rock-living or terrestrial 2

2. Toes without a serrated or fringed edge 3
   Toes with a serrated or fringed edge; subocular not bordering lip 8

3. Smooth or tubercular lamellae beneath the toes; subocular bordering lip 4
   Keeled lamellae beneath the toes 6

4. A distinct collar present; dorsal scales small, granular or flattened and not overlapping 5
   No distinct collar; dorsal scales large, strongly keeled and overlapping .... *Tropidosaura* (Mountain lizards, page 147)

5. Nostril pierced between 2-3 nasals and well separated from first upper labial; temporal scale rounded; terrestrial .... *Nucras* (Sandveld lizards, page 142)
   Nostril pierced between 2-4 nasals and first upper labial; temporal scale elongate; rock-living .... *Lacerta* (Mediterranean lizards, page 139)

6. A distinct collar present; dorsal scales small or granular; head shields smooth or slightly rough 7

   No collar present; dorsal scales large, keeled and overlapping; head shields striated and keeled .... *Ichnotropis* (Rough-scaled lizards, page 138)

7. Belly plates in 6 long rows; collar curved; an elongate temporal shield present .... *Heliobolus* (Bushveld lizards, page 137)

   Belly plates in 10 or more rows; collar straight; no upper temporal shield present .... *Pedioplanis* (Sand lizards, page 144)

8. A distinct collar present; femoral pores present .... *Meroles* (Desert lizards, page 140)

   No collar present; femoral pores absent; snout duck-billed .... *Aporosaura* (Shovel-snouted lizard, page 136)

SHOVEL-SNOUTED LIZARD
Aporosaura

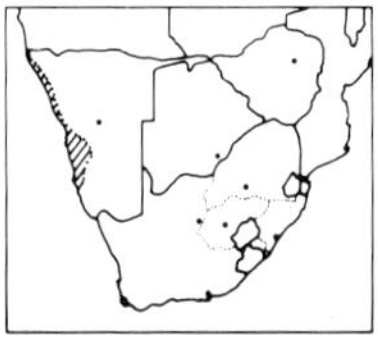

This unusual and highly specialized lizard has adapted to living on the shifting sands of the Namib Desert.

The genus contains a single species.

Shovel-snouted Lizard *Aporosaura anchietae* (Pl. 56)

10-11 cm; max. 12,5 cm

This is a small lacertid that is easily identified by its unusual, flattened snout, which has a sharp cutting edge. The nostrils are directed upwards and pierced between three nasals. The subocular does not border the lip. The lower eyelids are scaly. It lacks a collar and femoral pores. The dorsal scales are very small and granular. The hind legs are large and often splay outwards. The long toes have a conspicuous fringe of large scales. The tail has a broad base but tapers rapidly. The body is sand-coloured above, with network of black marks that extends on to the flanks and limbs. The head is paler and has a silvery sheen. A broken black stripe may extend along the backbone and on to the tail base. The belly is white. The tail may have a few black crossbands.

Biology and breeding: These lizards race at high speed over the loose sand dunes of the Namib Desert. Although they can withstand high body temperatures (up to 44 °C), they may overheat from the hot sand. To avoid this, they have an amusing 'thermal dance' in which they lift the tail and two of the feet high in the air. When disturbed, they dive into the dune face, swimming deep into the sand and out of danger. If they cannot escape they raise the body high and will jump and bite. They are territorial, particularly during the reproductive peak, when males establish and defend territories. A prime site is an eddy in a dune slipface where seeds collect. They feed on small beetles and other insects, but eat seeds during dry periods when insects are absent. All their moisture is obtained from their food. One or two large eggs (8-9 x 16-17 mm) are laid in a small chamber dug in firm sand in a dune slack. There is no fixed breeding season, and eggs are laid at irregular intervals, but with a peak during December-March. **Habitat:** Sparsely vegetated, unstable desert dunes. **Range:** Endemic to Namib Desert, from Lüderitz Bay to S. Angola.

USHVELD LIZARDS
Heliobolus

This small genus of African lacertids is closely related to the sand lizards, *Pedioplanis* (page 144), and desert lizards, *Meroles* (page 140), but has only six long rows of large belly scales. All are terrestrial and oviparous.

Four isolated species occur, widely distributed through Africa; *H. speckii* occurs in E. Africa, *H. quadrinasalis* in Chad, and *H. nitida* in W. Africa. One species is found in southern Africa.

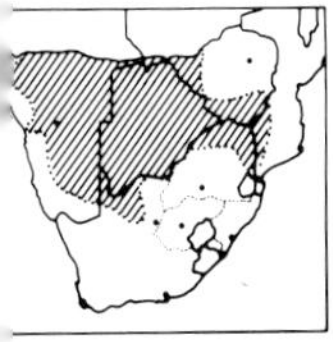

Bushveld Lizard *Heliobolus lugubris* (Pl. 57)

16-20 cm; max. 22 cm
A medium-sized lacertid with a small, elongate temporal scale bordering each parietal, a crescent-shaped tympanic scale above each ear opening, and a distinct collar. The subocular borders the lip. The lower eyelids are scaly. The dorsal scales are small and keeled. There are 12-18 femoral pores on each thigh. The toes are long and lack a fringe; there are 2-3 keels on the scales beneath the toes. Adults have a light grey-brown to red-brown back, with vague, dark crossbars and three pale dorsal stripes, the middle one dividing on the neck and extending on to the tail. The legs have white spots. The tail is usually pale brown. The belly is white. In contrast, hatchlings have a jet-black body, with broken, yellow-white lateral and dorsal stripes, and a sand-coloured tail.

Biology and breeding: These diurnal, terrestrial lizards can be commonly seen on sparsely vegetated, compacted, sandy plains, darting from bush to bush. They eat small insects, particularly termites. The hatchlings walk stiffly and jerkily and, with their very unusual colour pattern, resemble the distasteful oogpister ('eye squirter') beetle (*Anthia*), which can squirt an acidic, pungent fluid when threatened; it is probable that this mimicry protects the hatchlings from predators. The female lays 4-6 oval, soft-shelled eggs (6-7 x 11-12 mm) in a small chamber dug in loose sand. **Habitat:** Arid and mesic savannah. **Range:** Transvaal lowveld and SE. Zimbabwe, through Botswana, N. Cape and E. and central Namibia to S. Angola.

UE-TAILED TREE LIZARD
Holaspis

A unique and beautiful arboreal lizard. There is a single species in the genus, and it just enters the subcontinent.

laevis

Blue-tailed Tree Lizard *Holaspis guentheri* (Pl. 61)

10-11 cm; max. 12 cm
A very small, beautifully coloured lacertid with a flattened tail that has an unusual fringe of flat scales. It has a well-developed collar. The lower eyelids are scaly, and each has 3-5 enlarged, semi-transparent scales in the middle. The toes are long and have a fringe of flattened scales. A single, large preanal plate is present. There are 17-25 femoral pores on each thigh. The back is black. There is a broad cream stripe on the top of the head. Two dorsolateral cream stripes arise on the head and extend over the back before merging on the tail base. Another cream stripe extends from the upper lip to the tail base. The tail is blue below and black above, with a middle row of bright blue spots and yellow lateral scales. The throat, limbs and anal region are cream, and the belly is orange.

Biology and breeding: This lizard hunts on vertical tree trunks, feeding on spiders and ants. When threatened, it darts behind a trunk away from danger, and climbs into the canopy; it is very difficult to catch. It has been observed to jump between trees, and can glide quite long distances. It lays two relatively large eggs (5-6 x 9-11 mm). **Habitat:** Lowland forest. **Range:** Through tropical and coastal forests of central, E. and W. Africa, just entering the subcontinent in central Mozambique.

Subspecies: Two races are recognized, with only *H.g. laevis* entering the subcontinent; it is restricted to E. Africa, and is replaced in the rest of Africa by the typical race, which has two lateral stripes.

ROUGH-SCALED LIZARDS
Ichnotropis

This small genus of lacertids lacks a collar, and has rough head shields and large, spiny, overlapping scales on the back. The lower eyelids are scaly. The lamellae beneath the toes are strongly keeled. Femoral pores are prese The tail is long.

All are terrestrial and run around on sandy soil in savannah. They lay eggs and some species are 'annuals', maturing within 5-8 months of hatching, an dying soon after laying one (exceptionally two) clutch of eggs. When they occur together, the life cycles of the Cape and common rough-scaled lizard *I. capensis* and *I. squamulosa* (below and page 139, respectively) are staggered, so that the juveniles and adults of both species are found at different times, and therefore do not compete for the same food.

The genus contains seven species which are distributed through the savannahs of S. and central Africa. Three species, one of which is endemic, occur on the subcontinent.

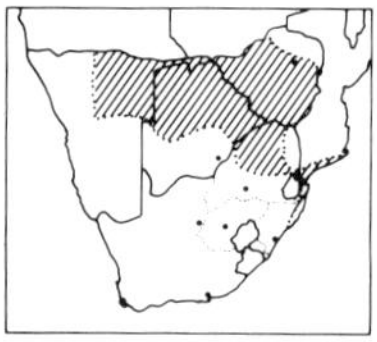

Cape Rough-scaled Lizard *Ichnotropis capensis* (Pl. 61)

16-18 cm; max. 20 cm
A medium-sized lizard with a narrow head. The large body scales are in 28- rows at midbody. The frontonasal is undivided. There are four upper labials front of the subocular, which borders the lip. The dorsal scales are strongly keeled and overlapping. There are 9-14 femoral pores on each thigh. Juveniles have a pale grey-brown back, with a white lateral stripe. In adults, the back becomes uniform grey to yellowish-brown or red-brown, with a narrow, white dorsolateral stripe that may be bordered above by a series of dark blotches. There is a broad, black stripe on the flank, bordered below b another white stripe. The belly is white. In breeding males, the white lateral stripe, chin and throat turn bright yellow, with an additional bright red stripe the lower flank.

Biology and breeding: These lizards are active hunters, searching throughout the day for termites and other insects. Mating occurs in October-December, and is followed almost immediately by egg-laying. The female digs an inclined burrow 10-20 cm long in soft soil, and lays 3-9 eggs (6,5 x 9,5 mm). Development is rapid, and they hatch in 56-77 days, the hatchlings (60-70 mm) emerging in January-March. Two clutches may be la by a female before she dies. Growth is rapid, and sexual maturity is reached 7-8 months (14-15 cm). By December adults begin to disappear, although some may be present until May; it is unusual for an individual to live for longe than 13-14 months. **Habitat:** Arid and mesic savannah. **Range:** N. Zululand, Transvaal and S. Mozambique, through Botswana, Zimbabwe and NE. Namibia, to Angola and Zambia.

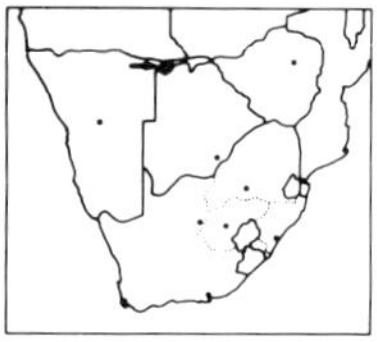

Caprivi Rough-scaled Lizard *Ichnotropis grandiceps*

(Endemic) 15-18 cm; max. 22 cm
A medium-sized species, similar to the Cape rough-scaled lizard, *I. capens* (above), but with a larger head, small body scales in 44-47 rows at midbody and without white borders to the dark dorsolateral stripe. The frontonasal is undivided. There are five upper labials in front of the subocular, which borde the lip. The dorsal scales are strongly keeled and overlapping. There are 13 femoral pores on each thigh. The back is pale grey-brown, with a few scattered, small, dark spots. A dark, broken dorsolateral band (which may b faint or absent) extends from the neck to the groin, where it breaks up into a line of lateral spots on the tail. The sides of the head and lower flanks are wh stippled with grey, and the belly is white.

Biology and breeding: It is doubtful that this species is an 'annual' as adults and juveniles have been collected together. It lives together with both other species of rough-scaled lizards. **Habitat:** Open mesic savannah. **Range:** Caprivi Strip and adjacent Botswana and NE. Namibia.

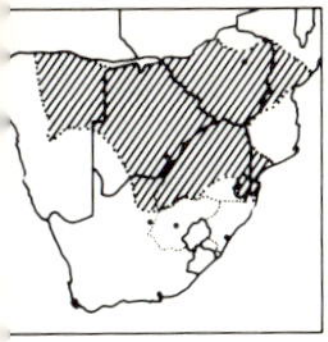

Common Rough-scaled Lizard
Ichnotropis squamulosa (Pl. 61)

17-20 cm; max. 23 cm
A medium-sized lizard with a small head and small body scales in 42-58 rows at midbody. The frontonasal is divided. The subocular does not border the lip. The dorsal scales are strongly keeled and overlapping. There are 11-18 femoral pores on each thigh. The back is pale buff to dark grey-brown, usually with narrow, broken, dark crossbands or blotches, and six long rows of pale spots. The belly is white to grey. In adult males, the lower labials and chin shields are mottled with black. **Biology and breeding:** These lizards are active hunters on sandy, flat clearings, feeding mainly on termites, but also catching grasshoppers, beetles, etc. They dig branching burrows in soft sand, usually at the base of acacia trees, that are sometimes shared by several individuals. The female lays 8-12 eggs (7 x 10-12 mm) in April-May (exceptionally as late as July), and hatchlings (70-80 mm) appear in October-November. The long incubation period (5-6 months) is due to cold winter temperatures. Growth is rapid, and sexual maturity is reached in 4-5 months. Adults die off during April-May. **Habitat:** Arid and mesic savannah. **Range:** N. Zululand, Transvaal, N. Cape, extreme S. and central Mozambique, through Botswana, Zimbabwe and E. Namibia, to Angola and Tanzania.

EDITERRANEAN ZARDS
certa

True *Lacerta* species include the beautiful common lizards of the Mediterranean region. They occur throughout most of Europe, even entering the Arctic Circle in Scandinavia. It is strange that two species endemic to South Africa should be placed in this genus, separated as they are from the other species by thousands of kilometres. It has been suggested that they are more closely related to the mountain lizards, *Tropidosaura* (page 147), and may even need to be placed in a new genus.

Southern Rock Lizard *Lacerta australis* (Pl. 62)

(Endemic) 14-19 cm; max. 22 cm
A medium-sized, graceful lizard with a well-developed collar and a tail that is twice as long as the body. The subocular borders the lip. The lower eyelids are scaly. The dorsal scales are very small, granular and smooth, in 67-68 rows at midbody. The ventral scales are in six long rows. The lamellae beneath the toes are smooth. There are 16-19 femoral pores. The back is dark olive-brown in adults (blue-green in juveniles) and has rows of small, pale yellow to white spots that fade on the rear of the body and are bright orange on the flanks. The top of the head is black, vermiculated with bright yellow spots. The throat and belly are bluish-white, with narrow, black edges to the scales. The tail is grey-green, tinged with blue at the tip, and with numerous black flecks. **Biology and breeding:** For many years this was South Africa's rarest lizard; only one specimen was collected between 1926 and 1973. It lives in mountainous country, in rock cracks and under large rock slabs on rugged sandstone outcrops. It often basks on the vertical faces and is locally common but relatively shy. It lays up to seven oval, soft-shelled eggs. **Habitat:** Fynbos vegetation. **Range:** Cedarberg, extending south to Hex River Mountains near Ceres.

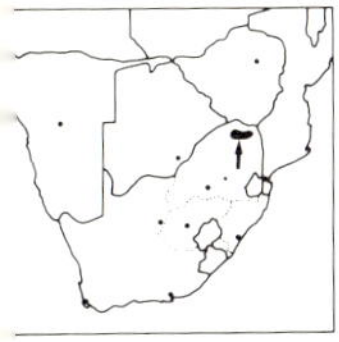

Soutpansberg Rock Lizard *Lacerta rupicola* (Pl. 62)

(Endemic) 11-13 cm; max. 15 cm
A small, graceful lizard with a well-developed collar. The subocular borders the lip. The lower eyelids are scaly, each with a few enlarged, elongate central scales. The dorsal scales are granular and smooth, in about 36 rows at midbody. The ventral scales are in six long rows. The lamellae beneath the toes are smooth. There are 15 femoral pores. The back is dark brown, with a

pair of rich reddish-brown, narrow vertebral stripes, and a white dorsolatera stripe that extends from the eye to the tail base. There may be a thin, pale lateral band that in adults is often broken into scattered white spots. The top the head is reddish-brown, the tail is pale brown, and the belly is bluish with dark-blue to black speckles. **Biology and breeding:** This diurnal lizard forages on exposed bedrock and rocky mountain slopes. (SA RDB, Restricted.) Its reproduction is poorly known; a female contained two eggs. **Habitat:** Sparsely vegetated mountains. **Range:** Endemic to Soutpansberg N. Transvaal, from Waterpoort to Lake Funduzi.

DESERT LIZARDS
Meroles

This is a small genus of lizards that have fine, granular dorsal scales and a well-developed collar. There is no window in the lower eyelid, and the subocular does not reach the lip. The feet are well developed, with long, usually fringed toes. Femoral pores are present.

These active, diurnal, terrestrial lizards are well adapted for desert life. Their reproduction is poorly known; all appear to be oviparous. They are camouflaged in pale colours, and none develops bright breeding colours.

There are six species in the genus, all occurring in the western arid regior of southern Africa; all but one are endemic.

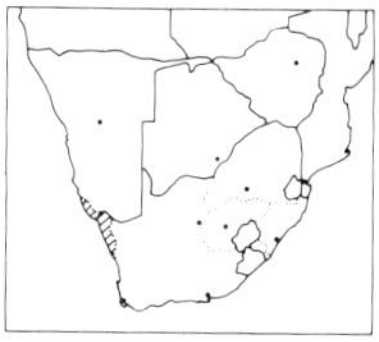

Smith's Desert Lizard *Meroles ctenodactylus* (Pl. 56)

(Endemic) 18-22 cm; max. 25 cm
A large desert lizard with a flattened, wedge-shaped snout and toes that ha a prominent, serrated fringe. Each ear opening is covered with a skin fold. T supranasals are in contact. There are 78-92 scale rows at midbody, 22-26 long rows of ventral plates, and 27-38 femoral pores on each thigh. The bac varies in colour from greyish-fawn to orange-brown or russet-brown, and is sometimes densely speckled with dark brown to black. A yellowish-white, dark-edged dorsolateral stripe runs from the ear to the tail base, with anothe broad, greyish to dark brown lateral stripe (that may have white and yellow spots) and a yellow lateral stripe, below. The limbs have large, yellow-white spots and the belly is white or pale yellow. **Biology and breeding:** Sit-and-wait hunters, these lizards shelter in the shade of sparse vegetation, dashin out to grab insects. They lay approximately six eggs (9-10 x 13-14 mm). **Habitat:** Vegetated coastal dunes and adjacent sandy plains. **Range:** Little Namaqualand and coast of S. Namibia.

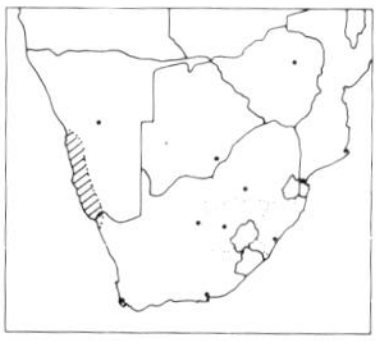

Wedge-snouted Desert Lizard *Meroles cuneirostris* (Pl. 56)

(Endemic) 13-15 cm; max. 16 cm
A small desert lizard with a flattened, wedge-shaped snout and toes that ha a prominent, serrated fringe. Each ear opening is covered with a skin fold. T nasals are swollen. The supranasals are not in contact. There are 90-110 sc rows at midbody, 24-30 long rows of ventral plates, and 18-24 femoral pore on each thigh. The back is greyish, sand-coloured or reddish-brown, with p spots (that may be absent) and dark flecks. Males have a more reticulated dorsal pattern. A pale dorsolateral stripe extends from the eye to the tail. Th lateral skin folds and upper lip may be yellow-orange. The belly is white.

Biology and breeding: The diet varies seasonally, depending on the availability of food. In summer, these desert lizards eat mainly adult beetles but take termites and other small insects during winter. They forage in the co morning and evening, and often sit next to ant paths, robbing the ants of their insect prey. They dive beneath loose sand to escape danger, and to sleep at night. There is no distinct breeding season. The female lays 2-4 eg (7-8 x 12-14 mm) in soft sand, up to 60 cm below the surface. Hatchlings measure approximately 56 mm. **Habitat:** Sparsely vegetated desert and coastal dunes. **Range:** Restricted to S. Namib Desert, from Richtersveld to Walvis Bay.

Knox's Desert Lizard *Meroles knoxii* (Pl. 57)

(Endemic) 15-20 cm; max. 23 cm
A medium-sized desert lizard with a rounded snout that lacks a sharp edge. Specimens from the Cape Peninsula are small (rarely longer than 18 cm). The lobed ear openings are visible. The supranasals are not in contact. The dorsal scales may be smooth (in northern populations) or weakly keeled, and are in 54-78 rows at midbody. There are 10-12 long rows of ventral plates, and 13-22 femoral pores on each thigh. The toes have a small fringe. Coloration is varied. Juveniles are dark brown to black, with five white stripes and scattered white spots. In adults, the back becomes red-brown (paler in coastal regions), with a dark dorsolateral stripe that may be broken into a series of partially white-edged, black spots. The flanks are grey, with a series of yellowish or brownish spots or circles. The limbs may have numerous pale spots. The belly is creamy white or bluish-grey. Adult breeding males are bright yellow on the lower head, throat and anal region.

Biology and breeding: Active, diurnal lizards, commonly seen dashing between bushes in sandy areas. They feed on insects, mainly beetles and flies. They shelter in a burrow dug in compacted soil at the base of a succulent bush or under refuse. Clutch size varies; small lizards from the Cape Peninsula lay only 2-3 eggs, while larger lizards from Namaqualand may lay up to six eggs (7,8 x 12-13 mm). **Habitat:** Varied; coastal dunes and succulent karroid veld. **Range:** Along W. Cape coast, from Cape Peninsula to coastal regions of S. Namibia, extending inland to Matjiesfontein and the Tanquwa Karoo.

Small-scaled Desert Lizard *Meroles micropholidotus* (Pl. 56)

(Endemic) 13-15 cm; max. 16 cm
A small desert lizard with a flattened, wedge-shaped snout and toes that have a serrated fringe. Each ear opening is covered with a skin fold. The nasals are not swollen. The supranasals are in contact. The dorsal scales are smooth and minute, in 126-138 rows at midbody. There are 24-28 long rows of ventral plates, and 17-20 femoral pores on each thigh. The back is light grey above, with a vague network of dark blue-grey that extends on to the limbs. The belly is yellowish-white. **Biology:** When disturbed, these lizards dive into coarse sand, and wriggle until they are completely covered; they also sleep under the sand. **Habitat:** Sparsely vegetated desert. **Range:** S. coastal Namib Desert, between Lüderitz Bay and Walvis Bay.

Reticulated Desert Lizard *Meroles reticulatus* (Pl. 56)

14-16 cm; max. 17 cm
This small desert lizard has a pointed snout with a weak edge along the upper lip. The lobed ear openings are visible. The supranasals are in contact. There are 50-56 scale rows at midbody, 16-18 long rows of ventral plates, and 19-23 femoral pores on each thigh. The toes have a prominent, serrated fringe. The back is grey to blue-grey, speckled with paler and darker grey tones that may form a dense network. The belly is white. **Biology and breeding:** These lizards burrow into mobile sand dunes, and feed on small insects. When disturbed, they run for some distance before diving just beneath the sand. Crows have been reported as major predators. A female contained four large eggs in October. **Habitat:** Sparsely vegetated coastal desert. **Range:** Coastal areas of N. Namib Desert, from near Walvis Bay to S. Angola.

Spotted Desert Lizard *Meroles suborbitalis* (Pl. 57)

(Endemic) 16-18 cm; max. 21 cm
A medium-sized desert lizard with a rounded snout that lacks a sharp edge on the upper lip. The lobed ear openings are visible. The supranasals are not in

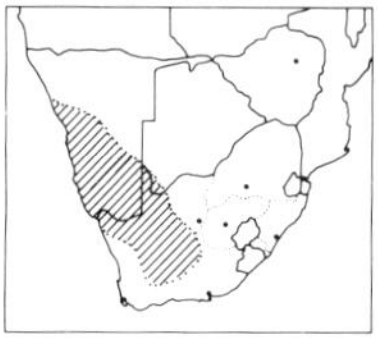

contact. There are 60-75 scale rows at midbody, 12-14 long rows of ventral plates, and 14-20 femoral pores on each thigh. The toes have a feeble fring Juveniles are pale yellow-white above, with four black stripes, and a black network enclosing large, pale spots on the limbs. In southern specimens th dark stripes develop light brown centres with age, and the back becomes grey-brown and covered with large, pale spots that may form rows. In the Namib Desert, adults are slate-coloured with a pinky sheen, and with irregu rows of dark, pale-edged spots. The belly is creamy white or bluish. The thr of males may be yellow. **Biology and breeding:** These sit-and-wait hunters live on flat gravel or sandy plains with scattered bush. They dash from the shade of low vegetation to grab bees, beetles, grasshoppers and termites. the central Namib Desert, breeding seems to be continuous and females m lay clutches of 3-7 eggs throughout the year. In the Kalahari, mating occurs early winter and females may lay two clutches, each of 4-8 eggs (7,5 x 12,5 mm), per season. Growth is relatively slow, and they may take up to thr years to mature. **Habitat:** Varied; arid savannah to desert. **Range:** Central Karoo to Little Namaqualand and S. Namibia, with an isolated population in central Namib Desert.

SANDVELD LIZARDS
Nucras

This is a small genus of lacertids that have a rounded snout, a cylindrical body, and a very long tail. The subocular borders the lip, and the lower eyeli are scaly. Each nostril is pierced between 2-3 nasals and is well separated from the upper labials. A distinct collar is present. The dorsal scales are sm smooth and non-overlapping. There are smooth lamellae beneath the toes, which lack a fringe. Femoral pores are present.

These secretive, diurnal, terrestrial lacertids may be locally common, but are rarely seen. They forage in the early morning and evening, and are ofter found gorging themselves on flying termites. They may store fat in the long, thick tail. All appear to be oviparous. Their predators include birds of prey a snakes.

They are restricted mainly to arid and mesic savannah on sandy soils, although some species enter fynbos and montane grassland. Seven specie are recognized from southern and E. Africa, with five entering the subcontinent; three of these are endemic. Identification is based mainly on coloration, as there is often a great overlap in scale counts between the species.

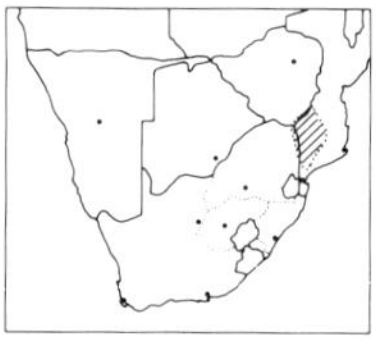

Blue-tailed Sandveld Lizard *Nucras caesicaudata* (Pl. 61)

(Endemic) 17-20 cm; max. 22 cm
A small, slender sandveld lizard with a brilliant blue tail. It has a series of six enlarged plates under the forearm. There are 40-54 scale rows at midbody. There are 14-15 femoral pores on each thigh. The back is dark brown, with seven thin, cream stripes, three of which extend on to the back of the head. The limbs are light brown with cream blotches, and the belly is pure white. **Biology:** Very little is known about these lizards (SA RDB, Peripheral). Like most other lizards, they eat termites. **Habitat:** Arid savannah on deep sand. **Range:** Restricted to Gazaland Plain of S. Mozambique, extending just into SE. Zimbabwe and N. Kruger National Park.

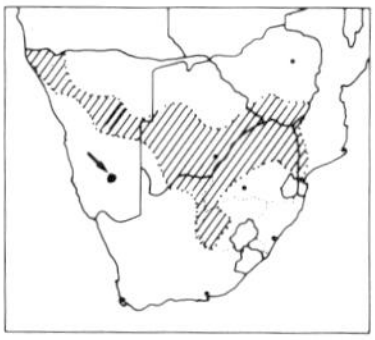

Spotted Sandveld Lizard *Nucras intertexta* (Pl. 60)

(Endemic) 22-26 cm; max. 28 cm
A large sandveld lizard, usually with a spotted back and an orange-brown ta It has 4-7 enlarged plates under the forearm, and 34-56 scale rows at midbody. There are 11-15 femoral pores on each thigh. The back is light brown to reddish-brown, with a series of pale spots that may be black-edge or fuse to form a mesh pattern or irregular, transverse bands. Sometimes a pale vertebral stripe is present. The belly is creamy white, with dark specks

the outer scale rows. Juveniles have darker backs, with rows of distinct cream spots; the sides of the neck are barred in pale yellow, and the tail is coral-red. **Biology and breeding:** Foraging widely in open dry savannah, these lizards search for slow-moving food such as spiders and scorpions. They lay about four eggs (8 x 12-13 mm). **Habitat:** Arid savannah, usually on Kalahari sand. **Range:** S. Mozambique and N. Transvaal, through Kalahari region to Kaokoveld in N. Namibia.

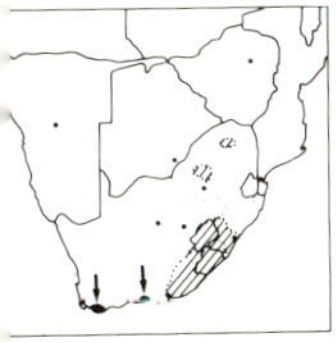

Delalande's Sandveld Lizard *Nucras lalandii* (Pl. 61)

(Endemic) 20-25 cm; max. 29 cm
A large, stout-bodied sandveld lizard that has a blunt snout and a thick, cylindrical tail. It lacks enlarged plates under the forearm. There are 34-42 scale rows at midbody. It has 10-15 femoral pores on each thigh. Juveniles are light red-brown, with 8-10 irregular rows of large, white, black-edged spots on the back, and black and white vertical bars on the sides of the head and neck. In adults, the back varies from grey-green to reddish-brown, and the black edges to the dorsal spots expand, often fusing and forming irregular black crossbands. The tail is olive-grey, heavily spotted in black, and the belly is creamy white, usually with numerous black spots.

Biology and breeding: These attractive lizards are usually found under stones in open grassland. They move slowly, scratching in loose soil around the base of grass clumps and boulders for spiders, beetles and larvae. The female lays 4-9 eggs (9-13 x 14-16 mm) in a chamber, often dug under a stone, in early summer. The hatchlings measure 70-90 mm. They do well in captivity and make fine, long-lived pets. **Habitat:** Montane and temperate grassland. **Range:** Distributed through the grassy uplands of E. Cape to Natal and adjacent OFS, with relict populations in central and N. Transvaal and S. Cape.

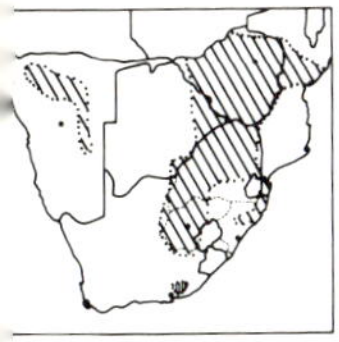

N.t. taeniolata
N.t. ornata

Ornate Sandveld Lizard *Nucras taeniolata* (Pl. 60)

20-30 cm; max. 35 cm
A beautiful sandveld lizard that has various colour patterns, but always has a striped back. The typical race from E. Cape rarely grows larger than 20 cm. There are 6-7 enlarged plates under the forearm and 36-64 scale rows at midbody. It has 10-21 femoral pores under each thigh. The typical race is dark brown above, with 8-11 fine, cream stripes that are fainter in the middle. The northern populations are grey or buff to reddish-brown above, with three well-marked, pale stripes along the back and 1-2 rows of spots on the flanks. The tail is light buff to coral-red, and the belly is creamy white. Juveniles are similar, but have a more distinct pattern. **Biology and breeding:** A secretive lizard, it may be locally common. None may be seen for 8-9 months, and then, following late summer rain when termite alates swarm, up to 20 may be found feeding avidly. They lay 4-5 eggs. **Habitat:** Varied; open grassland to mesic savannah and valley bushveld, usually in moist regions. **Range:** N. Cape and Natal, through Transvaal, Botswana and Zimbabwe, to Zambia and Malawi. There are isolated populations in E. Cape and Namibia. **Subspecies:** Two races occur, and both are found in the region. *N.t. taeniolata* has 4-5 pale stripes extending on to the back of the head; it occurs in the Albany district of E. Cape. *N.t. ornata* has no more than three pale stripes extending on to the back of the head; it is found over the rest of the species' range.

Striped Sandveld Lizard *Nucras tessellata* (Pl. 60)

20-25 cm; max. 31 cm
A slender, long-tailed sandveld lizard with vertical bars on the flanks and brilliant colours. There are 5-7 enlarged plates under the forearm, and 36-54 scale rows at midbody. It has 11-18 femoral pores on each thigh. Colour

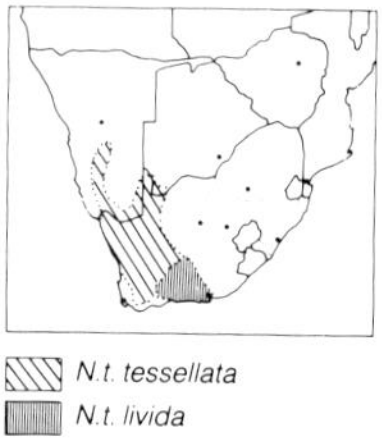

patterns vary. In the eastern race, the back is black, with six thin, cream stripes that fade towards the tail. The flanks are black, with small, white spot which may fuse to form vague, vertical bars. The typical race has a black ba with only four thin, cream stripes. The black flanks are brilliantly barred in black and white, and the rear of the body and tail are rich red-brown. The be is pure white. Some specimens are pale red-buff, with only two faint stripes the back.

Biology and breeding: They are specialist feeders on scorpions, and forage widely. The scorpions, and also beetles and spiders, are caught by digging them from their daytime retreats. The female lays 3-4 eggs. **Habitat:** Open arid savannah and karroid veld. **Range:** Karoo, through Namaqualand to S. Namibia and adjacent Botswana. **Subspecies:** Two rac are recognized, and both occur in the region. *N.t. tessellata* has 2-4 pale stripes on the neck, flanks with black and white stripes, and 36-48 scale row at midbody; it is found in W. Cape, through Namaqualand to S. Namibia. *N.t. livida* has six pale stripes on the neck, spotted flanks, and 44-54 scale rows at midbody; it occurs in central and E. Karoo, to S. Cape.

SAND LIZARDS
Pedioplanis

This is a group of small lizards that have a cylindrical body and a long tail. The subocular borders the lip. The lower eyelids may be scaly or have a transparent window. A distinct collar is present. The dorsal scales are usua small, smooth and non-overlapping. There are keeled lamellae beneath the long toes, which lack a fringe. Femoral pores are present.

These diurnal, terrestrial lizards are commonly seen dashing between sparse vegetation in the arid west. They are oviparous, laying small clutche soft-shelled eggs.

The genus contains at least eight species which are restricted mainly to southern Africa, with several species entering S. Angola. *P. benguellensis*, from S. Angola, is similar to *P. namaquensis* (page 146), but has a black-edged window in each lower eyelid. Seven species occur on the subcontine five of which are endemic.

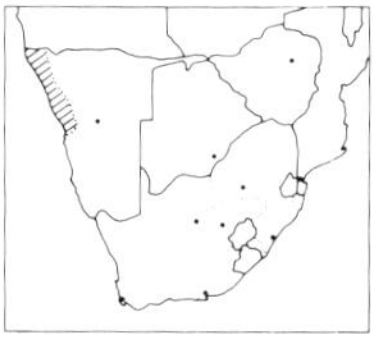

Short-headed Sand Lizard *Pedioplanis breviceps* (Pl. 58)

(Endemic) 12-14 cm; max. 16 cm
A small, slender sand lizard with a short head, scaly lower eyelids, no enlarged tympanic shield, four upper labials in front of the subocular, and 1 long rows of scales on the belly. The scales on the lower leg are small and n keeled. There are 11 femoral pores on each thigh. Males and females have different coloration. Hatchlings and adult females have five distinct, dark stripes on the back, with those on the flanks darker and more conspicuous. young males the bands begin to fade, and the backs of adult males are uniform light brown, with many grey spots that may form faint lines. The bell white and the tail is light brown, with thin, dark brown bands fading to red at the tip. **Biology and breeding:** These active lizards forage on open, sandy plains and in dry riverbeds. Eggs begin to develop in December, and 2-4 large eggs (6 x 11 mm) are laid in March-May. Hatchlings (50-80 mm) appear in May-July. **Habitat:** Desert. **Range:** Western areas of central and N. Namib Desert.

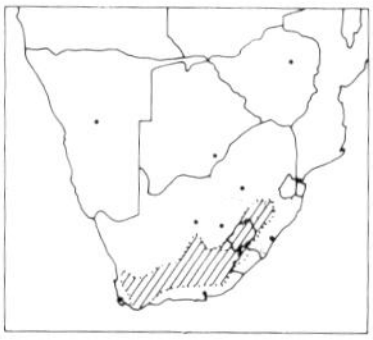

Burchell's Sand Lizard *Pedioplanis burchelli* (Pl. 59)

(Endemic) 14-16 cm; max. 19 cm
A small sand lizard with a relatively short tail and a slightly flattened head ar body. It has no enlarged tympanic or temporal shields. The lower eyelids ar opaque, each with 10-15 small scales across the middle. There is a faint gu fold. The nasals are usually in contact behind the rostral. Four to 13 small granules occur in front of the supraoculars. There are four pairs of chin shields. There are no lobes at the front of the ear openings. There are 62-75

scale rows at midbody, and 14 (rarely 16) long rows of ventrals. It has 13-16 femoral pores on each thigh. Coloration is extremely varied. Juveniles are black, with seven thin, white stripes on the body, and large, white spots on the limbs. The belly is white, and the tail is orange below. This coloration may continue in some adults, but usually the stripes fade or become broken into rows of white spots, and the back becomes buff, grey-brown or red-brown, usually with irregular black markings.

Biology and breeding: These lizards inhabit exposed bedrock with sparse vegetation and scattered rock slabs. They shelter in a small chamber excavated in soil under a flat rock. They are sit-and-wait hunters, dashing from behind a rock to catch small insects. In winter they lie dormant in their shelters. The female lays 4-6 oval eggs (7,5-10 x 9-13 mm) in moist soil under a rock slab in December-January. These hatch in 60-70 days (February-March); the hatchlings measure 66-78 mm. **Habitat:** Rocky montane grassland. **Range:** Cape Agulhas, through Cape fold mountains and inland Cape escarpment to W. OFS, Lesotho and SE. Transvaal.

Cape Sand Lizard *Pedioplanis laticeps* (Pl. 59)

(Endemic) 15-18 cm; max. 20 cm
This medium-sized sand lizard is very similar to Burchell's sand lizard, *P. burchelli* (page 144), but its nasals are not in contact behind the rostral and there is no gular fold. There are 13-30 small granules in front of the supraoculars, and five pairs of chin shields. It has 48-62 scale rows at midbody, and 16-18 long rows of ventrals. There are 12-18 femoral pores on each thigh. Coloration is extremely varied. Juveniles are similar to young Burchell's sand lizards, but have only five white stripes on the back. In adults, the lateral stripes fade and become broken into spots, and the back becomes brown with scattered pale spots, although three faint stripes remain. **Biology and breeding:** These lizards are similar in behaviour and habits to Burchell's sand lizards, except that they are found on sandy or hard soils down to sea level. **Habitat:** Coastal dunes and succulent karroid veld. **Range:** Central Karoo to W. Cape and Namaqualand.

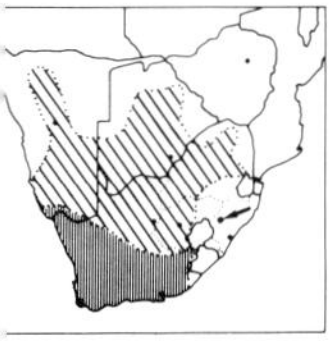

P.l. lineoocellata
P.l. pulchella

Spotted Sand Lizard *Pedioplanis lineoocellata* (Pl. 59)

(Endemic) 13-17 cm; max. 19 cm
A small sand lizard with a relatively short tail and a slightly flattened head and body. It has no enlarged tympanic or temporal shields, but has a window, formed by two black-edged, transparent scales, in each lower eyelid. There are 3-4 lobes at the front of each ear opening. There are 12-14 long rows of ventrals, and 10-17 femoral pores on each thigh. Coloration is extremely varied. The back ranges from buff to grey-brown or red-brown, usually with 2-4 rows of small, pale dorsal spots that may fuse into broken or continuous dorsolateral stripes. The flanks have a series of 4-7 large, pale blue-white spots that may be dark-edged. Specimens from around Lüderitz Bay have dull, dark grey bodies, occasionally with four faint dorsal stripes, and lack the large flank spots; they are sometimes treated as a separate race (*P.l. inocellata*). Some Namaqualand specimens are brightly coloured and have paired, black dorsolateral stripes. The hind limbs are usually spotted, and the tail is flecked with pale and dark dots. The belly is white, and males may have a blue-grey throat.

Biology and breeding: These lizards prefer flat, rocky veld. They are sit-and-wait hunters, grabbing small insects after a short dash from shaded cover. The diet varies with the season; they take mostly termites during winter, and beetles and locusts during summer. They are active during the day, even on warm winter days. They shelter in a small burrow dug beneath a flat rock. When chased, they dash from stone to stone, turning sharply. They are difficult to catch, unless pursued in the open, where they tire quickly. The female lays

4-8 eggs (9-10 x 12-13 mm) in November, in a small chamber dug in moist s beneath a rock. The eggs hatch after about 70-80 days (in February); hatchlings measure 50-55 mm. Sexual maturity is reached in 15-18 months, and they may live for 4-5 years.

Habitat: Very varied; karroid veld, valley bushveld and arid and mesic savannah. **Range:** Western half of the subcontinent, but absent from region of deep sand (eg. Namib Desert and central Kalahari). **Subspecies:** Two races are recognized, and both occur in the region. *P.l. lineoocellata* has smaller scales on its back than it does on its lower forelimbs, and they are slightly overlapping and keeled; it is found in the northern part of the range, i Transvaal, Botswana (where it is restricted to hard soil around the pans), OF N. Cape and N. Namibia. *P.l. pulchella* has small scales on its back, which a not overlapping and are almost smooth; it occurs in E. and W. Cape, through Karoo to Great Namaqualand.

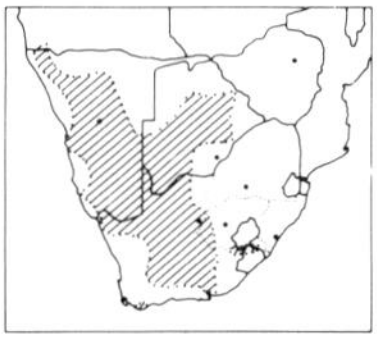

Namaqua Sand Lizard *Pedioplanis namaquensis* (Pl. 57)

14-17 cm; max. 18 cm
A small, slender sand lizard with a long tail. It has an enlarged tympanic shield. The lower eyelids are semi-transparent, each having 10-12 enlarged scales across the middle. A small group of 3-8 granules occurs in front of th supraoculars. There are no lobes at the front of the ear openings. There are 47-65 scale rows at midbody, and 12-14 long rows of ventrals. It has 10-16 femoral pores on each thigh. The juveniles are black, with four thin, white stripes; they have brown legs with white spots, a white belly, and a pinky-brow tail. This coloration may persist in adults in the Karoo on dark soils, but the dorsal stripes of adults from the western regions fade or become pale brow and the flanks may be irregularly barred. **Biology and breeding:** These amazingly fast lizards can be seen in the heat of the day, dashing over open sparsely vegetated sand and gravel flats. They forage widely, feeding on small insects. During winter they remain dormant in their burrows, which they dig i sand accumulated at the base of bushes. They lay 3-5 eggs in November. **Habitat:** Karroid veld, arid savannah and semi-desert. **Range:** E. Cape, through Karoo, Namaqualand, Namibia and Botswana, to S. Angola.

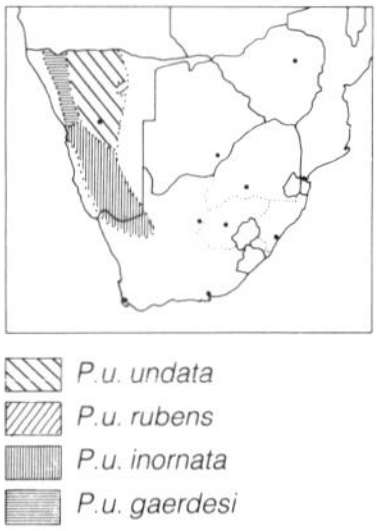

Western Sand Lizard *Pedioplanis undata* (Pl. 58)

14-18 cm; max. 20 cm
This small, slender species is very similar to the Namaqua sand lizard, *P. namaquensis* (above), but has a window composed of a few large scales each lower eyelid (see Subspecies). In addition, there are 57-75 scale rows midbody, usually 10 long rows of ventrals, and 6-15 small granules in front c the supraoculars. There are 8-15 femoral pores on each thigh. Coloration is varied. Juveniles are very similar to young Namaqua sand lizards, with 5-6 black stripes on the back. In adults of the typical race, the stripes remain, or become broken into a dense, reticulated pattern, and there is a series of pa bluish-green spots on the lower flanks. In other races, the stripes fade and tl tail may become brightly coloured (see Subspecies).

Biology and breeding: They are active hunters of small insects on gravel plains (the typical race) or on exposed rocky areas (other races). They mate November-January, and the young (55-75 mm) hatch in January-March. **Habitat:** Semi-desert, including rocky flats. **Range:** N. Cape, through inlanc areas of Namibia (occasionally nearing the coast), to S. Angola.

Subspecies: Four races are recognized (some may be full species). In *P.u. undata*, each lower eyelid has 2-4 large, semi-transparent scales, and t body usually has stripes; it occurs in S. Angola to N. and central Namibia. *P.u. rubens* has similar lower eyelids, but the head and forebody are red-brown, without dorsal markings, and the hind body and tail are bright brick-red; it is found on Waterberg. *P.u. inornata* is similar to *P.u. rubens*, but the body is grey-brown, becoming orange-brown on the tail, with a series of pale greenish spots on the flanks; it occurs in N. Cape and Great Namaqualand. *P.u. gaerdesi* has lower eyelids each with a large, undivided, transparent window, and a red-brown body that has a golden tinge and pale spots on the flanks, and faint or no dorsal stripes; it occurs from Erongo to Kaokoland.

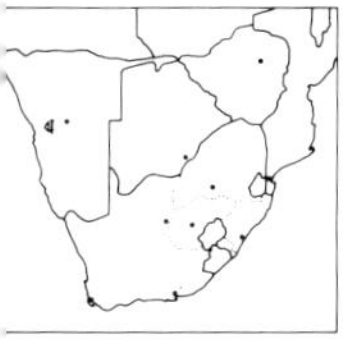

Husab Sand Lizard *Pedioplanis* sp. (Pl. 58)

(Endemic) 15-19 cm; max. 21 cm
This newly discovered species is very similar to the Western sand lizard, *P. undata* (page 146), but it has a smaller tympanic shield, and eight opaque scales in each lower eyelid. The back is reddish-brown, without stripes or lateral spots. **Biology and breeding:** A terrestrial species, they prefer expanses of flat rock on exposed bedrock. They are more predatory than the Western sand lizard, with which they live, and readily eat small lizards. Hatchlings emerge in April-June, and are 40-50 mm long. **Habitat:** Rocky desert. **Range:** Restricted to vicinity of Husab Mountains in central Namibia.

OUNTAIN
ZARDS
'opidosaura

These small lacertids have a short head, a cylindrical body, and a long, cylindrical tail. They lack a collar, but usually have a faint gular fold. The body is covered with rough, spiny, overlapping scales. Femoral pores are present. The body is camouflaged in brown and olive, but breeding males develop bright colours on the flanks.

They are terrestrial. Most live among rocks and heather on mountain summits, basking and foraging on flat bedrock and sheltering in short tunnels that they dig under rock slabs, or in soil-filled cracks. They hibernate in their retreats during winter, when snow often blankets the mountains. The females lay a small clutch of 2-7 eggs.

The four species in the genus are endemic to southern Africa. Two are restricted to the montane grassland of the Natal and Cape Drakensberg, and another to the fynbos of the Cape fold mountains. The remaining species is more widely distributed through the southern coastal regions.

Essex's Mountain Lizard *Tropidosaura essexi*

(Endemic) 11-13 cm; max. 15 cm
This small species is very similar in appearance to the common mountain lizard, *T. montana* (page 148). The head is short, and there is a well-marked gular fold. The subocular borders the lip. The scales on the sides of the neck are small, smooth and granular. The body scales are large, spiny and overlapping. There is a single large preanal plate, and 7-11 femoral pores. The back is olive to olive-green, with a dark, pale-edged stripe along the backbone. The flanks have dark-edged dorsolateral and lateral stripes. The head is bluish-green, spotted with brown. The belly is bluish, usually with dark edges to the scales. The tail is green-blue. **Biology and breeding:** These lizards rarely forage far from thick vegetation along mountain streams and at the base of rock faces, and quickly disappear when danger threatens. Two eggs are laid in summer. **Habitat:** Rocky montane grassland. **Range:** Summit slopes of Natal Drakensberg and adjacent OFS.

Cottrell's Mountain Lizard *Tropidosaura cottrelli* (Pl. 62)

(Endemic) 12-15 cm; max. 17 cm
This large mountain lizard has a flattened head and a well-developed gular fold. The subocular borders the lip, and the first upper labial is well separated from the nostril. The scales on the side of the neck are small, smooth and granular, while the body scales are large, spiny, overlapping and rhombic in shape. There are two large preanal plates and 11 femoral pores. Coloration is varied. Cape specimens have a brown back with scattered black and orange

scales. There is a row of yellow or yellow-tipped scales on the flanks, and a black, yellow-edged stripe on the neck that continues on to the lips. The tail i olive-brown above, with scattered, dark brown scales, and blue-white below The belly is bluish, with large, black blotches on the scales. Natal specimens are brown-black above, with numerous green and blue spots that form crossbands on the rear half of the body and tail. The flanks are pale blue, wit a black stripe. **Biology and breeding:** This terrestrial species hunts small insects among clumps of grass and heather. It shelters in a tunnel that it digs in deep soil. A female had four eggs in her oviducts. **Habitat:** Montane grassland. **Range:** Natal and Cape Drakensberg escarpment.

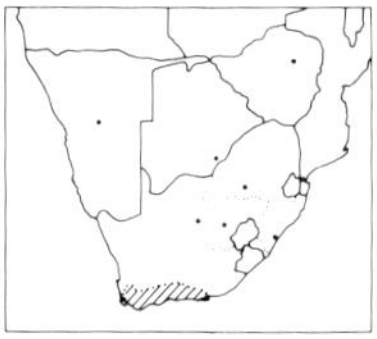

Cape Mountain Lizard *Tropidosaura gularis* (Pl. 62)

(Endemic) 14-18 cm; max. 20 cm
A large mountain lizard that has a distinct gular fold across the throat. The subocular borders the lip, and the first upper labial enters the nostril. The scales on the sides of the neck are small, smooth and granular, while the bo scales are large, spiny and overlapping. There are two large preanal plates, and 9-12 femoral pores. The back is dark brown to olive-brown, usually with pale olive-brown band along the backbone. The scales are black-edged, giving a speckled appearance. There are two white-yellow lateral stripes tha may be faint and/or broken up into spots, with a row of yellow spots below. The tail is blue-grey in juveniles and females, and pale olive with black flecks in adult males. In breeding males, the dorsal band becomes bright green, th yellow flank spots turn bright orange, and the head becomes black with yello spots. **Biology and breeding:** These lizards are common among the heathe and proteas of mountain summits, but are shy and difficult to catch. They readily climb on to rock faces and boulders, and feed on flies and bees attracted to the heather flowers. The female lays 4-8 eggs (7-8 x 11-12 mm) November in a small chamber under a flat rock. **Habitat:** Fynbos-covered mountain summits. **Range:** Cape fold mountains, from Matroosberg to Lady Slipper near Port Elizabeth.

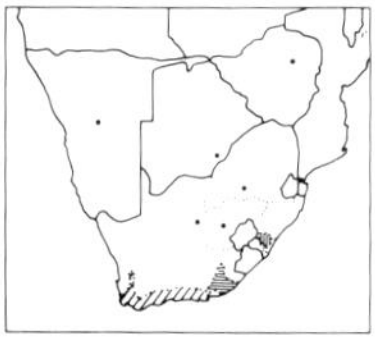

T.m. montana
T.m. rangeri
T.m. natalensis

Common Mountain Lizard *Tropidosaura montana* (Pl. 62)

(Endemic) 12-15 cm; max. 18 cm
A small mountain lizard with a short head, a faint gular fold and a long tail. The subocular borders the lip. The scales on the sides of the neck are keele and the body scales are large, spiny and overlapping. There is a single larg preanal plate, and 5-8 femoral pores (see Subspecies). The back is olive-brown to dark brown, with a dark streak along the backbone and a greenish-white to yellowish dorsolateral stripe. A broken white stripe extend from the upper lip on to the lower flank, with a series of large, pale yellow (bright orange in breeding season) spots below this. The belly is greenish-white, often with large, dark spots. The tail becomes faint blue-green in breeding season, and has black spots below.

Biology and breeding: These shy and secretive lizards often bask on vegetation, where they are perfectly camouflaged. They are active in the ea morning and late afternoon, and feed on small insects. The female lays 4-5 eggs (6,5-7 x 10-11 mm) in mid-November. They hatch in 33-34 days (at 28-30 °C), and the hatchlings are 56-61 mm long. **Habitat:** Fynbos and montane grassland. **Range:** Cape fold mountains, through E. Cape and Amatola Mountains, to Natal midlands and Drakensberg foothills. **Subspecies:** Three poorly defined races are recognized. *T.m. montana* usually has 7-8 femoral pores; it occurs in S. and W. Cape. In *T.m. rangeri*, t gular fold is usually absent and there are 5-6 femoral pores; it is found in E. Cape. *T.m. natalensis* has a longer head and frontal scale, the body scal are smaller, and there are only five femoral pores; it occurs in foothills of Nat Drakensberg and S. Natal midlands.

LATED LIZARDS, IRDLED LIZARDS ND THEIR ELATIVES
amily Cordylidae

These lizards have longish tails and well-developed legs; some grassland species have become snake-like with reduced limbs. The head has large, symmetrical head shields with osteoderms. Body scales are usually rectangular (plates), overlapping, arranged in regular rows (girdles), often heavily keeled, sometimes granular. The body is often depressed, and in most species is box-like in cross-section, with a lateral fold. The eyes and eyelids are well developed, external ear openings are visible, and the tongue is stout, roundly pointed and sometimes has a shallow notch. The femoral glands have conspicuous pores in most species, particularly in males.

This is the only lizard family restricted to Africa and Madagascar, with more than 50 species in 10 genera. True cordylid fossils are few, and are limited to the Late Eocene or Early Oligocene epoch of France (*Pseudolacerta*) and the Miocene epoch of Africa. There are two subfamilies; these were previously treated as separate families.

LATED LIZARDS
ıbfamily
errhosaurinae

This subfamily is the basal stock from which girdled lizards (cordylines) evolved, probably during the breakup of the southern supercontinent Gondwanaland, after the separation of Madagascar from Africa in the middle Cretaceous epoch (80-100 million years ago). They are diurnal, oviparous lizards, most having stout bodies and well-developed limbs, but including one genus of snake-like lizards that have reduced limbs. The tongue is moderately long and covered with scale-like papillae. The body scales are rectangular and have osteoderms. The granular lateral body fold is usually prominent. The clavicle (collar bone) is flattened and perforated.

There are six genera, with two occurring on Madagascar (*Tracheloptychus* and *Zonosaurus*) and four in subSaharan Africa, none of which is endemic.

ey to the southern frican genera in the errhosaurinae

1. Prefrontal shields present 2
   Prefrontal shields absent 3

2. Snout with sharp horizontal edge (spade-like); toes with lateral fringe .... *Angolosaurus* (Desert plated lizard, page 149)
   Snout not spade-like; toes without fringe .... *Gerrhosaurus* (Plated lizards, page 150)

3. Lower eyelids with transparent disc; small, and usually with bright blue tail .... *Cordylosaurus* (Dwarf plated lizard, page 150)
   Lower eyelids scaly; limbs short or rudimentary (some with forelimbs absent); colour dark grey-olive .... *Tetradactylus* (Seps and Plated snake-lizards, page 153)

ESERT PLATED ZARD
ngolosaurus

This is a medium-sized plated lizard with a spade-like snout, a relatively short, cylindrical tail, a cylindrical body, and long limbs with lateral fringes to the toes. These are all adaptations to the desert regions in which it lives. There is a single species in the genus.

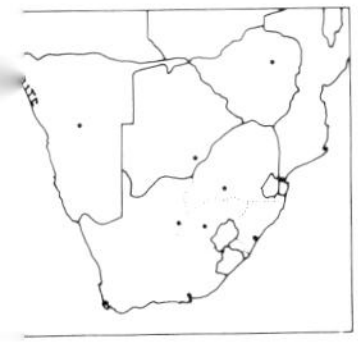

Desert Plated Lizard *Angolosaurus skoogi* (Pl. 64)

20-25 cm; max. 30 cm
A medium-sized plated lizard with a spade-like snout and a cylindrical body and tail. Prefrontals are present, and each nostril is pierced between two nasals and the first labial. The tympanic shield is large, covering most or all of each ear opening. The dorsal scales are smooth and small, in 32-35

longitudinal and 62-65 transverse rows. The lateral scales bordering the fold are granular. Ventrals are in eight longitudinal rows. The tail is short, tapering to a fine point, and is only slightly longer than the body. There are 12-13 femoral pores on each side (these are absent or rudimentary in females). The back is ivory, grey-white or light buff, often with numerous, scattered, small, rectangular, maroon-orange blotches. The flanks are paler and often have large, light blotches. The chin, throat and lower chest are black. The belly is white. Juveniles are sand-coloured. **Biology:** These active, diurnal lizards a often found in small colonies. They forage on loose wind-blown sand dunes, feeding on beetles and dry plant debris, including seeds and grass stems. When disturbed, they dash into *Narras* clumps or dive into dunes, disappearing beneath the sand with a swimming motion. They can spend up to 24 hours under the sand, sheltering from both danger and temperature extremes. **Habitat:** Scrub-covered sand dunes. **Range:** N. Namib Desert of Namibia and S. Angola.

DWARF PLATED LIZARD
Cordylosaurus

This very small, brightly coloured plated lizard has well-developed limbs, a prominent lateral fold and a longish tail.

There is a single species in the genus. It is restricted to the arid western region, from S. Angola to the Karoo.

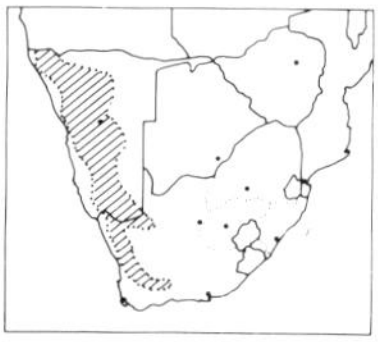

Dwarf Plated Lizard *Cordylosaurus subtessellatus* (Pl. 65)

12-14 cm; max. 16 cm
The head of this lizard is short, with large head shields and no prefrontals, ar each nostril is pierced between two nasals and the first labial. Each lower eyelid has a transparent disc. The dorsal scales are smooth to distinctly keeled, in 15 longitudinal and 52-55 transverse rows. The ventrals are in eigh longitudinal rows. There are 7-10 femoral pores on each side. The tail may b up to twice the length of the body. The back is dark brown to black, with distinct pale cream to light yellow dorsolateral stripes that become electric blue on the tail. The belly is off-white. **Biology and breeding:** This beautiful lizard forages among succulent vegetation on small rock outcrops and feeds on grasshoppers, flies, etc. It frequently stops to bask, resting on its belly an lifting its limbs off the hot sand. It will often tolerate very close approach befo slipping away between the rocks. If grabbed, it will readily shed its tail, whic continues to writhe for some time, attracting the predator's attention while its recent owner slips away to safety. Its breeding is poorly known; it is reported lay two eggs. **Habitat:** Succulent and karroid veld. **Range:** S. Angola, throug W. Namibia, Namaqualand and W. Cape, reaching Beaufort West in Karoo.

PLATED LIZARDS
Gerrhosaurus

These medium to large lizards have fully developed but relatively small limbs and a long tail. The body has a prominent lateral fold. Each nostril is pierced between two nasals and the first labial. The lower eyelids are opaque. The scales beneath the toes are smooth or have tubercles. Femoral pores are present.

Diurnal and terrestrial, most species live in savannah or sandveld, althoug some are rock-living. They dig holes in loose sand around bushes or excava leaf litter from large rock cracks or under boulders. Slow-moving, they eat large insects, snails, etc; larger species also eat some plant material. Shy ar solitary, they are rarely common and never form colonies. When foraging, th scrape away loose soil or leaf litter looking for hidden prey. They may slowly toboggan down gentle slopes on their smooth belly; this behaviour gives an indication of how the evolution of limb-loss in the snake-like gerrhosaurines (*Tetradactylus*) occurred. Plated lizards often bask resting on the belly with t limbs flexed upwards, off the ground. The tail is readily shed and can be full regenerated. Oviparous, they lay a few soft-shelled eggs in moist sites.

There are six species, all occurring on the subcontinent, with some extending into central and E. Africa. One is endemic.

Yellow-throated Plated Lizard
Gerrhosaurus flavigularis (Pl. 65)

25-35 cm; max. 44 cm
This medium-sized, graceful plated lizard has a slender, rounded body and a small head. The prefrontals are in narrow contact or are separated. The frontonasal and rostral are separated by the nasals. There are five supraciliaries. The dorsal scales are strongly keeled, in 22-24 longitudinal and 54-64 transverse rows. There are eight rows of ventral plates. Males have 11-17 femoral pores; females have no femoral pores. The back is dark red-brown to olive in colour, with a pair of prominent, dark-edged, yellow dorsolateral stripes. The flanks are darker, often with faint yellowish bars. The belly is cream. In breeding season, the chin, throat, chest and sides of the head of the males turn bright yellow, red or, in some regions, light blue (the taxonomic significance of this is unknown).

Biology and breeding: This lizard can be very common, even in urban areas. It lives in burrows which it digs at the base of bushes, under boulders, etc., but is also found under rubbish. It moves surprisingly quickly, catching grasshoppers, termites and millipedes. It is alert and difficult to catch without causing it to shed its tail, which is quickly regenerated. It bites readily when first captured, but tames easily and makes an interesting and long-lived (longer than 11 years) pet. The female buries her 4-6 oval, soft-shelled eggs (19-25 x 13-14 mm) in a small chamber dug in leaf litter or under a stone. Hatchlings, measuring about 100 mm, emerge in late summer.
Habitat: Varied; montane grassland, savannah, bushveld and low, open coastal forest. **Range:** Throughout the eastern part of the subcontinent; absent from karroid and western arid regions (with a relict population near Gobabis in Namibia). Elsewhere, through E. Africa to Ethiopia and Sudan.

Rough-scaled Plated Lizard *Gerrhosaurus major* (Pl. 64)

30-40 cm; max. 48 cm
A large, stout plated lizard, with a short head and large eyes. The tympanic shields are long and narrow. The prefrontals touch each other, as do the frontonasal and the rostral. The dorsal scales are rough, in 14-18 longitudinal and 31-33 transverse rows. There are 10 rows of ventral plates. Both sexes have 9-13 femoral pores. The back is straw-coloured to light brown; each scale is often dark-centred, creating a speckled to striped effect. It has a pale dorsolateral stripe. The chin and throat are light straw to cream, and the belly is smoky-grey to light brown. Males from Jos Plateau in Nigeria develop a blue throat (yellow in females), but this has not been noted elsewhere. **Biology and breeding:** This lizard regularly eats soft fruits and flowers, in addition to insects and millipedes; a large wild specimen ate a gravid Kirk's agama, and captive specimens will eat any small lizard they can catch and swallow. It lives in cracks in small, well-vegetated rock outcrops and also in old termitaria. It tames easily and settles well in captivity. A few (2-4) large, oval eggs (45-58 x 23-26 mm) are laid in moist soil. **Habitat:** Arid and mesic savannah. **Range:** Savannahs of Africa, from Togo in the west and Ethiopia in the north, to Zululand in the south-east. **Subspecies:** A number of races have been proposed (*G.m. grandis, G.m. zechi, G.m. bottegoi*), but all are based on small differences in colour pattern. Fossils are known from the lower Miocene epoch (10 million years ago) of Mfanganu Island, Lake Victoria in Kenya.

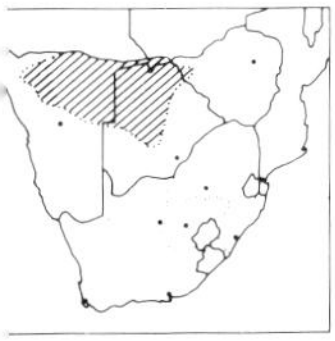

n. auritus

Kalahari Plated Lizard *Gerrhosaurus multilineatus* (Pl. 64)

30-35 cm; max. 49 cm
This large, thick-bodied, handsome plated lizard has a moderately sized head. The prefrontals are in broad contact, while the frontonasal and rostral are separated by the nasals. There are four supraciliaries. The tympanic

shields are broad and crescent-shaped. The dorsal scales are strongly keeled, in 26 longitudinal and 51-54 transverse rows. There are eight rows of ventral plates. Both sexes have 14-18 femoral pores. The body is pale brown to sepia, often with dark brown flecks on the head. The scales on the back and tail have a yellow spot on the inner edge, giving a flecked appearance. Sometimes there is a series of 3-4 narrow, dark-edged, pale dorsolateral stripes. The flanks are lighter, with dark brown spots. The limbs have dark-edged spots. The belly is creamy white. Juveniles are darker above, with more distinct spots. **Biology:** This lizard lives in holes burrowed among the roots of shrubs. Its diet consists of grasshoppers, beetles, termites and scorpions. **Habitat:** Bushveld and Kalahari sandveld. **Range:** N. Kalahari, north to Angola and Zaire. **Subspecies:** Two races are recognized, with only one occurring in the region. *G.m. multilineatus* has mucronate dorsal scales and spiny proximal caudals; this northern race occurs in Angola, Zaire and Zambia. *G.m. auritus* attains a maximum length of 44 cm, its dorsal scales are not mucronate, and its proximal caudals are not spiny; this southern race is found in northern regions of Namibia and Botswana.

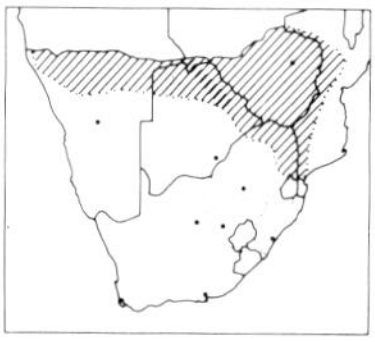

Black-lined Plated Lizard *Gerrhosaurus nigrolineatus* (Pl. 65)

35-45 cm; max. 56 cm
A large, thick-bodied plated lizard with a large, robust head. The prefrontals are in broad contact, while the frontonasal and rostral are separated by the nasals. There are four supraciliaries. The tympanic shields are narrow. The dorsal scales are strongly keeled, in 22-24 longitudinal and 54-64 transverse rows. There are eight rows of ventral plates. Both sexes have 16-18 femoral pores. The back is reddish-brown, with well-defined, black-edged, yellow dorsolateral stripes, and often with yellow streaks down the backbone. The flanks are vermillion or chestnut-brown, usually with irregular yellow spots. The belly is cream to yellow-white. Juveniles often have irregular, dark transverse bars on the back. **Biology and breeding:** This species is similar to the yellow-throated plated lizard, *G. flavigularis* (page 151), but prefers more open bushveld and lives in rodent and meercat burrows and old termitaria. Shy and elusive, it dashes to its burrow at the first sign of danger. The diet consists mainly of grasshoppers and beetles, but snails are also taken. It has many predators, including snakes, birds of prey and small carnivores. The female lays 4-6 eggs (30 x 22 mm) in rotting vegetation. Incubation takes 70-80 days. Hatchlings measure approximately 180 mm. **Habitat:** Savannah and bushveld. **Range:** E. and N. Transvaal, through Zimbabwe and adjacent regions to N. Namibia, reaching Gabon and lower Congo.

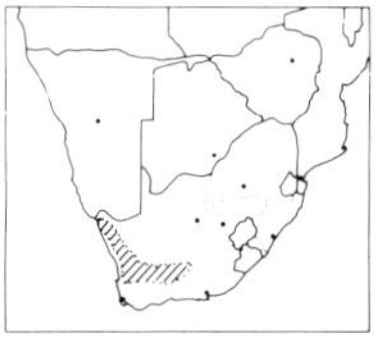

Namaqua Plated Lizard *Gerrhosaurus typicus* (Pl. 64)

(Endemic) 25-30 cm; max. 34 cm
This medium-sized plated lizard has a shortish head, large eyes and a rounded body. The tympanic shields are large and crescent-shaped. The prefrontals are separated (occasionally in short contact), while the frontonasal and rostral are in contact. The dorsal scales are keeled, in 22-24 longitudinal and 56-58 transverse rows. There are 10 rows of ventral plates. Both sexes have 15-18 femoral pores. The back is fawn to light purple-brown in colour, sometimes with a few scattered, dark spots. A pair of dark-edged, white dorsolateral stripes is present. The flanks are darker, with scattered, white, elongate spots. The limbs are pale pinky-brown. The tail is irregularly spotted, and the belly is white. The undersurfaces of the limbs and the base of the tail are carrot-red in breeding males. **Biology:** This species is rarely seen (SA RDB, Rare). It is similar in habits to the black-lined plated lizard, *G. nigrolineatus* (above), living in small burrows that it digs at the base of bushes. It is active in the early morning and evening. **Habitat:** Karroid succulent veld. **Range:** Little Namaqualand and central Karoo.

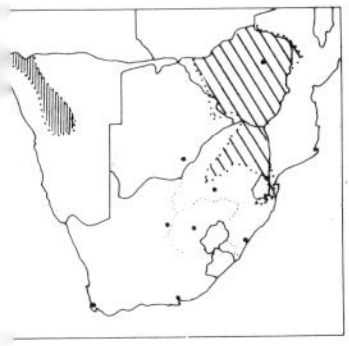

G.v. validus

G.v. maltzahni

Giant Plated Lizard *Gerrhosaurus validus* (Pl. 64)

40-60 cm; max. 69 cm
A large lizard with a flattened head and body. The tympanic shields are large and triangular. The prefrontals are in contact, as are the frontonasal and rostral. The dorsal scales are small, in 28-34 longitudinal and 52-55 transverse rows. There are 12-16 rows of ventral plates (see Subspecies), and 18-25 femoral pores in both sexes. In adults, the back is dark brown to black, with each head shield and dorsal scale spotted yellow, giving a speckled appearance; a pair of broad dorsolateral stripes is sometimes present.
The throat is dirty white, and the belly is light brown. Juveniles are black, with a distinct series of yellow spots on the back and bars on the flanks. In breeding season, the chin, throat and sides of the head of the males become tinged with pink-purple.

Biology and breeding: Rock-living, these lizards prefer the upper slopes of large granite koppies. They are shy and difficult to approach. They forage for invertebrates and vegetable matter (flowers, leaves, figs and other soft fruit), but will also eat small lizards, and even baby tortoises. They are very difficult to catch, and jam themselves into large cracks by inflating the body. The female lays 2-5 (usually four) large, oval eggs (44-46 x 24-26 mm) in soil-filled rock crevices in midsummer. Hatchlings measure 150-170 mm.

Habitat: Arid and mesic savannah. **Range:** Northern regions of the subcontinent (see Subspecies). **Subspecies:** Two races are recognized, with both occurring in the region. In *G.v. validus*, the subocular does not border the lip, and there are 14-16 ventral scale rows, and 18-24 lamellae under the fourth toe; this eastern race occurs from Zululand, through Transvaal lowveld, Zimbabwe and Mozambique, to Malawi and Zambia. *G.v. maltzahni* has its subocular bordering the lip, 12-14 ventral scale rows, and 15-17 lamellae under the fourth toe; this western race is found in central and NW. Namibia, extending into S. Angola.

EPS OR PLATED NAKE-LIZARDS *tradactylus*

These are small plated lizards, with species that include those having small but normal limbs, to snake-like forms lacking forelimbs and having only small, spiky hind limbs. The body scales are arranged in straight longitudinal and transverse rows. There is a prominent lateral fold. The head is small, with large head shields. The prefrontals are absent in southern African species. The lower eyelids are scaly. Femoral pores are present, except in the extralimital *T. ellenbergeri*. The tail is elongate.

Most of these lizards are snake-like and speedily hunt grasshoppers and other insects in grassland. They are diurnal, retreating at night into a grass tussock or beneath a stone. The vestigial hind limbs are minute, and are used to support the lizard when stationary and to assist small movements in long vegetation. The very long tail (up to three times the body length) is used for propulsion, rendering them almost impossible to catch when they 'swim' through grass. If grabbed, the tail is readily shed; regeneration is very rapid, as they are helpless without it. All are probably oviparous, although their breeding is poorly known.

Six species exist, five of which are endemic to the subcontinent. Extralimitally, *T. ellenbergeri* is distributed through the grasslands of E. Angola to SE. Tanzania.

T.a. africanus

T.a. fitzsimonsi

African Long-tailed Seps *Tetradactylus africanus* (Pl. 50)

(Endemic) 25-30 cm; max. 33 cm
This snake-like seps has minute, single-toed forelimbs and hind limbs (the forelimbs are sometimes absent; see Subspecies). Each nostril is pierced between two nasals and the first labial. The frontoparietals are just in contact. The dorsal scales are strongly keeled, in 14 longitudinal and 69-72 transverse rows. Ventrals are in six longitudinal and 50-52 transverse rows. There are 2-3

femoral pores on each side. The tail is more than three times the length of the body. The back is olive, with dark brown stripes running down the middle of the scale rows. The head is irregularly spotted with dark brown. The sides of the neck have a series of short, black bars. The belly is pale olive.
Habitat: Montane grassland. **Range:** Natal (Pondoland to Zululand) to S. Cape. **Subspecies:** Two races are recognized, and both occur in the region. *Tetradactylus africanus africanus* has minute forelimbs, and occurs in Natal. *Tetradactylus africanus fitzsimonsi* lacks forelimbs, and is found in S. and E. Cape.

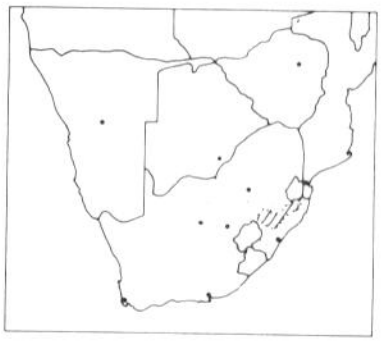

Breyer's Long-tailed Seps *Tetradactylus breyeri* (Pl. 51)

(Endemic) 18-22 cm; max. 24 cm
A snake-like seps with two minute toes on each forelimb and one on each hind limb. The three supraoculars touch the frontal. Each nostril is pierced between two nasals. The frontoparietals are rarely in contact. The dorsal scales are strongly keeled, in 12-14 longitudinal and 70-72 transverse rows. The ventrals are in eight longitudinal and 54 transverse rows. There are two (rarely three) femoral pores on each side. The tail is more than three times the body length. The back is olive-brown, with faint, darker dorsolateral stripes. The head is dark-spotted. The sides of the neck have short, dark bars. The belly is pale olive. (SA RDB, Rare.) **Habitat:** Montane grassland. **Range:** SE. Transvaal and adjacent Natal and OFS.

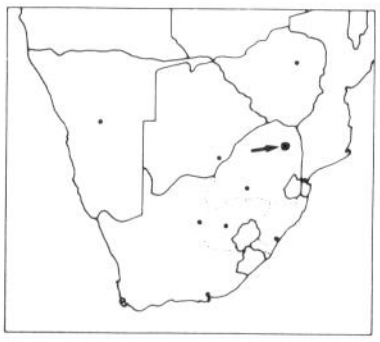

Eastwood's Long-tailed Seps *Tetradactylus eastwoodae*

(Endemic) 19-20 cm
This snake-like seps is similar in appearance to Breyer's long-tailed seps, *T. breyeri* (above). It has three toes on each forelimb and two on each hind limb. The two supraoculars touch the frontal. Each nostril is pierced between two nasals. The frontoparietals are band-like and not in contact. The dorsal scales are strongly keeled, in 12 longitudinal and 67-70 transverse rows. Ventrals are in 6-8 longitudinal and 50 transverse rows. There are three femoral pores on each side. The tail is more than twice the length of the body. The back has olive-brown, indistinct, dark longitudinal stripes. The head is dark-spotted, and the belly is grey-brown. (SA RDB, Vulnerable.)
Habitat: Unknown. **Range:** Woodbush Forest in N. Transvaal. It has not been rediscovered since its original description in 1913; it is possibly extinct, its habitat having been destroyed by pine plantations.

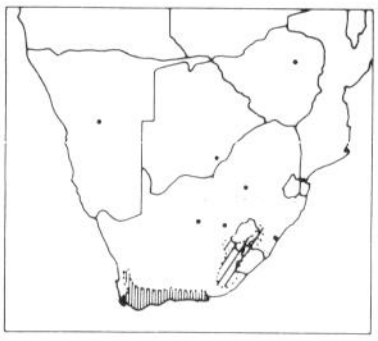

T.s. seps
T.s. laevicauda

Short-legged Seps *Tetradactylus seps* (Pls. 51 and 65)

(Endemic) 13-15 cm; max. 20 cm
A small, long-tailed plated lizard, with reduced but fully formed limbs. The lower eyelids are scaly. The dorsal scales are in 13 longitudinal rows. The tail is twice the body length. The body is uniform olive to reddish-brown, with slightly fainter flanks. The head has dark brown spots on top and dark-edged spots on the upper lip. The lower neck is white, with short bars. The belly is olive to bluish-grey.

Biology and breeding: This seps is rarely seen, but is locally common. It basks among thick vegetation in moist clearings. When alarmed, it 'swims' through the grass, using its long tail. It feeds on bees, grasshoppers, etc. The female lays 2-3 large, oval, creamy white eggs (14 x 8 mm) among rotting logs or leaf mould. **Habitat:** Coastal forests or mountain plateaus. **Range:** S. and W. Cape from Cedarberg along Cape fold mountains, through Amatola Mountains to Natal midlands. **Subspecies:** Two races are recognized, and both occur in the region. *T.s. seps* has keeled head shields and tail scales; it occurs in Cape fold mountains. *T.s. laevicauda* has smooth head shields and tail scales; it is found from Amatola Mountains to Natal midlands.

T.t. tetradactylus
T.t. bilineatus

Common Long-tailed Seps *Tetradactylus tetradactylus* (Pl. 51)

(Endemic) 18-24 cm; max. 29 cm
This thin, snake-like seps has minute, four-toed forelimbs and hind limbs. The frontoparietals are in contact. The dorsal scales are keeled, in 14 longitudinal and 59-62 transverse rows. Ventrals are in six longitudinal and 45-46 transverse rows. There are 4-5 femoral pores on each side. The tail is three times the body length. The back is olive, with a pair of dark brown dorsolateral stripes. The sides of the neck have short, black and white bars. The belly is pale olive. **Biology:** This seps is very quick and difficult to catch. It shelters at night in tufts of grass. **Habitat:** Montane grassland and fynbos. **Range:** S. and E. Cape, with relict populations in montane grassland of old escarpment (eg. Beaufort West and Cradock). **Subspecies:** Two races are recognized, both occurring in the region. *T.t. tetradactylus* has each nostril between two nasals and the labial, and short hind limbs that reach to the 6-7th tail scale row; it is found in Cape fold mountains and inland escarpment. *T.t. bilineatus*, which is poorly known, has each nostril between three nasals, long hind limbs that reach to the 9th tail scale row, and more conspicuous, dark brown dorsolateral stripes; it is found in E. Cape.

RDLED LIZARDS ID THEIR ELATIVES
bfamily Cordylinae

These lizards have a short tongue that is covered with long papillae. The body scales are usually arranged in regular girdles (although they are granular in some species) and lack osteoderms (except in *Cordylus*). The scales on the tail are arranged in regular rings and are spiny or strongly keeled. All except *Platysaurus* are viviparous.

There are four genera in southern Africa; one is endemic, and the others are restricted mainly to the subcontinent, but contain a few species that extend into central and E. Africa.

y to the southern ican genera in the rdylinae

1. Legs very reduced; body snake-like .... *Chamaesaura* (Grass lizards, page 155)
   Legs well developed; head with an obvious neck 2
2. Back scales granular and mostly of the same size; tail without spines; body strongly flattened .... *Platysaurus* (Flat lizards, page 164)
   Body scales large; tail with spines; body not strongly flattened 3
3. Dorsal scales large and with osteoderms; nasal shield pierced by nostril .... *Cordylus* (Girdled lizards, page 156)
   Dorsal scales small and without osteoderms; nostril pierced between nasal and upper labial .... *Pseudocordylus* (Crag lizards, page 169)

RASS LIZARDS
amaesaura

These very unusual lizards have extremely reduced limbs (little more than spikes) and a very long tail (3-4 times longer than the body). The body scales are very rough, strongly keeled, and arranged in regular rows.

The elongate shape of grass lizards allows them to move freely in long grass, through which they 'swim' with the speed and agility of snakes (although the generic name translates more correctly as 'creeping lizards'). They are not as mobile on smooth or sandy surfaces. Although the minute limbs appear useless, they give stability when the lizard is at rest. The tail is proportionately shorter in hatchlings; only the minimum needed for survival is shed, as without their tails they move very slowly and noisily, and thus succumb easily to predation or starvation. Regeneration of the lost segment rapid. Their rustic colours camouflage them in dried grass. Small invertebrates, particularly grasshoppers, are actively chased and eaten. Viviparous, they give birth to 5-9 babies (up to 150 mm) in late summer. Parturition may take 2-3 days, and the young often escape by themselves fro the mother's body.

There are three species in the genus, all occurring in southern Africa. Two have northern subspecies in the savannas of central and E. Africa, and the other is endemic.

Transvaal Grass Lizard *Chamaesaura aenea* (Pl. 50)

(Endemic) 30-35 cm; max. 40 cm
A slender grass lizard with five clawed digits on each of its four feet, and sma body scales (42-46 transverse body rows). Its coloration is drab, the head an back being dark brown, with three light yellow or grey-olive, black-edged stripes, the flanks straw-coloured, with 2-3 series of dark spots or a reddish-brown (the specific name means 'coppery') lateral stripe, and the belly off-white. **Habitat:** Grass-covered mountain slopes and plateaus. **Range:** Eastern escarpment grasslands of Natal and Transvaal, with an isolated, relict population on Amatola Mountains in E. Cape.

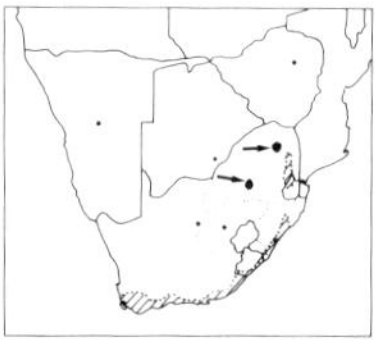

C.a. anguina

Cape Grass Lizard *Chamaesaura anguina* (Pl. 50)

35-40 cm; max. 49 cm
An elongate grass lizard with 1-2 clawed digits on each of its minute forelimb and hind limbs. The body scales are in 26-30 longitudinal and 36-40 transverse rows. In colour, it is similar to the Transvaal grass lizard, *C. aenea* (above), but the vertebral stripe is broader and the dorsolateral stripes (if present) are black. The flanks are straw-coloured, sometimes with a narrow, white lateral band. The belly is whitish or golden-yellow. **Breeding:** Gravid females have very swollen bellies. They give birth to 6-9 babies in April. **Habitat:** Grassy or fynbos-covered, gentle slopes. **Range:** Cape Town, through Cape fold mountains, E. Cape and Natal lowlands, and along the escarpment to Transvaal Drakensberg. **Subspecies:** Only the typical race, *C.a. anguina*, occurs on the subcontinent; isolated, relict populations occur Angola (*C.a. oligopholis*), and in upland grasslands (900-2 500 m) of E. Zair and E. Africa (*C.a. tenuior*).

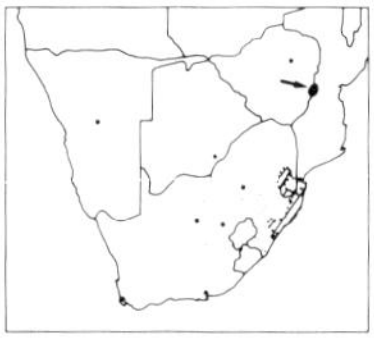

C.m. macrolepis

Large-scaled Grass Lizard *Chamaesaura macrolepis* (Pl. 50)

30-35 cm; max. 41 cm
Forelimbs are absent in this grass lizard, and the hind limbs are vestigial spikes, each having only one claw. The scales are enlarged, in 22 longitudin and 38-40 transverse rows. It is light brown, with two dark brown stripes (ofte broken into a series of elongate spots). The flanks are straw-coloured, and th belly is off-white. **Breeding:** The female gives birth to 6-8 babies in March. **Habitat:** Grassveld and mountain plateaus. **Range:** Natal and Zululand, through Swaziland to E. Transvaal, with an isolated population on Chimanimani Mountains in Zimbabwe. **Subspecies:** Only the typical race, *C.m. macrolepis*, occurs on the subcontinent; a relict race (*C.m. miopropus*) found in N. and E. Zambia.

GIRDLED LIZARDS *Cordylus*

These are the most characteristic lizards of southern Africa. The body is stocky, and the limbs are always well developed. The head is triangular and flattened on top, and covered with large shields that are fused to the skull. T eyes and eyelids are well developed. The eardrums are visible, but are part shielded by scales. The body scales have osteoderms and are overlapping usually keeled, and sometimes spiny. They are arranged in regular rows (girdles), sometimes separated by granular interspaces. The tail, which has whorls of spiny scales, can be shed and regenerated, albeit slowly and poorly.

They are diurnal and mainly rock-living (some species live on the ground or in trees). The thick scales with their bony plates protect them from abrasion against rough rock. To evade predators, many species jam themselves into rock cracks by inflating the body and shortening and thickening the skull, which has an unusual hinged structure. They eat a wide variety of large invertebrates, while some of the bigger species also eat small vertebrates and plant matter. All are viviparous, giving birth to a few (1-6) large babies each year. Some live in diffuse colonies, but they only rarely are territorial or have obvious social hierarchies. This is reflected in the drab colour of the adult males, who do, nonetheless, have active femoral pores. Sexual maturity is reached in 2-4 years, and they are long-lived (up to 25 years is known in captivity).

Their greatest diversity and probable origin is south of the Zambezi River, although a number of species extend into central and E. Africa (*Cordylus rivae* reaches Ethiopia). The taxonomy of some species is confused, particularly regarding the races of *C. warreni* and *C. cordylus*. Twenty species are recognized at present, with 19 in the region (many with local races), 17 of which are endemic.

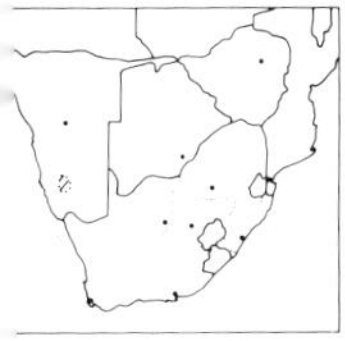

Campbell's Girdled Lizard *Cordylus campbelli*

(Endemic) 15-17 cm; max. 18 cm
A small girdled lizard similar in appearance to the Herero girdled lizard, *C. pustulatus* (page 162). It has a flattened body. The nasals, which touch each other, are not or are only feebly tubular, and separate the rostral and the frontonasal. Each lower eyelid has a semi-transparent disc. There are 6-8 occipitals, the middle ones often being small. The dorsal scales are keeled on the back and moderately spiny on the flanks, in 27-31 rows. The ventrals are mostly smooth and in 16-18 longitudinal rows. There is a pair of enlarged preanal plates. Five large femoral pores are present on each thigh. The back and tail are chestnut to light brown, with irregular dark brown crossbars, and sometimes with a dark central band. There are numerous small, yellow specks along the back and on the head. The belly is off-white, with rusty infusions on the chin and throat. **Biology:** This lizard lives in rock cracks on arid mountain slopes, feeding on termites and beetles. **Habitat:** Rocky, arid savannah. **Range:** Vicinity of Helmeringshausen, Great Namaqualand in Namibia.

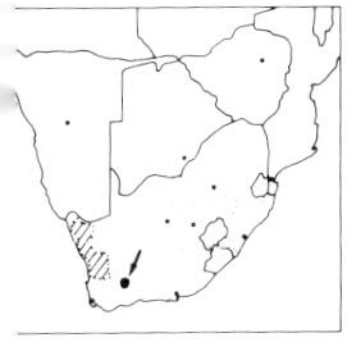

Armadillo Girdled Lizard *Cordylus cataphractus* (Pl. 68)

(Endemic) 12-16 cm; max. 21 cm
This thick-set, flattened girdled lizard has a broad head. The scales on the sides of the neck have sharp spines. The nasals are in contact, separating the rostral and the frontonasal. It has six large, keeled occipitals. There are 15-17 broad, sometimes asymmetrical, bands of spiny scales around its body. The tail is ringed with large spines. Males have prominent femoral pores (13-16 on each side). The back is plain, dirty yellowish-brown. The upper lips are dark brown. The throat is yellow (especially in males) or violet, with dark brown blotches. The belly is yellow, with darker infusions, particularly on the inner surfaces of the limbs. **Biology and breeding:** This heavily armoured, lovable lizard is too often illegally collected (SA RDB, Vulnerable). It lives in large cracks in low rock outcrops, and feeds on the insects attracted by the floral splendour of Namaqualand. In an unusual behaviour for a lizard, it forms family groups that inhabit the same rock cracks. It is very wary and retreats at the first sign of danger. It is very difficult to extract from cracks because of its spiny scales, but if caught in the open, it will bite its tail and roll into a tight ball (hence its common name), making it too spiny for predators to eat. One or two large babies are born in late summer. **Habitat:** Dry, succulent, karroid veld. **Range:** W. Karoo, from Little Namaqualand to Matjiesfontein.

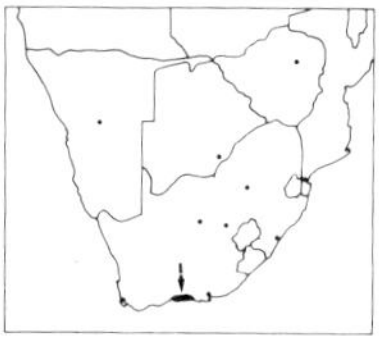

Blue-spotted Girdled Lizard
Cordylus coeruleopunctatus (Pl. 71)

(Endemic) 13-16 cm; max. 20 cm
A graceful girdled lizard. Its nasals are in contact, separating the rostral and the frontonasal. It lacks occipitals. The scales behind the head are small and granular. The dorsal scales are small, with no interspaced granules, in 40-43 transverse and 20-22 longitudinal rows. There is a pair of enlarged preanal plates, and 12-16 femoral pores on each thigh. The tail has regular whorls of keeled scales. The back is greyish-olive, with numerous black streaks that fuse towards the rear. There is a reddish or orange-yellow band on the flanks, and often on the snout and the side of the head. The throat and chest are greenish-yellow to orange (especially in adult males). It characteristically has scattered enamel-blue spots, particularly on the sides of the head, hence its common and scientific names. **Biology and breeding:** It is common in suitab moist habitat, on coastal cliffs and small rock outcrops, where it forages for invertebrates. It may excavate a small tunnel in loose soil in a rock crack or beneath a boulder. It gives birth to 3-4 young in midsummer. **Habitat:** Fynbo and forest fringes. **Range:** S. Cape, from Mossel Bay to Witelsbos.

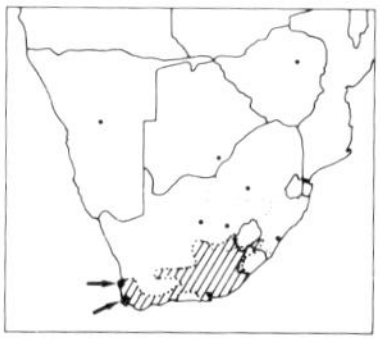

C.c. cordylus
C.c. niger

Cape Girdled Lizard *Cordylus cordylus* (Pls. 66 and 67)

(Endemic) 13-19 cm; max. 21 cm
This small girdled lizard has a flattened body and a triangular, flattened head with slightly rough head shields. The nasals are usually in contact, separatin the rostral and the frontonasal, which is in contact with the loreals. The anteri parietals are smaller than the posterior ones. There are six small, non-spiny occipitals. The dorsal scales, which are large and faintly keeled (with the kee arranged in parallel rows) are in 22-30 transverse and 16-22 longitudinal row The ventrals are smooth (but faintly keeled along the flanks) and in 10-14 longitudinal rows. There is a pair of feebly enlarged preanal plates. There are 4-11 femoral pores on each thigh. The tail has whorls of large spines. Coloration is varied. It is usually dull brown to reddish-brown, with irregular darker markings, and sometimes with a pale cream, irregular vertebral stripe The belly is dirty yellow to dull red-brown. Melanistic populations occur along the coast and in the mountains of W. Cape.

Biology and breeding: These girdled lizards often live in dense colonies (up to 300 specimens per hectare) where there are suitable rock cracks in which to shelter. Adults are aggressive, and form social hierarchies with dominant males. In territorial disputes, males circle one another, bobbing th heads and arching their backs. The weaker male usually signals submission by moving its tail. Fights may ensue if dominance is not resolved by these rituals. They are active in the early morning and evening (and all day on overcast days), foraging for insects in the veld and dashing back to cover when danger threatens. Predators include snakes, small carnivores and sma birds of prey (including owls). They mate in spring and give birth to 1-3 youn in January-February. Hatchlings measure 69-80 mm. **Habitat:** Diverse; coastal cliffs, rock plateaus in fynbos and montane grassland, and shale bands in valley and Fish River bushveld. **Range:** Coastal regions of the Cap from Saldanha Bay to East London, but absent from George to Witelsbos, where it is replaced by the blue-spotted girdled lizard, *C. coeruleopunctatus* (above). Inland, through Cape fold mountains and montane grassland of NE. Cape and SE. OFS, and through inland Transkei to Natal midlands.

Subspecies: Two races are recognized at present. *C.c. cordylus* is dark brown to olive-buff in colour, the subocular does not reach the lip, the prefrontals are usually in contact, and there are 5-9 femoral pores; it occurs over most of the range. *C.c. niger* is uniform jet-black in colour, the subocula reaches the lip, the prefrontals are not in contact, and there are 7-10 femora pores; it is found on Cape Peninsula and at Saldanha Bay.

Giant Girdled Lizard or Sungazer
Cordylus giganteus (Pl. 65)

(Endemic) 20-35 cm; max. 40 cm
A very large girdled lizard. The nasals are separated by the rostral and the frontonasal. There are four very large, spiny occipitals. The dorsal scales are spiny, in 22-25 transverse and 10-12 longitudinal rows. There is a pair of enlarged preanal plates, and 10-12 femoral pores on each thigh. The tail has whorls of very large spines. The back is yellow to dark brown, and often extensively clouded with dark brown. The sides of the head, flanks and belly are pale yellow, sometimes with grey-brown infusions on the chest. Juveniles are more intensely marked, with irregular crossbars of red-brown on the back, and black and yellow bands with many orange to red scales on the tail.

Biology and breeding: These terrestrial girdled lizards live in colonies in burrows that they dig in silty, fine soil. The burrows, which measure approximately 42 cm deep and 180 cm long, end abruptly, without an enlarged chamber, but with the end nearer the surface so that they do not become completely flooded. Each burrow is usually occupied by a single individual, although females will often share their burrow with juveniles; three species of small frog also hibernate in winter in these burrows with the lizards. If it is threatened by a predator entering the burrow, the sungazer will retreat backwards, lashing its spiny tail from side to side. If grasped, it will jam its occipital spines into the tunnel roof. Sungazers are long-lived (longer than 20 years in captivity). They are often seen during the day, basking at the entrance to their burrows or on a termite mound, staring at the sun – hence the common name. They feed mainly on invertebrates (beetles, grasshoppers, millipedes, termites and spiders), although they will take small vertebrates if the opportunity arises. Their numbers are declining (SA RDB, Vulnerable) due to habitat destruction (maize and sunflower farming) and, to a limited extent, because of collecting for the pet trade. One or two babies, measuring 130-150 mm, are born in February-March, possibly only every alternate year. **Habitat:** Flat or gently sloping *Themeda* grassland, or transitional zones. **Range:** Small, scattered populations in NE. OFS, extreme W. Natal and SE. Transvaal.

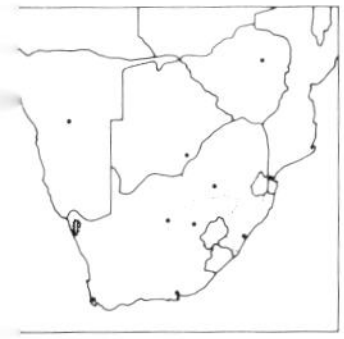

Lawrence's Girdled Lizard *Cordylus lawrenci* (Pl. 68)

(Endemic) 13-15 cm; max. 16 cm
A small, thin girdled lizard with a flattened body and a triangular, flat head with rough head shields. The nasals are slightly swollen and in contact, separating the rostral and the frontonasal, and the subocular is separated from the lip. The lower eyelids are opaque. There are six occipitals, the middle pair of which is enlarged. The dorsal scales are keeled on the back and moderately spiny on the flanks, in 24 transverse and 22 longitudinal rows. The ventrals are mostly smooth, and in 12 longitudinal rows. There is a pair of enlarged preanal plates, and eight large femoral pores on each thigh. The tail has whorls of large, spiny scales. The back, head and tail are dark brown, with irregular black infusions and a few yellow specks on the head and forebody. The belly is off-white to grey-brown, with a black-flecked throat. The tail often has a dark band. **Biology:** This species is similar in habits and behaviour to Campbell's girdled lizard, *C. campbelli* (p. 157). (SA RDB, Restricted.) **Habitat:** Succulent karroid veld. **Range:** Richtersveld in Little Namaqualand.

Large-scaled Girdled Lizard *Cordylus macropholis* (Pl. 66)

(Endemic) 10-13 cm; max. 15 cm
A small, round-bodied girdled lizard. The nasals are in contact, separating the rostral and frontonasal. There are six strongly keeled occipitals. The nostril pierces the lower part of the nasal. The dorsal scales are very large, strongly

keeled and in 16-18 transverse and 14-18 longitudinal rows. The ventrals ar keeled and in 10 rows. There is a pair of enlarged preanal plates, and 10-12 femoral pores on each thigh. The tail has whorls of very large spines. The ba and flanks are grey to olive-grey, with irregular dark markings. The belly is p grey. **Biology and breeding:** Terrestrial, this girdled lizard is found among dead wood or *Euphorbia* bushes in sand dunes, under debris at the strand line, or sometimes in soft limestone cracks. It is very shy, scuttling off when danger. It is well camouflaged among twigs and dead leaves. One or two young are born in April-May. These measure 60-70 mm. **Habitat:** Coastal dunes and strand line. **Range:** W. Cape coast, from Yzerfontein to Kleinzee

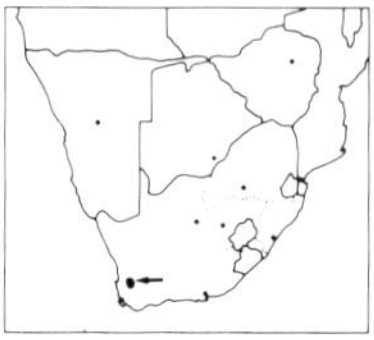

McLachlan's Girdled Lizard *Cordylus mclachlani* (Pl. 67)

(Endemic) 10-13 cm; max. 15 cm
This small girdled lizard has a very flattened body and a triangular head wit rough head shields. There are only two supraciliaries. The supranasals are contact, separating the rostral and frontonasal. The nasals are small and slightly tubular. The anterior parietals are larger than the posterior ones. The are eight spiny occipitals, and no enlarged chin shields. The dorsal scales a small, obliquely keeled, and in 28 transverse and 22 longitudinal rows. The laterals are spiny and larger than the dorsals. The ventrals are smooth, in 12 longitudinal rows. There is a pair of feebly enlarged preanal plates, and 8-1 femoral pores are present on each thigh. The tail has whorls of very large spines. The back, tail and limbs are olive-brown, with numerous, irregular black markings and a few yellow speckles. The flanks are reddish-brown. T head is dark brown to blackish, with a faint, dark line from the snout through the eye. The belly is creamy white. **Biology:** This species has very recently been described (1986). (SA RDB, Restricted.) Specimens are found individually in narrow cracks on low sandstone outcrops. The tail is easily shed, and many specimens have incomplete or regenerated tails. **Habitat:** Succulent karroid veld. **Range:** Koue Bokkeveld in W. Cape.

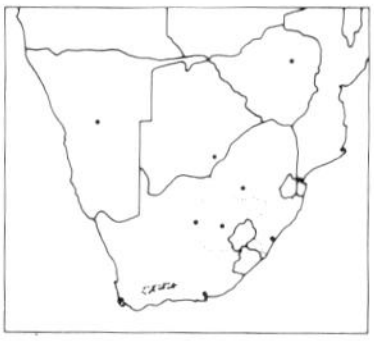

Dwarf Girdled Lizard *Cordylus minor* (Pl. 67)

(Endemic) 10-13 cm; max. 16 cm
This small girdled lizard is similar to McLachlan's girdled lizard, *C. mclachla* (above). The body is flattened and the head is triangular, with rough head shields. There are three supraciliaries. The supranasals are in contact, separating the rostral and frontonasal. The nasals are small and not tubular. The anterior parietals are larger than the posterior ones. There are six occipitals, and no enlarged chin shields. The dorsal scales are small, obliquely keeled, and in 26-28 transverse and 24-26 longitudinal rows. The laterals are spiny and larger than the dorsals. The ventrals are smooth, 12-14 longitudinal rows. There is a pair of feebly enlarged preanal plates, a both sexes have 4-6 femoral pores on each thigh. The tail has whorls of larg spines. The back, tail and limbs are dirty brown in colour, with diffuse, irreg black markings. The head is dark brown to blackish. The belly and upper labials are dirty white. **Biology:** This shy species lives in scattered colonies It chooses small, vertical rock cracks in low, north-facing bands of Dwyka tillite. **Habitat:** Karroid veld. **Range:** Known from only three isolated populations, at Matjiesfontein, Prince Albert, and near Meiringspoort.

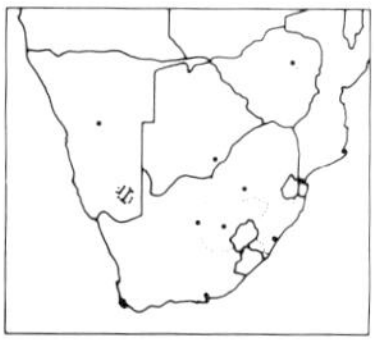

Namaqua Girdled Lizard *Cordylus namaquensis* (Pl. 68)

(Endemic) 15-17 cm; max. 18 cm
A small girdled lizard with a flattened body and a triangular, flat head with rough shields (but with smooth second and third supraoculars). The nasals are slightly tubular and in contact, separating the rostral and frontonasal. Th lower eyelids are opaque. There are eight occipitals, the middle ones often

being small. The dorsal scales are smooth on the back and moderately spiny on the flanks, in 27-31 rows. The ventrals are mostly smooth, in 16-18 longitudinal rows. There is a pair of enlarged preanal plates, and both sexes have 9-10 femoral pores on each thigh. The tail has whorls of large, spiny scales. The back is chestnut to light brown in colour, usually with dark brown to black mottling. The side of the head has two blackish streaks. The lower labials are black-edged, and the belly is dirty white to pale brown.
Biology and breeding: This girdled lizard lives in rock cracks and crevices in mountains, feeding on termites and beetles. Two or three young are born in January-February, after a four- to five-month gestation period.
Habitat: Semi-desert. **Range:** Great Karasberg district in S. Namibia.

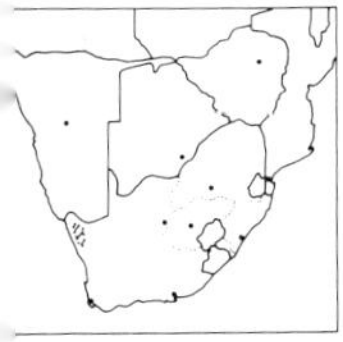

Peers's Girdled Lizard *Cordylus peersi* (Pl. 68)

(Endemic) 15-17 cm; max. 19 cm
This small, thin girdled lizard has a flattened body and a triangular, flat head with rough shields. The nasals are swollen and in contact, separating the rostral and frontonasal. The subocular borders the lip. The lower eyelids are opaque. There are six occipitals, the middle pair of which is often small. The dorsal scales are keeled on the back and moderately spiny on the flanks, in 22-24 transverse and 16 longitudinal rows. The ventrals are mostly smooth, and in 12 longitudinal rows. There is a pair of enlarged preanal plates, and both sexes have 9-12 large femoral pores on each thigh. The tail has whorls of large, spiny scales. The back and tail are jet-black. The belly is very dark purple-brown, with pale yellow femoral pores and glandular scales.
Biology: This girdled lizard is very visible when basking on light-coloured rocks, but quickly retreats into a deep crack when approached. It prefers the upper surfaces of large, shattered boulders on small, rocky outcrops. It captures large insects and caterpillars. **Habitat:** Succulent karroid veld.
Range: Little Namaqualand.

Karoo Girdled Lizard *Cordylus polyzonus* (Pl. 69)

(Endemic) 20-25 cm; max. 26 cm
A large, graceful girdled lizard with a flattened body. The supranasals are in contact, separating the rostral and frontonasal. The nasals are small and slightly tubular. The lower eyelids have transparent discs. There are two occipitals. The dorsal scales are small, smooth on the back and keeled on the flanks, in 38-46 transverse rows. There is a pair of enlarged preanal plates, and 10-19 femoral pores on each thigh. There are two rows of large, spiny scales in each tail whorl. Coloration is varied. In juveniles, the back is yellow-brown, chequered with dark brown and pale cream, and the tail is banded in dark brown. In adults, the body may retain the juvenile coloration (in specimens from S. Karoo), or become dark brown or black (those from central Cape and Little Namaqualand coast), uniform olive (those from S. OFS), olive-brown with vivid orange-red flanks (those from NW. Cape), or even blue-green (those from SW. Cape). All populations have a characteristic black blotch on the side of the neck, and some individuals retain irregular black spots on the back.

Biology and breeding: This is one of the most common lizards in the central karroid regions of the Cape, and is often found in diffuse colonies, living in sun-split rocks of small rock outcrops and lower mountain slopes. During the heat of the day it perches on a boulder, basking in the sun and making short forays to grab beetles or grasshoppers in the veld. During winter it may hibernate in a deep tunnel which it digs in soil beneath a large boulder. Very alert, at the first sign of danger it retreats into a rock crack and curls its tail over its head. Two (exceptionally 3-4) large babies (100-110 mm) are born in late summer. **Habitat:** Karroid regions. **Range:** Central and W. Cape, extending into S. OFS and S. Namibia.

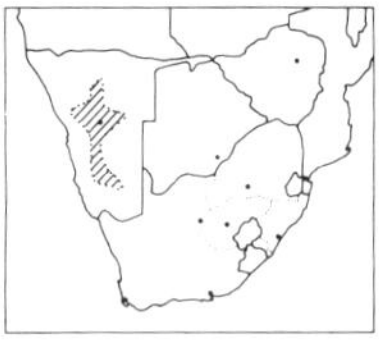

Jordan's Girdled Lizard *Cordylus jordani* (Pl. 69)

(Endemic) 20-24 cm; max. 28 cm
This species is similar to the Karoo girdled lizard, *C. polyzonus* (page 161), but grows larger and is more robust in build. It has 5-8 femoral pores. There a single row of scales in each tail whorl. The body is uniform olive-brown in adults, and there is no dark blotch on the side of the neck. Juveniles are buff-coloured, with a typical, dark, chequered pattern. **Biology and breeding:** This species is similar in habits and behaviour to the Karoo girdle lizard. **Habitat:** Rocky hillsides. **Range:** Central Namibia, north of Aus.

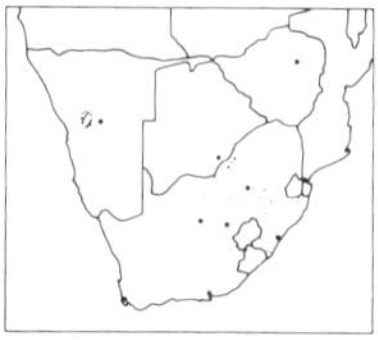

Herero Girdled Lizard *Cordylus pustulatus* (Pl. 67)

(Endemic) 15-18 cm; max. 20 cm
This smallish, finely scaled girdled lizard has a very flattened body and a triangular head with rough head shields. The nasals are in contact, separati the rostral and frontonasal, which is broader than it is long and touches the loreals. There are six non-spiny occipitals. The dorsal scales are small and smooth down the back, and in 30-32 transverse and 27-29 longitudinal rows The laterals are keeled. The ventrals are smooth (except the outermost rows in 14 longitudinal rows. There is a pair of feebly enlarged preanal plates. Th tail has whorls of large spines. The back is either uniform olive-brown to dar brown, or, more usually, has numerous small, scattered, yellow spots that fo a vague vertebral band. The belly is yellowish-brown. **Habitat:** Arid savann **Range:** Hereroland in Namibia.

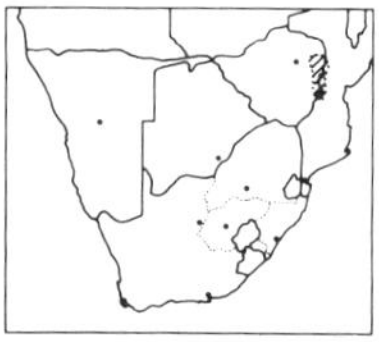

Zimbabwe Girdled Lizard *Cordylus rhodesianus* (Pl. 67)

(Endemic) 13-16 cm; max. 18 cm
A small girdled lizard with a very flattened body and a triangular head that h smooth or slightly rough head shields. The nasals are usually in contact, separating the rostral and the frontonasal, which is four-sided and separate from the loreals. There are six non-spiny occipitals. The dorsal scales are small and slightly keeled down the back, in 25-29 transverse and 20-26 longitudinal rows. The laterals are keeled. The ventrals are smooth (except t outermost preanal plates) and both sexes have 5-8 femoral pores on each thigh. The tail has whorls of large spines. The back is olive-brown, with irregular darker markings and sometimes with paler blotches on the upper flanks. The belly is yellowish or greyish-white. **Biology:** This girdled lizard is long-lived and slow-growing. It lives under stones and in rock cracks on roc outcrops, and feeds on beetles and grasshoppers. Predators include the be adder. **Habitat:** Montane grassland. **Range:** Eastern Highlands of Zimbabw and adjacent Mozambique.

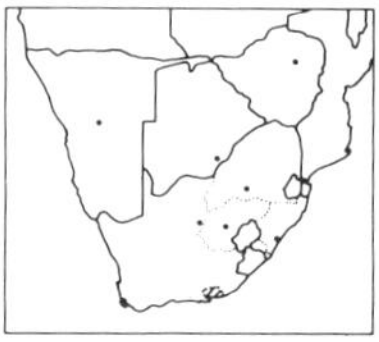

Tasman's Girdled Lizard *Cordylus tasmani* (Pl. 66)

(Endemic) 13-15 cm; max. 17 cm
A small girdled lizard with a rounded body and a triangular head that is not obviously flattened and has slightly rough head shields. The nasals are usu in contact, separating the rostral and the frontonasal, which is in contact wit the loreals. There are six small, non-spiny occipitals. The dorsal scales are large, slightly keeled and in 22-30 transverse and 16-20 longitudinal rows. T ventrals are smooth, and in 10-12 longitudinal rows. There is a pair of feebly enlarged preanal plates, and both sexes have 4-6 femoral pores on each thigh. The tail has whorls of large spines. The back is reddish-brown to mahogany, with irregular darker markings, sometimes with a pale cream, irregular vertebral stripe. The belly is dirty yellow to dark red-brown. **Biology and breeding:** It lives under the 'apron' of dead leaves on tall aloe or on dead aloe stems lying on stony slopes; it also occurs under the bark o

trees or in piles of rotting spekboom trunks, and will occupy cracks in limestone or sandstone outcrops. It feeds on small invertebrates. It is usually shy and difficult to approach. Two or three young are born in late summer. **Habitat:** Succulent valley bushveld. **Range:** Algoa Basin in E. Cape.

Tropical Girdled Lizard *Cordylus tropidosternum* (Pl. 66)

13-15 cm; max. 17 cm
This small, round-bodied girdled lizard does not have a flattened head, and its head scales are rough and keeled. The nasals are in contact or are separated by the rostral and frontonasal. There are six occipitals. The dorsal scales are very large, strongly keeled, and in 22-27 transverse rows. The lateral scales are separated by granular interspaces. The ventrals are smooth (with the outer rows keeled in the typical race), and in 12-14 rows. There is a pair of enlarged preanal plates, and both sexes have 6-9 femoral pores on each thigh. The tail has whorls of large spines. The back and flanks are a dirty straw colour, grey-brown or dark brown, and sometimes uniform but more often irregularly blotched in dark brown, with off-white flecks along the backbone. There is a dark brown to black lateral band from the neck to the groin. The belly is off-white to straw-yellow. The femoral pores are yellow. **Biology and breeding:** This lizard lives under loose bark, in hollow logs of trees and in tree stumps. It is usually very shy and secretive. It feeds on moths and spiders, and is very fond of winged termites. It lays down fat reserves to tide it over the dry winter season. Usually two (but sometimes up to four) young (60-70 mm) are born in midsummer. **Habitat:** Dry lowveld, particularly mopane savannah. **Range:** E. African lowlands, south to Transvaal lowveld and adjacent regions. **Subspecies:** There are two races, and both occur in the region. *C.t. tropidosternum* is larger (up to 9 cm snout-vent length), and its 3-4 outer rows of ventrals are keeled; it is found in E. Africa, reaching its southernmost limit in E. Zimbabwe. *C.t. jonesi* is a small race (up to 7,5 cm snout-vent length), and all its ventrals are smooth; this southern race occurs in N. Transvaal, E. Botswana and S. Mozambique.

Transvaal Girdled Lizard *Cordylus vittifer* (Pl. 67)

14-17 cm; max. 18 cm
A small girdled lizard with a very flattened body and a triangular head with rough head shields. The first row of dorsals is elongated. The nasals are in contact, separating the rostral and frontonasal. There are six occipitals. The dorsal scales are small and strongly keeled, in 16-18 transverse and 14-18 longitudinal rows. The laterals are spiny. The ventrals are smooth, in 16-18 rows (but 14-16 in some populations). There is a pair of enlarged preanal plates. Usually, both sexes have 6-8 femoral pores on each thigh, but these are fewer, and are sometimes absent, in females of some populations. The tail has whorls of very large spines. The back is either uniform straw-coloured, yellow-brown to dark brown, or has irregular, darker spots that are sometimes arranged to form dorsolateral bands. A pale cream-white vertebral stripe is often present. The flanks are orange-brown. The head has a few scattered, yellow spots. The belly is dirty white to light brown. **Biology and breeding:** This girdled lizard lives in cracks in small rock outcrops, feeding on beetles, grasshoppers, etc. One to three young are born in late summer. **Habitat:** Grassland. **Range:** There are two disjunct populations: in Transvaal (and adjacent SE. Botswana, Swaziland and Zululand), and S. Angola (and adjacent Kaokoveld). **Subspecies:** Two races are recognized, and both occur in the region. *C.v. vittifer* has 20-24 longitudinal rows of dorsal scales and a large rostral; it is found in N. OFS and Transvaal, through Swaziland to Zululand. *C.v. machadoi* has 26-28 longitudinal rows of dorsal scales and a small rostral; it occurs in S. Angola and Kaokoveld. The taxonomy of specimens from N. OFS is confused.

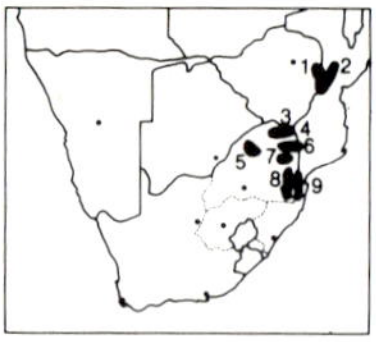

1. *C.w. regius*
2. *C.w. mossambicanus*
3. *C.w. depressus*
4. *C.w. laevigatus*
5. *C.w. breyeri*
6. *C.w. vandami*
7. *C.w. perkoensis*
8. *C.w. barbertonensis*
9. *C.w. warreni*

Warren's Girdled Lizard *Cordylus warreni* (Pls. 70 and 71)

(Endemic) 20-30 cm; max. 33 cm
This large girdled lizard has a flattened body. The nasals are separated by th rostral and frontonasal. There are 4-8 occipitals (see Subspecies). The dors scales are spiny, in 22-46 transverse rows (see Subspecies). There is a pair enlarged preanal plates, and both sexes have 7-14 femoral pores on each thigh. The tail has whorls of large, spiny scales. The back is dark brown to black, with varying degrees of yellow spotting and/or barring (see Subspecies). **Biology and breeding:** This lizard is found on rocky mountain slopes, favouring deep cracks in large boulders that are sheltered by trees. It is very shy and difficult to approach. The diet includes large invertebrates (beetles, grasshoppers, etc.), small land snails, small lizards and even frogs Two to six young (90-130 mm) are born in late summer. **Habitat:** Montane, well-wooded rocky outcrops. **Range:** Eastern escarpment from Ubombo Mountains in Zululand, through Transvaal Drakensberg, Soutpansberg, and E. Zimbabwe to Gorongoza Mountains in Mozambique.

Subspecies: The taxonomy is confused and is under investigation. Nine races are recognized at present; some of these are doubtful, while others ma be full species (eg. *C.w. laevigatus*). They all occur in the region. *C.w. regius* has 6-8 occipitals, 34-40 dorsal scale rows, 14 ventral scale rows, a dark brown back and bright orange or yellow flanks; it occurs in E. Zimbabwe. *C.w. mossambicanus* has 6-8 keeled occipitals, the loreal not entering the nostril, 35-46 dorsal scale rows, and a dark brown back with a few scattered whitish spots; it is found on Gorongoza Mountains in Mozambique to lower slopes of Chimanimani Mountains in Zimbabwe. *C.w. depressus* has six larg spiny occipitals, 24-28 dorsal scale rows, 14 ventral scale rows, and a dark brown back with irregular, scattered, large yellow spots or bars; it occurs on Soutpansberg and environs. *C.w. laevigatus* has six small occipitals, the scales on the back in about 20 rows and separated by granular interspaces and a dark brown back with a few scattered yellowish spots; it is found on E. Soutpansberg. *C.w. breyeri* has six spiny occipitals (the outer two being th largest), 24-26 dorsal scale rows, 10-12 ventral scale rows, and a uniform da brown head, back and tail (with some yellow spots in juveniles); it occurs on Waterberg in NW. Transvaal. *C.w. vandami* has four large, spinose occipitals 26-28 dorsal scale rows, 14-16 ventral scale rows, 10-12 femoral pores, a da brown back with up to six fragmented yellow crossbars, and a dark head; it i found on Drakensberg of NE. Transvaal, mainly north of Olifants River, and t Mozambique border. *C.w. perkoensis* is similiar to *C.w. vandami*, except tha has 12-14 femoral pores, and a dark brown back with a few scattered yellow scales; it occurs on E. Transvaal Drakensberg, south of Olifants River. *C.w. barbertonensis* has six spiny occipitals (the middle two being the smallest), 32-38 dorsal scale rows, 14 ventral scale rows, and a very dark brown back with bright yellow bars or large spots; it is found in Barberton in E. Transvaal, and S. Swaziland. *C.w. warreni* has six pointed occipitals, the loreal entering the nostril, 34-42 dorsal scale rows, and a dark brown back w small, yellow, black-edged spots forming vague bands; it occurs on Ubomb Mountains in Zululand.

FLAT LIZARDS
Platysaurus

These unmistakable and bizarre creatures are some of our most beautiful lizards. The body is very flattened and covered with granular scales. The leg and tail are well developed and often have scattered spiny scales. The eyes and eyelids are well developed, and the eardrums are visible. There are femoral pores in both sexes (these appear as small pits in females).

The flattened shape of these 'platys' (or 'flatties') permits them to squeeze under thin rock flakes where they are safe from predators. Up to 12 individua may squeeze into the same crack, although it is unusual to find adult males together. They are restricted to certain types of rock (eg. granite, gneiss and some sandstones), and are therefore found in isolated populations. Sociabl they form dense colonies. Prime territories on the rock faces are defended by dominant males during the breeding season. These depressed dandies are clothed in Jacobean splendour, the colours varying from species to species. They are most vivid on the belly, where their intensity of colour is hidden from predators. Females and juveniles have black backs, usually attractively marked with three pale, longitudinal, dorsal stripes. Males grow slightly larger than females. When confronting an intruder, the male raises his head and forebody on straightened forelimbs, revealing the bright coloration of his throat and chest. They mature at the end of their second or third year. Unlike other cordylines, they are oviparous and lay only two eggs, usually in November-December. The eggs are large, elongate (7-10 x 17-22 mm) and soft-shelled, and are laid in deep cracks, usually in damp leaf mould. Numerous females may nest in the same crack, where as many as 26 eggs may be laid. Most platys feed on small invertebrates (flies, beetles and larvae), although some species and subspecies (*P. guttatus*, *P. ocellatus* and *P. intermedius wilhelmi*) also eat plant material (flower petals, young leaves and seeds). They are relatively long-lived, and have lived for longer than 14 years in captivity.

This genus has speciated explosively in Zimbabwe and adjacent areas, where 10 species (many with local races) have evolved. An isolated, primitive species occurs in NW. Cape. Nine species, seven of which are endemic, occur in the region. Species and subspecies are often identified by the male's breeding colours, which makes the identification of females and juveniles difficult. However, few species are sympatric and if the locality is known there is little difficulty identifying the species. The taxonomy of specimens from central and N. Transvaal (eg. Blouberg and Waterberg) is confused and is under investigation.

Cape Flat Lizard *Platysaurus capensis* (Pl. 74)

(Endemic) 18-21 cm; max. 23 cm
The lower eyelids of this platy are opaque, and each is divided into a series of vertical septa. The supranasals are fused with the nasals. The middle row of gulars is not very enlarged. The scales on the sides of the neck are flattened and not enlarged. The ventrals are in 18-22 longitudinal rows. Males have 13-19 femoral pores. Females and juveniles have a dark brown back, with three broad, cream stripes that may be broken up into spots, or there may be spots between the stripes; the tail is straw-coloured, and the belly is white with a blackish patch in the middle. In adult males, the head and most of the body are Prussian blue to blue-green (northern specimens have numerous pale spots), and faint dorsal stripes may be present. The rear of the body is red-brown to pale brown, and the tail is brown. The throat is blue and lacks a collar. The chest is dark blue, and the belly is black in the centre. The hind limbs, tail and rear of the body are red below. **Biology:** These beautiful lizards are common on the smooth granite walls of Augrabies Falls, where they are tolerant of the thousands of tourists. Their major predator is the rock kestrel. **Habitat:** Succulent veld. **Range:** Lower Orange River valley below Augrabies Falls, extending along Fish River into S. Namibia and south to Garies in the Cape.

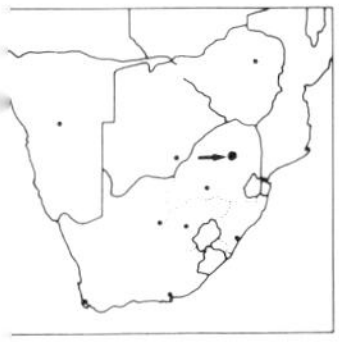

FitzSimons's Flat Lizard *Platysaurus fitzsimonsi* (Pl. 74)

(Endemic) 15-20 cm; max. 23 cm
A large platy with a transparent 'window' in each lower eyelid. The supranasals are fused with the nasals. The middle row of gulars is enlarged. The scales on the sides of the neck are conical and enlarged. The ventrals are in 20-22 longitudinal rows. There are 20 femoral pores in males. Females and juveniles have a black back with three broad, cream stripes, and 1-2 faint spots between the stripes. The throat is white, with three dark stripes. The belly is white with black spots. Adult males have a dark green head with three pale

stripes. The body is green, with a few faint light spots. The tail is orange or re above and yellow below. The throat is black, flecked with blue, and has a broad black collar. The belly and limbs are ultramarine-blue. **Habitat:** Arid savannah. **Range:** Sekhukhuni Mountains in E. Transvaal.

Dwarf Flat Lizard *Platysaurus guttatus* (Pl. 74)

(Endemic) 13-16 cm; max. 17 cm
A small platy with a transparent 'window' in each lower eyelid. The supranasals are fused with the nasals. The middle row of gulars is enlarged. The scales on the sides of the neck are flattened and enlarged. The ventrals are in 16-20 longitudinal rows (see Subspecies). There are 13-20 femoral pores in males. Females and juveniles have a dark brown back, with three narrow, broken, pale stripes and numerous pale spots between the stripes. The throat is blue-white, and the chest and belly are white. Adult males have varied coloration (see Subspecies). **Habitat:** Arid and mesic savannah. **Range:** Isolated populations in N. and W. Transvaal. **Subspecies:** Two race are recognized, and both occur in the region. *P.g. guttatus* has 20 ventral rows, spiny scales on the heels and spiny lateral caudals. Adult males have a green to blue-green back, with numerous pale spots, three pale stripes on th head, a bright orange tail, a pale green throat with black specks, no collar, a light blue chest and a dark blue belly. This subspecies occurs in N. and W. Transvaal. *P.g. minor* has 16-18 ventral rows, and the scales on its heels and the lateral caudals are not spiny. Adult males have a dark red-brown ba with numerous small spots, three pale stripes on the head, a dull red tail, a blue throat and chest, and an orange belly with a blue centre. It is found on Waterberg in N. Transvaal.

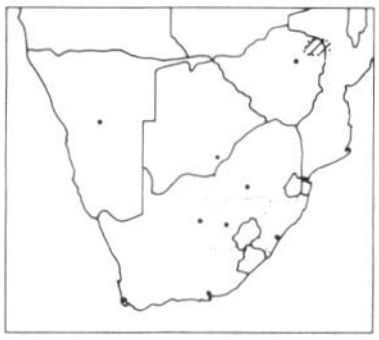

Emperor Flat Lizard *Platysaurus imperator*

(Endemic) 25-35 cm; max. 40 cm
A magnificent giant platy that is closely related to the common flat lizard, *P. intermedius* (below), but is immediately distinguishable by its great size a striking coloration. The lower eyelids are opaque, and each is divided into a series of vertical septa. The supranasals are fused with the nasals. The midd row of gulars is not very enlarged. The scales on the sides of the neck are conical and enlarged, and those on the flanks are no larger than those on the back. The ventrals are in 22-28 longitudinal rows. There are 17-24 femoral pores in males. Females and juveniles are black, with three light cream strip (yellow on the head), the middle stripe being narrow and broken towards the rear. The tail is straw-coloured. The throat is white, and the belly largely blac Adult males have an ochre-yellow head, a crimson body with numerous larg pale spots anteriorly, and are ochre-yellow towards the rear and on the tail. The limbs are black. The throat is brick-red, with a broad black collar. The chest is reddish or yellowish, the belly is black, and the tail is orange to light yellow below. **Biology and breeding:** This platy is the largest in the genus. It lives on massive boulders on top of gneiss hills, and feeds mainly on beetl It has lived for longer than 14 years in captivity. **Habitat:** Mesic savannah. **Range:** NE. Zimbabwe and adjacent Mozambique.

Common Flat Lizard *Platysaurus intermedius* (Pl. 75)

18-28 cm; max. 33 cm
A medium to large platy with opaque lower eyelids, each divided into a serie of vertical septa. The supranasals are fused with the nasals. The middle row gulars is not very enlarged. The scales on the sides of the neck are conical and enlarged, and those on the flanks are no larger than those on the back. The ventrals are in 16-22 longitudinal rows (see Subspecies). There are 12-2 femoral pores in males. Females and juveniles are black, with three buff

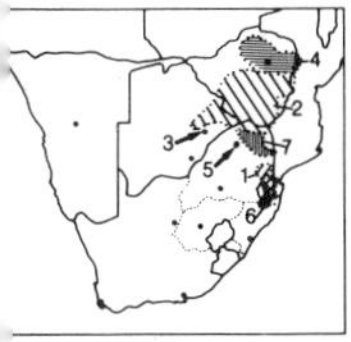

P.i. wilhelmi
P.i. rhodesianus
P.i. nigrescens
P.i. subniger
P.i. parvus
P.i. natalensis
P.i. intermedius

stripes that sometimes have light spots between them. The belly is brownish in the centre and white at the edges. Adult males have varied coloration (see Subspecies). **Biology:** This is the most widely distributed and common platy. It lives under exfoliating rock flakes on smooth outcrops of granite, gneiss or sandstone, and may form dense colonies (eg. at Rhodes' Grave in the Matopos in Zimbabwe). **Habitat:** Varied; mesic and arid savannah. **Range:** Throughout most of Zimbabwe and N. and E. Transvaal, with peripheral races in S. Malawi, Swaziland and N. Natal.

Subspecies: Eight races are recognized, seven of which occur in the region. *P.i. wilhelmi* has different-sized back scales, those covering the pale dorsal stripes being the largest. The ventral scales are in 16-18 rows. It has a dull olive-green to brown back, with a few scattered, pale spots. The tail is red (the colour sometimes extends on to the flanks), and the belly is black, except for the red tail. This race occurs from SE. Transvaal to Zululand. All other races have dorsal scales that are uniform in size.

A group of three subspecies with 20-22 ventral scale rows occurs north of Transvaal Soutpansberg and south of Zambezi River; the first two have the nasals in contact and four upper labials in front of the subocular. *P.i. rhodesianus* has a blue-green to yellow-green head, with three faint stripes. Its back is blue-green anteriorly (but red in specimens from Mozambique), and red (in the west) or green (in the east) posteriorly. The tail is greenish or yellowish. Two ventral colour phases occur: the 'red' phase has a blue throat with a black collar, and a terracotta chest; the 'green' phase has a yellow throat with a black collar, and a blue or green chest. This subspecies occurs in Zimbabwe (except in the north-east), E. Botswana and N. Transvaal. *P.i. nigrescens* has a black head and body, with faint yellow spotting on the rear. The tail is bright orange. The chin and throat are black, with an irregular yellow patch. The chest and belly are black, with scattered yellow scales. It occurs in the vicinity of Shoshong Hills in NE. Botswana. *P.i. subniger* has five upper labials in front of the subocular, and the nasals are separated. It has a dark green back, becoming brown or black posteriorly (but uniform red in males from Trelawney in Zimbabwe), with numerous pale spots. The tail is orange. The throat is orange, yellow or white, and the chest and belly are black.

A group of three subspecies with 16-18 ventral scale rows occurs south of Transvaal Soutpansberg; two have the occipital touching the parietal, and a bright green head and body. *P.i. parvus* has a dark green back anteriorly (and red-brown posteriorly), with numerous pale spots. The tail is dull orange. The throat is pale blue, with the collar reduced to a black blotch on either side of the neck. The chest and belly are blue. This race is found on Blouberg in N. Transvaal. *P.i. natalensis* has a grass-green back, with numerous pale spots and three faint, pale stripes. The tail is bright orange. The throat is yellow or pale blue, usually with black blotches, and a black collar. The chest is light blue, and the belly Prussian blue. It occurs in Swaziland and Natal. *P.i. intermedius* has the occipital separated from the parietal. The head and body are dull green to brownish above, with faint stripes and numerous pale spots. The tail is reddish to yellowish. The throat is blue and lacks a collar. The chest is blue, and the belly blue-black in the centre. This subspecies is found in E. Transvaal.

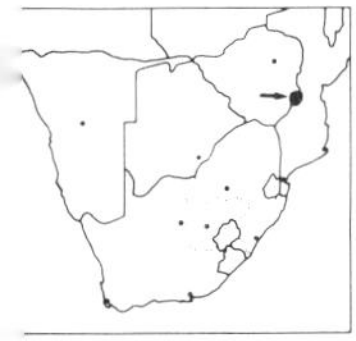

Ocellated Flat Lizard *Platysaurus ocellatus*

(Endemic) 18-22 cm; max. 25 cm
This medium-sized platy has similar scalation to the Cape flat lizard, *P. capensis* (page 165), but has different coloration, lacking the three dorsal stripes. It has opaque lower eyelids, each divided into a series of vertical septa. The supranasals are fused with the nasals. The middle row of gulars is very enlarged. The scales on the sides of the neck are spiny but not enlarged, and the collar is straight and composed of large plates. The ventrals are in

12-14 longitudinal rows. Males have 13-18 femoral pores. Females and juveniles have a bronze back, with numerous pale, ill-defined spots. The tail i: blackish. The throat is white, speckled with grey. The belly is cream, and yellow under the base of the tail. Adult males have a uniform black-brown head, and an olive-brown back with sulphur-yellow, dark-edged spots. The tε is brown above and orange to yellow below (but orange to yellow above and below in very large specimens). The belly is pale green to blue. **Habitat:** Low miombo woodland. **Range:** Lower western slopes of Chimanimani Mountains in Zimbabwe and adjacent Mozambique.

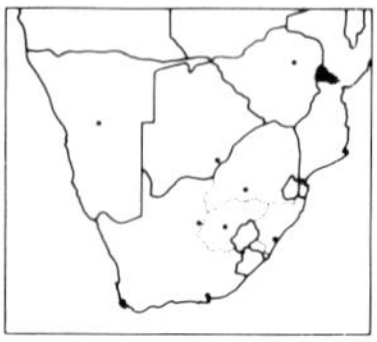

Pungwe Flat Lizard *Platysaurus pungweensis*

(Endemic) 16-20 cm; max. 24 cm
A medium-sized platy that resembles the common flat lizard, *P. intermedius* (page 166), but is smaller and has only 14-16 longitudinal rows of ventrals. It has opaque lower eyelids, each divided into a series of vertical septa. The supranasals are fused with the nasals. The middle row of gulars is not very enlarged. The scales on the sides of the neck are spiny and enlarged, and those on the flanks are no larger than those on the back. Males have 13-20 femoral pores. Females and juveniles have a black back with three buff stripes, the middle one often being broken into spots or ending on the neck; there are no spots between the stripes. Adult males have varied coloration (see Subspecies). **Habitat:** Mesic savannah. **Range:** E. Zimbabwe and adjacent Mozambique. **Subspecies:** Two races are recognized, and both occur in the region. *P.p. pungweensis* has an occipital. Adult males of this race have a brown back with numerous yellow spots, a tail that is brown abov and red on the sides, a blue throat blotched with black, a black collar, and a blue chest and belly; females have a bluish-white belly. It occurs in Eastern Highlands of Zimbabwe and adjacent Mozambique. *P.p. blakei* has a small occipital, or lacks one. Adult males have a dark green back, a red tail, a grey throat that lacks a collar, and a purple to black chest and belly; females have blackish belly. It is found on S. Manica Platform in central Mozambique.

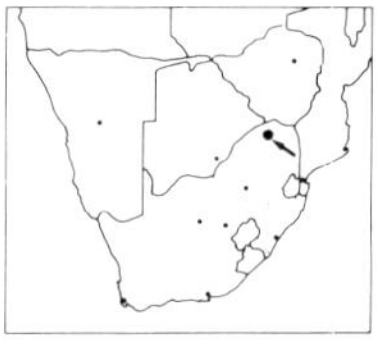

Relict Flat Lizard *Platysaurus relictus* (Pl. 74)

(Endemic) 13-16 cm; max. 16 cm
This small platy has opaque lower eyelids, each divided into a series of vertical septa. The supranasals are fused with the nasals. The middle row of gulars is not very enlarged. The scales on the sides of the neck are flattened and enlarged, and those on the flanks are no larger than those on the back. The ventrals are in 18-20 longitudinal rows. There are 18-20 femoral pores in males. Females and juveniles have a dark brown back with three cream stripes and a few spots between the stripes. The tail is straw-coloured. Adult males have a dark green back and limbs, with pale yellow-green marks on th head, a faint yellow-green stripe along the backbone, and numerous yellow-green spots. The tail is bright orange. The throat is blue-white, with a black collar. The chest and belly are dark blue. **Biology:** This flat lizard lives on sandstone outcrops. (SA RDB, Restricted.) **Habitat:** Arid savannah. **Range:** Waterberg and northern side of Soutpansberg in N. Transvaal.

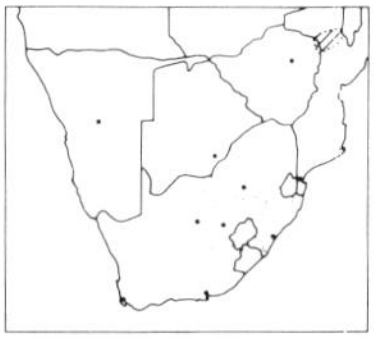

Striped Flat Lizard *Platysaurus torquatus*

15-18 cm; max. 20 cm
This medium-sized platy is the only species in the genus in which males reta the pale longitudinal stripes, and females and juveniles have uniform Cambridge-blue tails. It has opaque lower eyelids, each divided into a series of vertical septa. The supranasals are fused with the nasals. The middle row gulars is not very enlarged. The scales on the sides of the neck are flattened and enlarged, and those on the flanks are larger than those on the back. The

ventrals are in 16-20 longitudinal rows. Males have 15-22 femoral pores. Females and juveniles have a blackish-brown back, with three buff stripes with a few or no spots between them. The tail is blue, with a dark median stripe. The throat and chest are white, and the belly and base of the tail are orange. Adult males have a dark brown back with three buff stripes. The flanks and tail are bright orange. The limbs are grey-brown. The throat is white, with a black collar. The chest is orange or yellow, suffused with bright green, and becoming Prussian blue on the belly. **Habitat:** Mesic savannah. **Range:** NE. Zimbabwe, extending into west-central Mozambique and S. Malawi.

RAG LIZARDS
eudocordylus

These beautiful tyrants are some of the largest cordylines, surpassed in size only by the sungazer, *Cordylus giganteus* (page 159). They are very similar in appearance to the girdled lizards, *Cordylus* (page 156), differing only in that the neck and back are usually covered with granular scales, the body scales lack osteoderms and the tail is less heavily spined.

Rock-living, crag lizards usually inhabit a large fissure in a shattered boulder that commands a good vantage point from which to spot potential danger and forage for food. Unlike the flat lizards, *Platysaurus* (page 164), they are not sociable, and a crack is usually occupied by a single individual. They may, however, aggregate in diffuse colonies in good habitat. They eat a wide range of large invertebrates, including beetles, crickets and grasshoppers, and also take small vertebrates, particularly other lizards. They have tremendously strong jaws and an unusual, pliable skull structure that allows the shape of the head to thicken if the jaws are clammed shut. This is effectively used as a defence: the lizard wedges its head into a narrow part of a crack, and as long as it clenches its jaws, the top of the head and lower jaw will be tightly jammed against the rock walls. To prevent damage to the lizard, the scales on the top of the head are thickened with bony osteoderms. Their bulldog-like tenacity makes it almost impossible to pull them from their retreats. Although large and aggressive, they are very wary, and quickly retreat into their cracks at the first sign of danger. Viviparous, the females give birth to 1-5 young in late summer. These disperse and are usually found in marginal habitats, on the lower mountain slopes. Later, as they grow and mature, they seek more permanent homes in more prominent positions.

There are five species. The small graceful crag lizard, *P. capensis* (below) is sometimes grouped with the girdled lizards, and illustrates the close relationship between the two genera. All are endemic to the region. They are distributed in a wide arc in the old mountainous escarpment of the subcontinent, from the Cedarberg in the west, through the Cape fold mountains and the inland mountains of E. Cape, the Natal and Transvaal Drakensberg, to the Soutpansberg in N. Transvaal.

P.c. capensis
P.c. robertsi

Graceful Crag Lizard *Pseudocordylus capensis* (Pl. 71)

(Endemic) 18-22 cm; max. 26 cm
This gracile species has long toes and a thin tail. The flanks are entirely covered with granular scales. The nasals are in contact, separating the rostral and frontonasal. Each nostril pierces the lower part of the nasal. The temporal scales are small, in three rows. There are 15-18 femoral pores on each thigh. The tail is spinose and considerably longer than the body. The body, head and tail are blue-black in colour, with yellow blotches and vermiculations, particularly on the top of the head; these are fainter on the back. The belly is uniform slate-grey, and sometimes paler in the centre. The throat has a rust-red suffusion in some populations. **Biology and breeding:** This species may be found together with the Cape crag lizard, *P. microlepidotus* (page 171), but prefers more vertical cliff faces. It is agile and runs at speed over the smooth rocks. Very alert, it retreats at the first sign of danger. They form small, diffuse colonies, and several specimens (usually a male and a

female) may be found in the same crack. They feed on insects, particularly bees and wasps. One to three babies are born in December-January. **Habitat:** Fynbos. **Range:** Isolated populations from Cedarberg in the north, through Cape fold mountains to Kamanassieberg. **Subspecies:** Two poorly defined races occur. *P.c. robertsi* has two rows of enlarged scales down the backbone, and the dorsal scales in 45-54 transverse and 12-18 longitudinal rows; it is found in the northern part of the range (Cedarberg to Bokkeveld mountains). *P.c. capensis* has no enlarged scales down the backbone, and the dorsal scales in 34-36 transverse and 10-12 longitudinal rows; it occurs i the Cape fold mountains.

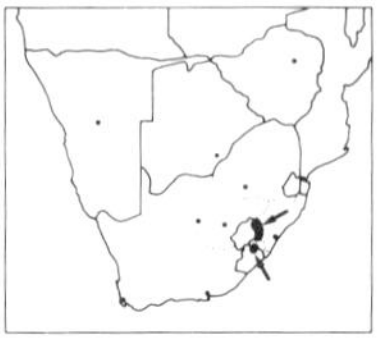

Lang's Crag Lizard *Pseudocordylus langi*

(Endemic) 18-22 cm; max. 26 cm
This medium-sized crag lizard is similar in appearance to the Drakensberg crag lizard, *P. melanotus* (below), but breeding males lack the yellow flanks. has small, granular scales on its flanks. There is a single row of 4-6 elongate temporal scales. The nasals are in contact, separating the rostral and the frontonasal, which is undivided. There are usually four upper labials anterior the subocular, and five lower labials. This species lacks occipitals. Both sex have 11-17 very small femoral pores. The back and head are olive-grey, and heavily blotched and streaked in black. The back has numerous pale grey-green blotches that form irregular crossbars. There are two large black blotches on the side of the neck, followed by a series of 1-6 bright sky-blue blotches. The belly is slate-grey, with dark blotches. The throat has a large, dark brown patch, flanked by three narrow, brown stripes. **Biology:** This cra lizard is found in small colonies and may live in the same crack as the Drakensberg crag lizard. It eats beetles and flying insects, as well as large amounts of the leaves and flowers of everlasting daisies and other plants. (SA RDB, Restricted.) **Habitat:** Rock cracks at high altitudes. **Range:** Summ of Natal Drakensberg (2 600-3 000 m).

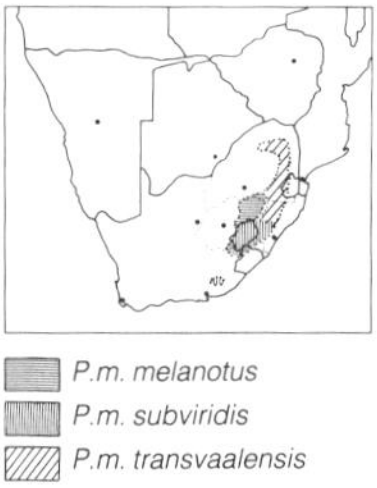

Drakensberg Crag Lizard *Pseudocordylus melanotus* (Pl. 73)

(Endemic) 20-28 cm; max. 34 cm
A medium-sized crag lizard that has the scales on its flanks separated by granular interspaces. There is usually a single row of 4-6 elongate temporal scales (and a lower row of small scales in *P.m. transvaalensis*). The nasals a in contact, separating the rostral and the frontonasal, which is undivided anc touches the anterior loreals. There are usually four upper labials anterior to th subocular, and six lower labials. Occipitals number up to 13, but may be absent. There are 5-13 femoral pores (but these are absent in *P.m. melanotu* females). Coloration is complicated; regional differences occur and breedin males are more colourful (see Subspecies). In females, the back is greyish t olive-brown, with extensive, irregular pale spots. The sides of the head, neck and body are yellowish-green, yellow or orange, often with dark dorsal coloration extending as bars on to the flanks. There are 1-2 black spots on th side of the neck. The belly is off-white, suffused with pale orange at the edge

Biology and breeding: These lizards are found in large, diffuse colonies i suitable habitat, but rarely with more than a single individual in a rock crack. They feed on small beetles and flying insects. One to four babies are born in late summer, after a three- to four-month gestation period. **Habitat:** Rock outcrops on mountain plateaus and in rolling grassland. **Range:** Escarpmer mountains, from Amatola Mountains to Transvaal Drakensberg, extending o to OFS highveld.

Subspecies: Three races are recognized, and all occur in the region. *P.m. melanotus* has a divided frontonasal. Females lack femoral pores. Breeding males having 1-17 glandular femoral scales; a broad dark brown t black band on the back (sometimes with small, pale flecks); bright orange o the flanks and sides of the neck; a red-brown temporal region; and a diffuse blue-grey patch on the throat. This race is found in E. OFS. *P.m. subviridis* does not exceed 25 cm in length, has an undivided frontonasal, and the lateral scales are larger than the spaces between them. Females have femoral pores. Breeding males have a black back with numerous large, pale olive-grey blotches; an olive-grey head and temporal region, heavily suffused with black; and orange on the flanks and sides of the tail. This race is found on Natal Drakensberg, through Lesotho, with an isolated population on Amatola Mountains. *P.m. transvaalensis* (the taxonomy of this race is confused, and the status of the different populations is at present under investigation) has an undivided frontonasal, and the lateral scales are smaller than the spaces between them. Females have femoral pores. Breeding males have similar coloration to those of the typical race. It occurs in Natal midlands, through Swaziland to Transvaal Drakensberg.

Spiny Crag Lizard *Pseudocordylus spinosus* (Pl. 71)

(Endemic) 15-18 cm; max. 21 cm
This small crag lizard has rough scales. The lateral scales are spiny and are not separated by granules, and there is a single row of elongate temporal scales. The nasals are in contact, separating the rostral and the frontonasal, which is undivided and small and does not touch the anterior loreals. There are usually 3-4 upper labials anterior to the subocular, and six lower labials. This species lacks occipitals. There are 3-5 femoral pores in both sexes. The back is dull brown, with pale, elongate spots forming irregular rows. The head is dark brown to black, with pale labials. The belly is pale brown. There is a pair of parallel grey stripes on the throat. Breeding males develop bright orange flanks and spots. **Biology and breeding:** This lizard is wary and difficult to approach. (SA RDB, Restricted.) It prefers scattered boulders in open grassland, where it feeds on small beetles, cockroaches, etc. One to four babies are born in late summer. **Habitat:** Mountain slopes. **Range:** Lower slopes (1 500-2 500 m) of Drakensberg, from Giant's Castle in Natal to Golden Gate in OFS.

P.m. microlepidotus
P.m. namaquensis
P.m. fasciatus

Cape Crag Lizard *Pseudocordylus microlepidotus* (Pl. 72)

(Endemic) 25-30 cm; max. 35 cm
This very large crag lizard has the scales on its flanks surrounded by granules. The temporal scales are in 1-3 rows, sometimes with the upper row elongated (in *P.m. namaquensis*). The nasals are in contact, or are separated by the rostral and frontonasal (in *P.m. microlepidotus*). The frontonasal is undivided. There are usually five upper labials anterior to the subocular. Males have 5-6 femoral pores on each thigh. Males are beautifully coloured; the back is dull reddish-brown to dark brown, usually with 7-8 large, irregular, yellowish (bright lemon-yellow in *P.m. namaquensis* and orange in *P.m. fasciatus*) crossbars. (S. Transkei populations are uniform dark brown, with numerous thin, irregular, pale yellow bars.) The heavy jaw muscles are deep red-brown, and the throat is suffused with grey-blue (the pattern and extent of which vary in different populations). The underside of the limbs and tail, the sides of the belly and the lips are yellow (again, orange in *P.m. fasciatus*). The upper surface of the tail has 10-13 irregular, yellow and dark brown bars.

Biology and breeding: This lizard selects large rock cracks, often partly filled with soil, within which it excavates a chamber. It forages for food on adjacent flat rocks, eating large grasshoppers, beetles, etc., and will readily kill small lizards (agamas and geckos). It bites readily and painfully, and holds on with a bulldog-like tenacity. It hibernates in winter in a deep retreat. Very pugnacious, males and females maintain exclusive territories. Three to six (usually four) young are born in January-February. **Habitat:** Mountain plateaus and upper slopes in fynbos or montane grassland. **Range:** Throughout Cape fold mountains, inland mountain ranges of old Cape escarpment, and S. Transkei. **Subspecies:** Three subspecies are recognized, and all occur ir the region. The taxonomy of Transkei populations is under investigation at present. In the typical race, *P.m. microlepidotus*, the frontonasal and rostral are in contact, and the dark crossbars on the back extend on to the flanks; it occurs in Cape fold mountains, from Cedarberg to Port Elizabeth. *P.m. namaquensis* has the frontonasal and rostral separated, the dark crossbars not extending on to the flanks, the last lower labial having a ridge, and the throat with a figure-of-eight-shaped, dark bluish mark; it is found on Nuweveldberg from Sutherland to Beaufort West. *P.m. fasciatus* also has the frontonasal and rostral separated, the dark crossbars on the back extending on to the flanks, the last lower labial lacking a ridge, and the throat a uniform bluish colour; it occurs on inland mountains of E. Cape

MONITORS (LEGUAANS)
Family Varanidae

This small family of about 30 living species contains the world's largest lizard They are closely related to the unusual earless monitor from Borneo (family Lanthanotidae), and the North American Gila monster and Mexican beaded lizard (family Helodermatidae), which are the world's only poisonous lizards; monitors are not poisonous. Very large fossils (approaching 6 m) are known from the Pleistocene epoch of S. Asia and Australia. All living monitors are placed in the genus *Varanus*.

MONITORS
Varanus

All monitors are similar in appearance, having well-developed limbs and strong claws; a long tail that is usually laterally compressed and cannot be shed; a long and flexible neck; small, polygonal, non-overlapping, bead-like scales that lack osteoderms; a single pair of preanal pores; and a long, smooth, retractile tongue, similar to a snake's. Size ranges from 20 cm to mo than 2 m, with the Komodo dragon (*V. komodoensis*) of Java attaining 3 m.

Most varanids are semi-aquatic, which has allowed them to colonize much of the East Indies; they are, however, absent from Madagascar and other Indian Ocean islands. All are predatory. Small species eat insects, while larg species take anything they can overcome. The Komodo dragon is capable o killing deer and small water buffalo, and may rarely take humans. The food is swallowed whole or is torn to bits with the strong claws. Monitors are shy lovers, and mating and egg-laying are rarely observed. They are oviparous, laying large, soft-shelled eggs in holes or termite nests. Males of some species engage in ritualized wrestling contests to determine the dominant males and their territories. The skins are attractive, and many thousands of varanids are slaughtered each year for fashion. All varanids are now protecte

Their distribution is restricted to the Old World, with the greatest diversity occurring in Australia. Only three of the 30 species occur in Africa, with two reaching the subcontinent.

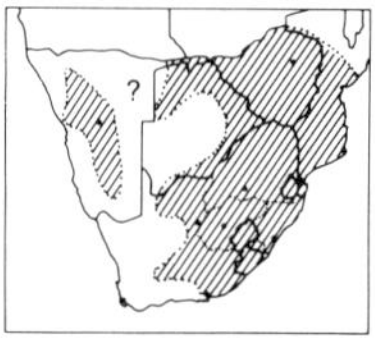

V.e. albigularis

Rock or White-throated Monitor
Varanus exanthematicus **(Pl. 63)**

70-110 cm; max. 132 cm
This very large, stout lizard has strong, stocky limbs and sharp claws. The sk is tough and covered with small, bead-like scales, in 150-165 rows at midbody. The head has a bulbous snout, with the nostrils slit-like and nearer the eyes than the end of the snout. The tail is longer than the body, and cylindrical at the base but compressed towards the tip. The back is dark grey-brown above, with 5-6 pale yellow, dark-edged blotches. The top of the head and neck are dark brown. The limbs are spotted with pale yellow, and the tail is banded in dark brown and off-white. The belly is dirty yellow, with scattered spots. Juveniles are more intensely marked.

Biology and breeding: This monitor lives in a tunnel that it digs under rock overhangs, or in a disused animal burrow, a hole in a tree or a rock crack. It is usually solitary and hibernates, semi-dormant, in its retreat in winter. Its skin i usually dulled with dirt and grime, and sullied with patches of unshed skin. It is also well-adorned with ticks in the soft skin around the eyes, nostrils and limb joints. Its scientific name *exanthematicus* alludes to its leprose appearance ('exanthema' is Greek for a skin eruption). The diet consists mainly of invertebrates (millipedes, beetles, grasshoppers and land snails), although it will kill and eat any animal small enough to swallow, and also scavenges on carrion; baby tortoises are frequently eaten. In defence, it adopts a side-on posture and lashes its tail. It will bite and hold on like a bulldog; if held behind the head, it usually ejects its cloacal contents, and finally it may sham death, hanging limp (but still keeping its eyes open). If this ruse works and it is released, it scampers to safety at the first opportunity. The martial eagle is the main predator on adults. It rarely tames in captivity, and, because of its size, needs a large enclosure. The flesh is reported to taste like chicken, but this monitor is protected by Provincial legislation (CITES, Appendix II). Mating occurs in August-September. The female may occasionally lay her eggs in a live termite nest, as does the Nile monitor, *V. niloticus* (below), but normally uses a hole dug in soft, moist soil. She may dig several 'test holes' before selecting a suitable spot. Between eight and 37 eggs (53-61 x 35-39 mm, 32-46 g) are laid, depending on the size of the female. The eggs hatch in 120 days in captivity, but may take up to 12 months in the wild. Many clutches are eaten by the banded mongoose.

Habitat: Savannah and arid, karroid areas. **Range:** Throughout savannah and semi-desert regions of the subcontinent, but absent from W. Cape. Elsewhere, throughout savannahs of subSaharan Africa. **Subspecies:** Two well-defined races are recognized, with only one occurring on the subcontinent. The typical race, *V.e. exanthematicus*, has a more uniform colour pattern, lacking the white throat, and enlarged scales on the back of the neck; it occurs from Senegal to Ethiopia. The southern race, *V.e. albigularis*, has a more contrasting colour pattern, with a lighter belly and throat, and smaller scales over the back of the neck; it occurs on the subcontinent and north to Ethiopia. Several ill-defined races from central Africa have been decribed, based on minor differences in scalation (*V.e. microstictus* and *V.e. angolensis*), or the juvenile colour pattern (*V.e. ionidesi*).

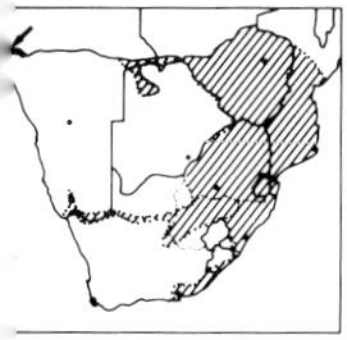

. niloticus

Nile or Water Monitor *Varanus niloticus* (Pl. 63)

100-140 cm; max. 200 cm

This is the largest African lizard. It has a stout body with powerful limbs and strong claws. The skin is tough and covered with small, bead-like scales. The head has an elongate snout, and the nostrils are round and situated midway between the eyes and the end of the snout. The tail is much longer than the body, and is laterally compressed, with a low dorsal crest. Adults are greyish-brown to dirty olive-brown on top of the head and back, with scattered darker blotches and light yellow ocelli and bands on the head, back and limbs. The belly and throat are paler, with black bars. Juveniles are beautifully patterned in black and bright yellow.

Biology and breeding: The Nile monitor is common in major river valleys, foraging for food in the marginal vegetation. It is an excellent swimmer, using its long, oar-like tail. It often basks on rock outcrops or tree stumps. In temperate regions, it may hibernate communally in a large rock crack on a rocky cliff or koppie bordering a river. The diet is varied. Adults forage in freshwater pools for crabs and mussels, but will also take frogs, fish, and birds and their eggs; they also excavate and eat the eggs from terrapin, sea turtle and unattended crocodile nests. Juveniles rarely enter deep water but shelter in marginal reed beds, where they hunt for frogs and insects. Adults' teeth are rounded and peg-like (ideal for crunching crabs), unlike the sharp, recurved teeth of the juveniles. When disturbed, they dive into the water and swim underwater to the safety of the reed beds. If cornered, they bite and lash the tail in defence, as does the rock monitor, *V. exanthematicus* (page 172).

Crocodiles and pythons are major predators on the adults. The flesh of these lizards is edible, and the fat is used for tribal medicine, but it is protected by Provincial legislation (CITES, Appendix II). After the spring rains (August-September), the female excavates a hole in a living termite nest anc lays 20-60 eggs (54-64 x 30-40 mm). This may take 2-3 days to complete. Tl termites then repair their nest, and the eggs develop inside it, incubated at a constant temperature and humidity. It may take up to a year before the youn emerge, although in captivity (at 30 °C), they hatch in 129-175 days. The young (200-320 mm, 23-32 g) emerge together, digging themselves out of t rain-softened nest the following summer.

Habitat: Rivers, pans and major lakes. **Range:** SubSaharan Africa, extending along Nile River to Egypt. On the subcontinent, it occurs through t eastern part of the region, extending along Orange River to Atlantic Ocean and Fish River Canyon, and along the south coast to Gamtoos River valley, b is absent from W. Cape. **Subspecies:** Two races are recognized, with only t typical race (*V.n. niloticus*) occurring on the subcontinent; it is replaced in th rain forests of W. Africa and Zaire by *V.n. ornatus*.

AGAMAS
Family Agamidae

These small to large lizards have either a cylindrical or a flattened body, well-developed limbs and a tapering tail. The head is large, with a distinct neck. The prominent eyes have movable eyelids. The pineal eye, which is visible as a small depression on the crown of the head, is often well developed. The ear openings are rarely covered with scales. The head scal are small and irregular, while those on the body are overlapping, keeled anc often drawn into spines. The tail is usually long, and cannot be shed or regenerated.

These diurnal reptiles are mainly terrestrial, although there are some specialized rock-living and arboreal forms. Many feed mainly or exclusively ants, while others are primarily herbivores. They often form social groups, ar display well-developed territorial behaviour; males are often brightly coloure and/or adorned with frills, crests or throat fans to enhance their displays. Mo are oviparous, although a few Asian species are viviparous.

This large family comprises about 300 species in 30-40 genera, which are distributed throughout most of the Old World, with the greatest diversity occurring in Asia and Australia. It is replaced in the New World by the Iguanidae, and is absent from Madagascar. At present, this family is represented by a single genus on the subcontinent.

The fragmented head shields of an agama

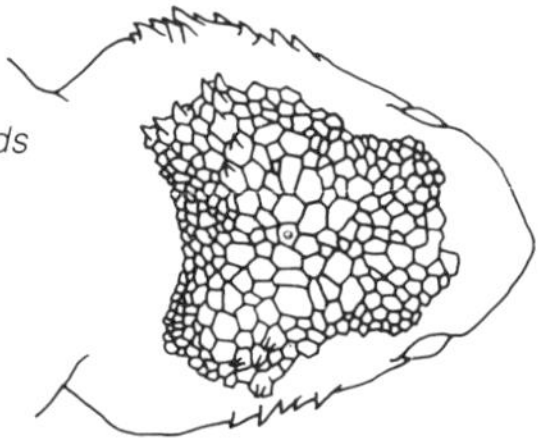

AGAMAS
Agama

The agamas are plump, short-bodied lizards with thin tails and triangular heads. The head is covered with small scales, and sometimes has an enlarged occipital scale under which lies the pineal eye. Two fang-like teeth are found in the upper jaw. The large eyes have scaly eyelids and round pupils. The legs are long, with thin toes. Preanal pores are present in males.

These active, diurnal and territorial lizards occupy a variety of habitats; some species are rock-living, others are mainly terrestrial, and a few climb trees. They feed almost exclusively on ants and termites, although all will supplement their diet with beetles and other insects. Males develop vivid breeding colours and engage in territorial displays, maintaining exclusive home ranges and chasing off vanquished opponents. Females and juveniles are more cryptically coloured to match the soil or lichen-covered rocks. All lay relatively large clutches of soft-shelled eggs in a hole that is dug in the ground.

This wide-spread genus is distributed throughout Africa and the near East. At present, it contains approximately 60 species (although these may be divided into six genera in the future). Nine species occur on the subcontinent, three of which are endemic.

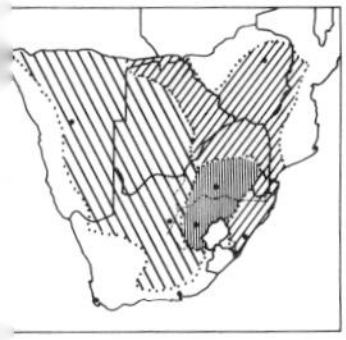

A.a. aculeata
A.a. armata
A.a. distanti

Ground Agama *Agama aculeata* (Pl. 78)

15-22 cm; max. 26 cm
A medium-sized agama with a triangular head and a relatively rounded snout. The earholes are large, and the tympanums are visible. The fat body has sturdy limbs. The scales are strongly keeled and directed obliquely inwards towards the backbone. The enlarged spines are arranged more or less in regular rows along the back. There are 75-115 scale rows at midbody. The ventrals are smooth (but sometimes keeled in *A.a. armata*). A dorsal crest is present but is weakly developed, and sometimes extends on to the tail. The tail is usually longer than the head and body in both sexes (but shorter in the female *A.a. distanti*). Males have a single row of 9-14 preanal pores. The body is olive to reddish-brown (and sometimes grey to yellowish), usually with a distinct grey-yellow dorsal streak which sometimes has a thin central black line. There are 4-5 paired, darker blotches on the back that continue as 10-13 irregular bars on the tail. There are often two bars or chevrons between the eyes. The belly is creamy white to pale dirty pink. Males have three parallel, blue-black lines that border a central network, on each side of the throat; females have only the central network, which may be faint or even absent. Breeding males develop blue sides to the head. **Biology and breeding:** Although they are terrestrial, these agamas will often climb into low scrub to bask. A short hole dug in loose soil at the base of a bush is usually shared by a pair. The diet is almost exclusively termites or ants, and they can often be seen browsing on a stream of ants passing to and from a nest. When disturbed, they run at top speed with the tail curved upwards, and then stop abruptly, hugging the ground and relying on their superb camouflage to escape detection. The female lays 10-18 eggs (13-15 x 9-10 mm) in a hole in sandy soil; the hole is often dug under a stone or at the base of a bush.

Habitat: Semi-desert and savannah. **Range:** Throughout most of the subcontinent, absent only from moister coastal regions and true desert. Elsewhere, to S. Angola and Tanzania. **Subspecies:** Three races occur, but their status is problematic and some may prove to be full species. All occur in the region. *A.a. aculeata* has smooth dorsal head shields, and the fourth toe longer than the third; it is found in the western regions, through Cape, Namibia and W. Botswana to S. Angola. *A.a. armata* has rough dorsal head shields, the third toe longer than the fourth, and keeled ventrals; it occurs in central Transvaal, through Zimbabwe and NE. Botswana to Tanzania, and through Natal to Transkei. *A.a. distanti* has rough dorsal head shields, the third toe longer than the fourth, and smooth ventrals; it is found in OFS and S. Transvaal.

Anchieta's Agama *Agama anchietae* (Pl. 78)

15-20 cm; max. 25 cm
A smallish agama with a broad head and a short, rounded snout. The scales on the back are strongly keeled and directed obliquely inwards towards the backbone, with a few scattered spines. There are 90-105 scale rows at midbody. The ventrals are keeled. A dorsal crest is present, but is weakly developed on the nape. The tail is about half again as long as the head and body, and has a serrated crest in males. There is a single row of 10-12 preanal pores in males. Breeding adult males are pinkish-brown above, with fine grey

and yellow flecks and 4-5 dark brown to black crossbands that are sometim separated by a pale dorsal streak. The flanks are reddish-brown. The tail is banded. The head is bright blue-green, with a large, dark blue spot on the throat. Females and non-breeding males have the head the same colour as the body, both being mottled in brown and grey. The belly is white to pale yellow, and there are dark bluish to grey-black wavy stripes on the throat. **Biology and breeding:** This species is similar in habits and behaviour to th Southern rock agama, *A. atra* (below), but does not form dense colonies. Th female lays 10-12 soft-shelled eggs (15-16 x 10-11 mm). **Habitat:** Semi-des and arid savannah. **Range:** NW. Cape, through Namibia and Angola to S. Zaire.

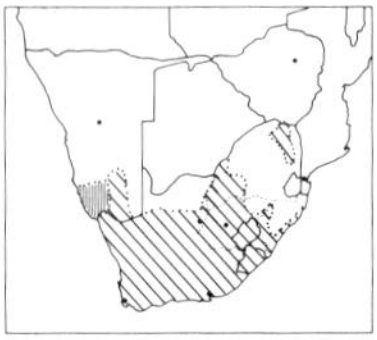

A.a. atra
A.a. knobeli

Southern Rock Agama *Agama atra* (Pl. 76)

(Endemic) 20-25 cm; max. 32 cm
A large agama with a flattened body and large limbs. The scales on the bac are very small, with scattered, enlarged spines (but females lack enlarged spines and their scales are distinctly keeled). There are 90-180 scale rows a midbody. The ventrals are smooth. A dorsal crest is present and extends we on to the tail (but occurs only on the nape of adult *A.a. knobeli* males). The ta is half again as long as the head and body, and is compressed in males and cylindrical in females. There is a single row (and occasionally two) of 9-16 preanal pores in males. Breeding adult males are olive-green to red-brown above, marbled with dark maroon to black, and with scattered, pale-centred spots. An orange-yellow to whitish vertebral streak extends from the neck to the tip of the tail. The flanks often have red blotches. The head and forelimbs are blue to greenish-blue. The throat is intense ultramarine to purple-blue, w irregular dark stripes that may extend on to the belly. The tail is greyish-white to yellow, usually with dark crossbands. Females and non-breeding males a mottled in tan, cream and dark brown above, sometimes with red blotches c the flanks. The belly is off-white, and there is a bluish network on the throat.

Biology and breeding: This species usually lives on rock outcrops and mountain plateaus, but may also be found on rocky plains and may even shelter under the bark of dead trees. Along the Cape coast they forage amc driftwood on the beach. They feed almost exclusively on ants and termites, although coastal populations may eat various intertidal arthropods. They ma form dense colonies (up to 165 specimens per hectare). Both males and females form hierarchies and maintain territories. Male territories are larger (approximately 90 m^2), and contain those of several females. There is alway single dominant male and female; she is 'loyal', while he may mate with any female. A dominant male perches on the highest point of his territory, noddi his brightly coloured head as a signal for lesser males to pay due respect ar stay clear. When danger threatens, they hug the rock; the bright head fades and becomes camouflaged against the lichen-covered rock. If this fails, the scamper off at top speed, leaping from boulder to boulder, to shelter in a de crack. Nonetheless, many fall prey to rock kestrels and other predators. Juveniles are frequently eaten by snakes (particularly cross-barred sandsnakes and spotted house snakes). Egg development starts in August-September. Gravid females, bulging with eggs, bask to speed the formation of yolk from the fat stores laid down in the previous autumn. A first clutch of 7-12 oval, soft-shelled eggs (15-18 x 10-12 mm) is laid in October-November in a shallow hole dug in damp soil; a second clutch is usually laid in January-February. Incubation takes 2-3 months. Hatchlings measure 53-59 mm.

Habitat: Semi-desert to fynbos, from sea level to mountain tops. **Range:** Throughout Cape (absent only from northern sandy areas) and nort to S. Namibia, Zululand and Transvaal escarpment. **Subspecies:** Two race are recognized, although these may prove to be separate species. *A.a. atra* has its vertebral crest weakly developed on the neck, and the scale keels o the flanks pointing outwards; it occurs over most of the range. *A.a. knobeli* has a well-developed vertebral crest over the neck, and the scale keels on the flanks pointing towards the backbone; it is found in Namaqualand, north and south of lower Orange River.

Etosha Agama *Agama etoshae* (Pl. 78)

(Endemic) 12-14 cm; max. 16 cm
A small agama with a broad head and a rounded snout. The earholes are small, and the tympanums are not visible. The back scales are strongly keeled and directed obliquely inwards towards the backbone. Enlarged spines are scattered on the flanks, sometimes in irregular rows. The ventrals are feebly keeled. A weak dorsal crest extends on to the tail base. The tail is longer than the head and body in males, and shorter in females. Males have a single row of 10 preanal pores. The back is pale yellowish-white to rust-red, usually with four dark, pale-ringed blotches. There are two dark bars between the eyes. The tail has 10-12 dark, paired, semicircular marks. The belly is cream. The gular has a black pentagonal spot that is surrounded by short black streaks. **Biology:** Terrestrial, these agamas forage for beetles and termites in sandy, flat country. At night they shelter in a hole that they dig at the base of a bush. They may be solitary or found in pairs, but do not form colonies. They shuffle into loose sand both to conceal themselves and to avoid the heat. They occasionally protrude their eyes to clear sand grains from around the eyelids, and to aid the shedding of old skin. They can often be seen displaying their gular pattern from the top of small rock piles, but are shy and very difficult to approach. **Habitat:** Sandveld. **Range:** Vicinity of Etosha Pan in N. Namibia.

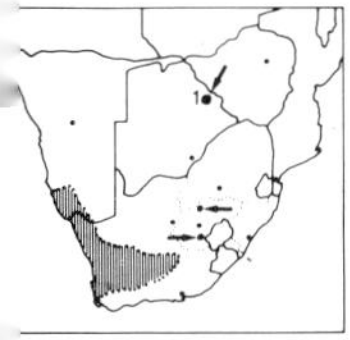

A.h. hispida
A.h. makarikarica

Spiny Agama *Agama hispida* (Pls. 77 and 78)

(Endemic) 12-20 cm; max. 24 cm
A small to medium-size agama (see Subspecies) with a broad head and a rounded snout. The earholes are small, and the tympanums are not easily seen. The back scales are strongly keeled and directed obliquely inwards towards the backbone. Enlarged spines are arranged in regular rows along the back. There are 70-90 scale rows at midbody. A dorsal crest is present but is weakly developed and does not extend on to the tail. The tail is longer than the head and body in males, and shorter in females. Males have a single row of 10-12 preanal pores. Coloration is varied (see Subspecies).

Biology: Terrestrial, these agamas live in a short tunnel dug at the base of a bush, and forage for ants and beetles in open, sandy veld. Large *A.h. hispida* males may display from boulders. They do not form colonies. **Habitat:** Arid semi-desert, coastal dunes and salt pans. **Range:** W. Cape and adjacent regions, with an isolated race in NE. Botswana. **Subspecies:** Two races are recognized, and both occur in the region. *A.h. hispida* is large (max. 24 cm), the gular region has irregular pale blotches, and the ventrals are keeled. Breeding males have a vivid, almost metallic, yellow-green head and body, with indistinct darker and paler blotches. The belly is bluish-grey to blue-green, darker on the chin and throat, and with an obscure gular pattern. Females and juveniles of this race are olive to brown above, with 4-5 darker crossbars that extend more faintly on to the limbs and tail. The belly is yellow-white to pale green, with a dark network on the throat that sometimes extends backwards. It is found in W. Cape and S. Karoo, extending into S. Namibia. *A.h. makarikarica* is small (max. 15 cm), the gular region has an extensive pale network, and the ventrals are feebly keeled. Both sexes are light to dark grey-brown above, usually with four pairs of dark rectangles on the back, and sometimes with a yellowish dorsal streak. There is a distinct X-shape on the snout and two dark chevrons between the eyes. The head is blue in breeding males. The belly is whitish, with a blue-black reticulation on the throat. It is restricted to Makarikari Pan in NE. Botswana.

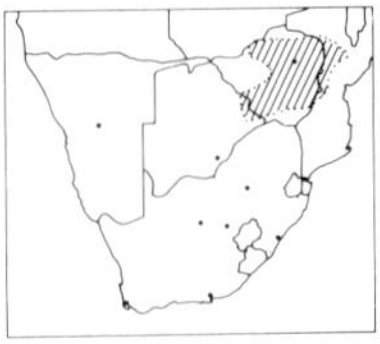

Kirk's Rock Agama *Agama kirkii* (Pl. 77)

18-25 cm; max. 34 cm

A large agama with a flattish body and long limbs. The back scales are abou the same size as those on the belly, are strongly keeled (although less so in females), and directed obliquely inwards towards the backbone. There are 99-114 scale rows at midbody. The ventrals are smooth. A dorsal crest, whic is well developed on the nape of adult males, is present and extends on to th tail. The tail is almost twice as long as the head and body. There is a single ro (and occasionally two) of 10-14 preanal pores in males. Breeding males hav a vermilion to yellow head, and a purple body with a whitish vertebral crest and adjacent scales. The hind limbs and tail are light blue-green, with narrow, white rings. Breeding females have a blue-green head, and a maroo body with blue-grey blotches. The limbs and tail are grey. In winter, juveniles and adults are camouflaged in mottled grey-black. **Biology and breeding:** Rock-loving, Kirk's rock agama lives on granite and paragneiss outcrops. The diet is almost exclusively ants and termites, although it also takes the occasional beetle. They are often seen stationed at the base of a fi tree, browsing off the constant stream of ants that are attracted to the fruit. Th female lays about 10 soft-shelled eggs. **Habitat:** Arid and mesic savannah. **Range:** E. Botswana, through Zimbabwe and Mozambique to Malawi, Zamb and S. Tanzania.

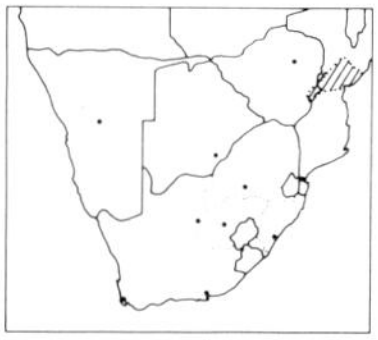

Mozambique Agama *Agama mossambica*

18-25 cm; max. 33 cm

This large agama is very similar to Kirk's rock agama, *A. kirkii* (above), but th scales on the back are larger than those on the belly. There are 69-94 scale rows at midbody. The ventrals are keeled (but only faintly so in some populations). Their coloration differs from that of Kirk's rock agama. Breedin males have a pinkish or grey-brown body with a blue head and a broad, blue-white vertebral band. The belly is suffused with pink, and the throat has large blue patch. Breeding females are brilliant orange on the back, and are otherwise similar to Kirk's rock agama. In winter, adults have an olive to dark brown head with a reddish bar across the forehead, a large reddish spot just above the shoulder, and sometimes a row of 4-5 brick-red to blackish spots c the flanks. In juveniles, the dorsal spots are distinct and interspersed with smaller yellow spots, and the vertebral streak is yellowish. **Biology:** Equally a home in trees and on the ground, these agamas will climb rapidly up the nearest tree when disturbed. The diet is mainly ants, supplemented with beetles, other insects and millipedes. **Habitat:** Lowland savannah and forest fringe. **Range:** E. Zimbabwe, extending through adjacent Mozambique to Malawi and Tanzania.

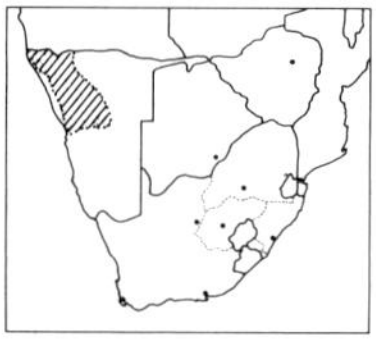

A.p. planiceps

Namibian Rock Agama *Agama planiceps* (Pl. 77)

25-30 cm; max. 33 cm

A large, graceful agama with a small, flat head and a pointed snout. The ear openings are large, and the tympanums are visible. The body is flattened, wi long limbs and toes. The large, feebly keeled back scales are directed obliquely inwards towards the backbone. There are 63-76 scale rows at midbody. The ventrals are small and smooth. The dorsal crest is weakly developed on the nape and absent on the back, but forms a crest on the tail males. The tail is longer than the head and body. Males have a single row of 8-14 preanal pores. Breeding males have a dull blue-purple, almost metallic back. The head, neck and throat are coral to orange-red, sometimes with scattered dark and yellow spots. The tail is olive-yellow at the base, changin to orange-red at the tip. The belly is dull purple to blue-black. Females and juveniles have a greyish to olive-brown body, with pale blotches and dark

purple-brown infusions. There is a bright orange blotch behind the shoulder, and a pale dorsal streak is sometimes present. The head has symmetrical lemon-yellow blotches. **Biology:** This species is very similar in habits and behaviour to Kirk's rock agama, *A. kirkii* (page 178), which it also resembles physically. They live on rock outcrops and boulder-strewn hills, and are very active and difficult to approach. The diet includes leaves and seeds, as well as beetles and insects. **Habitat:** Semi-desert and arid savannah.
Range: Damaraland and Kaokoveld in N. Namibia. **Subspecies:** Two races are recognized, but only the typical race, *A.p. planiceps*, occurs in the region. It is replaced in Angola by *A.p. schacki*.

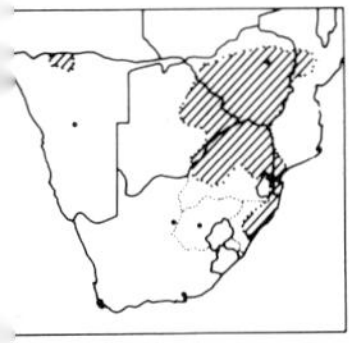

a. atricollis

Tree Agama *Agama atricollis* (Pl. 76)

20-30 cm; max. 39 cm
A very large agama with a broad head (particularly in males) that lacks an occipital scale. The ear openings are larger than the eyes, and the tympanums are visible. The scales on the body are small, rhomboidal and keeled, with those along the back larger and mixed with scattered, enlarged, spiny scales. There is no vertebral crest. There is a strong fold across the throat. The ventrals are smooth. There are two (and sometimes three) rows of 10-12 preanal pores in males. Breeding males have a dull blue to bluish back, with bright blue (anteriorly) to straw-yellow (posteriorly) spines. The head is coppery-green to brilliant ultramarine on top, blue-green on the sides and peacock-blue on the throat. There is a large black spot on each side above the shoulder, and a broad, blue-green to yellowish vertebral stripe. The tail is dull green to olive-brown. Females and non-breeding males are olive to green-brown, marbled with black above, with a black shoulder spot. Juveniles have a similar ground colour, with dark X-shapes surrounded by white blotches along the sides. The tail is banded with dark brown-black.

Biology and breeding: These beautiful lizards are frequently seen nodding their heads in display while clinging to a tree trunk. They come to the ground only to cross to another tree, and occasionally to eat (especially flying ants and termites). They will supplement their diet with grasshoppers and beetles. When threatened, they retreat around the tree trunk, always keeping the trunk between themselves and danger. They will gape the mouth widely, showing the bright orange mouth lining, and will deliver a painful bite if caught. Contrary to popular belief, they are not poisonous. They sleep at night in a hollow branch or under peeling bark. The female lays 8-14 oval, soft-shelled eggs (10-18 x 22-27 mm) in a hole dug in moist soil. They hatch after 90 days. Hatchlings measure 70-80 mm; they triple in size in their first year, but growth slows thereafter. They become sexually mature in their second year.
Habitat: Open savannah, particularly with *Brachystegia* and *Acacia* trees.
Range: Ethiopia, extending through E. Africa to Natal, Transvaal lowveld, E. Botswana (Lobatsi) and Ovamboland. **Subspecies:** There are possibly up to six races in E. and central Africa, but all are poorly defined. Only the typical race, *A.a. atricollis*, occurs on the subcontinent.

HAMAELEONS
amily
hamaeleonidae

Chamaeleons are unmistakable, and look unlike any other lizards. Their scales are small, do not overlap and lack bony plates. The head and body are compressed and the neck is not defined. Limbs are always present, with the toes usually bound together and opposed. The tail is prehensile and cannot be shed or regenerated. The protruding eyes are independent. The hearing is very poor, the middle ear cavity and external opening being absent. The tongue is telescopic and can be shot further than the body length to capture prey. Preanal and femoral pores are absent. Males are characterized by their bright coloration, and often have horns and other ornamentation on the head.

Chamaeleons are primarily arboreal, which explains many of their unusual features. Their colour, compressed bodies, crests and casques serve as camouflage; the prehensile tail and opposable toes, bound in uneven bundles, clasp swaying branches; the turreted eyes move independently, scanning for food and danger; and the unique tongue captures prey at a distance in flimsy foliage. Social displays and sexual dimorphism are well developed. Both sexes are asocial, maintaining exclusive territories by ritualized combat which includes head-butting, inflating the throat pouch, ar rapid colour changes. They rarely fight, but will bite if an intruder fails to withdraw. Colour varies to match the surroundings. The most dramatic darkening occurs when they are stressed, and the colour is most intense wh they are aroused, either in territorial disputes with other males or when courting females. Most are oviparous, laying large clutches of soft-shelled eggs, but some montane and temperate species are viviparous.

This family has 70-100 species in four genera, and is restricted mainly to Africa and Madagascar. At present, there are three genera with at least 16 species in the region; 14 species are endemic.

Key to the southern African genera in the Chamaeleonidae

1. Toes with single claw; tail prehensile; soles of feet smooth	2
Toes with bicuspid claws (small, secondary cusp directed downwards); soles of feet spiny; tail feebly prehensile; lungs simple; oviparous	*Rhampholeon* (Leaf chamaeleons, page 187)
2. Parietal narrow and compressed; lungs with elongate sacs; oviparous	*Chamaeleo* (Common chamaeleons, page 185)
Parietal broad, with lateral processes; lungs simple; viviparous	*Bradypodion* (Dwarf chamaeleons, page 180)

DWARF CHAMAELEONS *Bradypodion*

These are small chamaeleons. The tail is prehensile. The parietal forms a lar plate over the temporal region. The lungs are simple and lack diverticula. Th skin has scattered large tubercles, particularly on the flanks and limbs. Gula and dorsal crests are present and are usually composed of scaly flaps; vent crests, tarsal spurs and occipital flaps are absent. The head lacks protuberances, but often has a well-developed casque, and adult males hav a larger gular crest. Adult males develop intense breeding colours, while juveniles and females are camouflaged in mottled brown and grey-green.

They are often found in small, localized colonies that seem to undergo regular cyclic changes, with populations exploding and crashing in number Fecundity is high, and the females give birth to 2-3 clutches, each of 5-20 minute babies, every season. Growth is rapid, and sexual maturity is reache in 1-2 years. Adult males become intensely coloured when defending their territories against other males or when courting females. They engage in we developed ritual displays; the head is bobbed and the gular region expande to reveal the brightly coloured interstitial skin. All are arboreal, although som small species live in low vegetation. They prey on small insects. Water is required regularly and is licked from dew or raindrops on foliage. They have many predators, particularly snakes (eg. the boomslang and spotted bush snake) and birds (eg. the fiscal shrike). They climb into the top of vegetation the morning to bask in the sun, and retreat at night into bushes, where they turn pale grey-white (rendering them easy to locate with a torch). *Bradypodi* means 'slow foot', which describes well their deliberate, jerky gait.

Previously considered endemic to the subcontinent, this genus is now known to extend into E. Africa. Locally, the taxonomy is very confused, as th scattered populations are characterized by slight differences in scalation ar male breeding colour. Some conservative taxonomists treat all different forms as subspecies of a single, wide-ranging species, but this is an over-simplification. Conversely, treating all forms as full species confuses their relationships, as some are probably regional variants (eg. *B. gutturale* and *B. karroicum* are probably best considered races of *B. ventrale*). Given these limitations, 13 species are recognized at present, with at least six new species under investigation (see Plate 3). Two species groups can be recognized. The first, composed of the larger species, is associated with montane forests in Natal (eg. *B. thamnobates*), Cape (eg. *B. damaranum*) and Transvaal (eg. *B. transvaalense*), but occurs in a wider variety of habitats in the Cape (eg. *B. pumilum* and *B. ventrale*). The second group is restricted to coastal or fynbos vegetation and consists of small species (eg. *B. taeniabronchum*, *B. setaroi* and *B. melanocephalum*). Species are often difficult to distinguish; there are few diagnostic features, and these are often fully developed only in adult males. However, few species are sympatric, so if the locality is known, they should be easy to identify. All are endemic to the subcontinent.

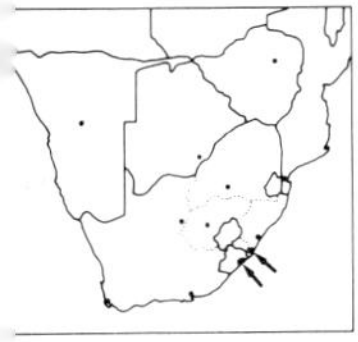

Transkei Dwarf Chamaeleon *Bradypodion caffrum* (Pl. 94)

(Endemic) 10-12 cm; max. 14 cm
A medium-sized dwarf chamaeleon. The flanks have rough granules, with a few scattered tubercles that may form an irregular row on the upper flanks. The head has an obvious, elevated, pointed and laterally compressed casque and moderate cranial crests. The gular region has 2-4 shallow throat grooves. There is a well-developed gular crest, with the anterior scaly flaps broader than they are long, and overlapping. The dorsal crest consists of large, compressed tubercles and extends well on to the tail in males, but fades out over the lumbar region in females. The tail is slightly longer than the body in males, and equal to or shorter than the body in females. The body is mottled brown, olive and cream, with an irregular, pale cream lateral stripe that extends on to the temporal region and may intensify to orange in display. The gular region is pale cream, with white throat grooves. **Habitat:** Low coastal forest. **Range:** Immediate vicinity of Port St Johns in Transkei, with an isolated population in Oribi Gorge in Natal.

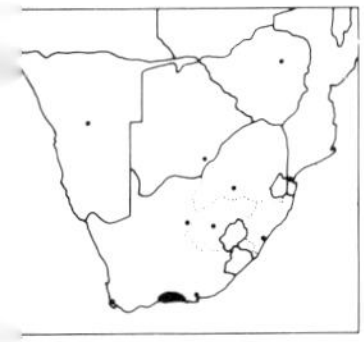

Knysna Dwarf Chamaeleon *Bradypodion damaranum* (Pl. 94)

(Endemic) 12-16 cm; max. 18 cm
A large dwarf chamaeleon. The flanks have granular scales with a few scattered, flattened tubercles and two longitudinal rows of enlarged tubercles. There is an oval patch of smooth skin around each forelimb. The head has a well-developed, narrow casque that is elevated and somewhat recurved posteriorly. The cranial crests, which are prominent, consist of large, convex tubercles that surround large, polygonal, smooth temporal scales. The gular crest consists of 8-13 large, scaly flaps that are broader than they are long and overlap strongly. The dorsal crest consists of 35-50 low, interrupted tubercles that extend almost to the end of the tail and are largest over the base of the tail. The tail is longer than the head and body in both sexes.

Males are brilliant emerald-green, often with a dark blue-green, blue-maroon or rust patch on the flanks, bordered with smooth yellow skin around the base of each forelimb. The cranial crests are yellow to dark blue, with the temporal scales pale blue-grey and sometimes light maroon when the animal is excited. The interstitial skin of the gular region is pale yellow, with light blue tubercles. Females, juveniles and unaroused males are mottled in green and yellow-brown. **Biology:** This chamaeleon may climb high into the canopy, and often sleeps in the centre of tree ferns, where its long, coiled tail resembles a young fern frond. **Habitat:** Wet coastal forest, entering low secondary bush in clearings and gardens. **Range:** S. Cape, from George to Tsitsikamma; despite its scientific name, it is not found in Damaraland.

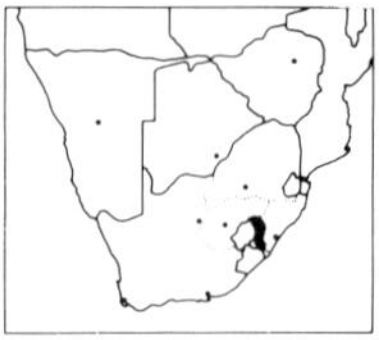

Drakensberg Dwarf Chamaeleon
Bradypodion dracomontanum (Pl. 93)

(Endemic) 12-14 cm; max. 15 cm
The flanks of this chamaeleon have fine, granular scales and a few small, scattered, convex tubercles. The head has a medium-sized, recurved casqu and distinct cranial crests. The gular crest is composed of 10-18 small, abutting scaly flaps that are longer than they are broad. The dorsal crest has 39-49 small, conical scales that extend on to the tail, which is longer than the head and body. Males have a green-blue body with dark blue blotches on th anterior flanks, and white tubercles. The belly and gular region are infused with bright yellow. The female is mottled in green, brown and white, sometime with a faint, pale lateral stripe. **Habitat:** Evergreen kloof forests and alpine ve between 1 500 and 2 500 m. **Range:** S. and central Natal Drakensberg.

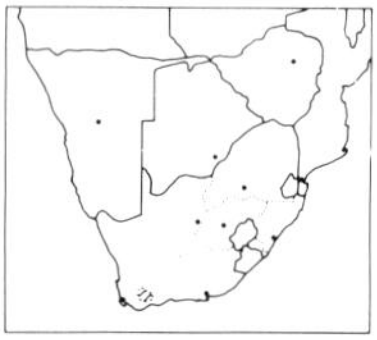

Robertson Dwarf Chamaeleon *Bradypodion gutturale* (Pl. 94

(Endemic) 11-14 cm; max. 15 cm
A medium-sized dwarf chamaeleon. The flanks are rough, and have fine granules and scattered, very large tubercles that form one (and sometimes two) obvious row on the sides. The head is covered with very rough tubercles and well-developed cranial crests, and has a prominent, pointed casque. Th gular region has irregular grooves and scattered, large tubercles. The gular crest is prominent, with numerous scaly flaps that are longer than they are broad and that overlap at the front. The dorsal crest is well developed, with 36-55 elongate tubercles that are largest over the shoulder region and extenc over half the length of the tail. The tail is much shorter than the body. This chamaeleon is blue-grey, with rust-coloured lateral tubercles and a dorsal crest of the same colour, becoming light orange-red anteriorly and on the cranial crests. The gular region has light orange scales and blue-grey interstitial skin. **Habitat:** Low montane fynbos scrub. **Range:** Vicinity of Worcester and Robertson in SW. Cape.

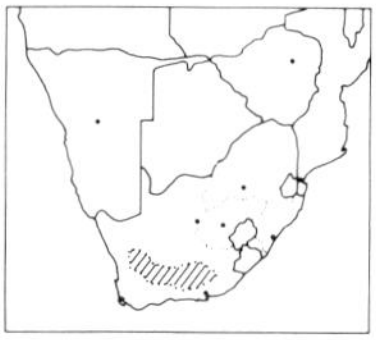

Karoo Dwarf Chamaeleon *Bradypodion karroicum* (Pl. 94)

(Endemic) 10-13 cm; max. 14 cm
A medium-sized dwarf chamaeleon. The flanks have granular scales and scattered, convex tubercles that form 2-4 irregular, longitudinal rows on the upper flanks. The head has a well-developed casque that is feebly elevated, and shallow cranial crests. The gular region has 3-5 irregular throat grooves and a row of small tubercles. The gular crest is composed of elongate, and sometimes pointed, scaly lobes that are longer than they are broad and do n overlap. The dorsal crest is composed of prominent conical tubercles, interrupted by smaller tubercles and granular skin, and extends over most of the tail. The tail is shorter than the head and body in both sexes. The body is mottled in ash-grey and cream-brown, tinged wih olive-green, and with 2-3 diffuse, dark-edged blotches. One or two rows of rust-orange flank stripes ar chrome-yellow interstitial skin in the gular region may develop. **Biology and breeding:** This species is similar in habits and behaviour to the Southern dwarf chamaeleon, *B.v. ventrale* (page 185). **Habitat:** Sparse thorn bushes along river courses; adapting to urban gardens. **Range:** S. and E. Karoo.

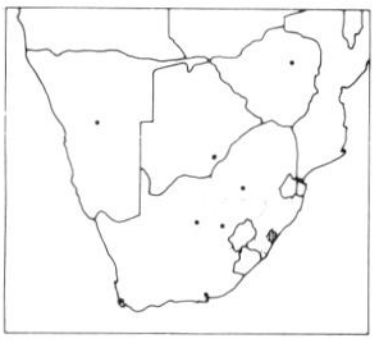

Blackheaded Dwarf Chamaeleon
Bradypodion melanocephalum (Pl. 95)

(Endemic) 9-11 cm; max. 12 cm
A tiny dwarf chamaeleon. The flanks are covered in granules, with scattered tubercles that may form 2-3 irregular rows. The head has a feeble, narrow, slightly recurved casque and a weak cranial crest. The gular region has 2-3

shallow throat grooves and a weak crest of 12-22 simple, small, conical tubercles. The dorsal crest is composed of 31-55 small, compressed tubercles and extends well on to the tail. The tail is slightly longer than the body in males, and shorter in females. The body is brown-green, with a pale brown area on the flanks that extends on to the temporal region and is interrupted or constricted by 2-3 dark-edged, irregular, brownish blotches. There are two brighter orange regions between the cranial crests. The gular region is off-white, with white throat grooves. The colour is intensified during displays. **Breeding:** Up to 12 tiny young are born in summer. **Habitat:** Coastal bush and reed beds around vleis. **Range:** Restricted to N. Natal coast (below 30 m), from Mount Edgecombe to Nqwazi River, with isolated populations around Pietermaritzburg and Greytown.

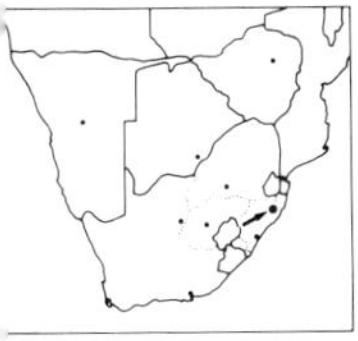

Zululand Dwarf Chamaeleon *Bradypodion nemorale* (Pl. 95)

(Endemic) 10-14 cm; max. 14,5 cm
A medium-sized dwarf chamaeleon. The flanks have fine, granular scales and a row of slightly enlarged tubercles. The head has a recurved, prominent casque and moderate cranial crests. The gular crest is composed of 8-10 scaly, non-overlapping flaps that are longer than they are broad. The dorsal crest consists of 10-29 well-spaced, triangular scales and usually extends on to the tail. The tail is shorter than the body. Males have dirty green flanks with large, irregular, rust-coloured blotches that intensify during display. The tubercles on the flanks are orange, and those on the dorsal crest are reddish-brown. The gular region is pale green, with white throat grooves. (SA RDB, Restricted.) **Habitat:** Forest. **Range:** Qudeni and Nkandla forests in Zululand.

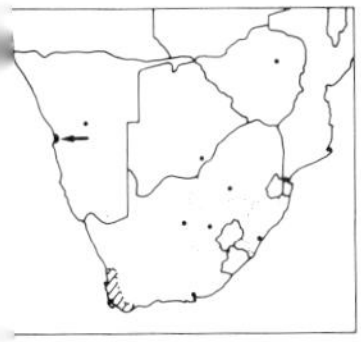

Cape Dwarf Chamaeleon *Bradypodion pumilum* (Pl. 93)

(Endemic) 13-16 cm; max. 18 cm
A large dwarf chamaeleon. The flanks are covered with granules and have 1-2 rows of enlarged, rounded, convex tubercles. The head has a moderate, narrowed and pointed casque and well-developed cranial crests. The gular region has 3-4 shallow grooves that are not separated by the enlarged tubercles. The gular crest is composed of 12-17 elongate, scaly flaps that do not overlap. The dorsal crest exends along two-thirds of the tail and is composed of small, separate, conical tubercles. The tail is longer than the body. Coloration is uniform leaf-green, usually with an orange to rust-red lateral stripe. The temporal region is red-orange, sometimes with wine-coloured cranial crests. The gular region is pale green to yellow. **Biology and breeding:** This species may be less territorial than other species, and can reach high densities (200 specimens per hectare), with 5-6 adults inhabiting a single bush. They live in low coastal bushes and reed beds. Females may have up to four litters per year; litters born in December consist of more babies (5-12) than those born in April (3-6). The foetal membranes surrounding the young stick to the foliage, whereupon the young wriggle free. Like all lizards, they fend for themselves from birth. **Habitat:** Around vleis and along river courses. **Range:** SW. Cape, with isolated populations around Clanwilliam and Bredasdorp.

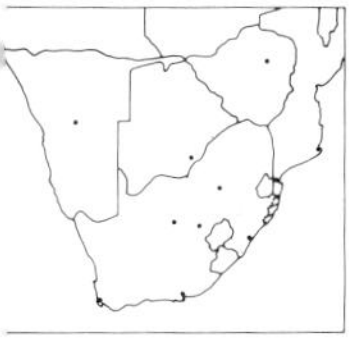

Setaro's Dwarf Chamaeleon *Bradypodion setaroi* (Pl. 95)

(Endemic) 9-12 cm; max. 13 cm
A small dwarf chamaeleon. The flanks have fine granular scales and an irregular series of enlarged, flattened tubercles bordering the backbone. The head has a pronounced, elevated, narrow casque and moderate cranial crests. The gular crest is composed of 13-23 weakly developed, irregular scaly flaps that are longer than they are broad. There are 2-4 shallow throat grooves. The dorsal crest is reduced, is composed of 10-18 low tubercles,

and does not reach the hind body. The tail is longer than the body in males, and shorter in females. The body is light grey-brown, with an irregular orange stripe on the flanks, which may have 2-3 darker orange blotches. The lateral stripe is bordered with vermiculated grey-green that also occurs as bars on the tail and speckles on the head. The gular region is light green, with white throat grooves. (SA RDB, Restricted.) **Breeding:** Eight babies (39-42 mm) ar born in late April. **Habitat:** Low coastal dune forest. **Range:** N. Zululand.

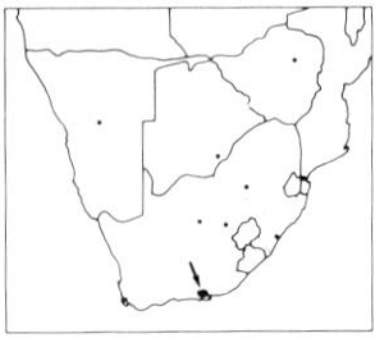

Smith's Dwarf Chamaeleon
Bradypodion taeniabronchum (Pl. 95)

(Endemic) 8-11 cm; max. 12 cm
A small dwarf chamaeleon. The flanks are finely scaled, with a few scattered, flattened tubercles that may form an irregular row on the dorsal surface. The head has a small, pointed, indistinct casque and weakly developed cranial crests. The gular region has two deep grooves. The gular crest is small, with 18-22 tiny, conical tubercles in an interrupted series. The dorsal crest is composed of 30-45 small conical tubercles and reaches to the base of the ta The tail is shorter than the body. Coloration ranges from grey-green to pale green (sometimes with a pale lateral stripe). The belly and gular region are paler. The throat grooves are black. Occasional specimens are rust-red in colour, with deep maroon throat grooves. **Biology and breeding:** It lives in Protea bushes, feeding on the insects that are attracted to the flowers. It shelters at night among the dead flower heads. Much of its habitat has been destroyed f pine plantations (SA RDB, Endangered). It gives birth to up to 13 minute babies (35-40 mm) in summer. An unreceptive female will reject the male's advances b taking an open-legged stance, gaping her mouth to reveal the bright yellow-orange lining and black throat grooves, and rocking from side to side. **Habitat:** Mountain summits. **Range:** Vanstadensberg in E. Cape.

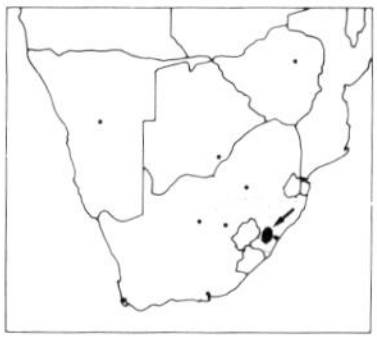

Natal Midlands Dwarf Chamaeleon
Bradypodion thamnobates (Pl. 93)

(Endemic) 15-18 cm; max. 19 cm
A large dwarf chamaeleon. The flanks have granular scales and 1-2 rows of enlarged, convex tubercles on the upper surface, and similar tubercles scattered over the lower sides and tail. The head has a strongly elevated, recurved casque and very distinct cranial crests composed of horn-coloured conical tubercles. The gular crest has 10-18 scaly lobes, usually overlapping and longer than they are broad. The pronounced dorsal crest is composed o 26-43 elongate, conical tubercles, and extends on to the tail, which is longer than the head and body in males, but shorter in females. Males are dark blue green on the back, with an elongate, cream, bright yellow or red-brown latera patch, and a light blue-green belly and limbs. The tubercles on the limbs and flanks are reddish-cream to horn-coloured. The cranial crests are horn-coloured, sometimes with dark blue-black blotches. There are scattered red tubercles on the temporal region and the eyes. The chin, throat and base of the forelimbs are white to cream, with white interstitial skin in the throat grooves. Females and juveniles are mottled in grey and brown, and some individuals are uniform green. (SA RDB, Restricted.) **Habitat:** Lowland forest and bush. **Range:** Natal midlands from Mooi River to Howick and Bulwer.

Transvaal Dwarf Chamaeleon
Bradypodion transvaalense (Pl. 94)

(Endemic) 14-18 cm; max. 20 cm
A large dwarf chamaeleon. The taxonomy of the many recently discovered populations is very confused. The flanks have granular scales and scattered tubercles over the back; these are sometimes confined to 1-2 longitudinal

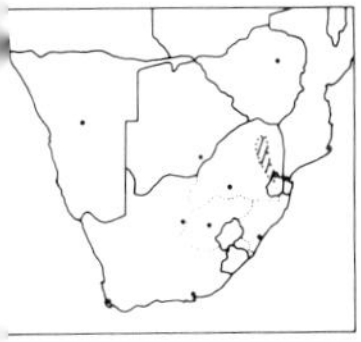

rows. The head has a moderate casque that is curved slightly upwards, and prominent cranial crests. The gular crest has 9-13 well-developed, scaly flaps, the anterior ones overlapping and broader than they are long. The dorsal crest is prominent and extends on to the first third of the tail. The tail is slightly longer than the head and body. Males have a prominent, elongate, pale yellow-mustard stripe that is bordered with dark maroon, and extends from the temporal region on to the flanks. The gular is cream-coloured, and the belly often infused with rust. Females and juveniles are mottled in brown and cream. The back is dark and the flanks cream-coloured, with a number of large, dark-edged, irregular blotches. The edge of the casque is bright mustard. **Habitat:** Wet forest of escarpment kloofs. **Range:** Isolated population along Transvaal escarpment, extending into adjacent Swaziland.

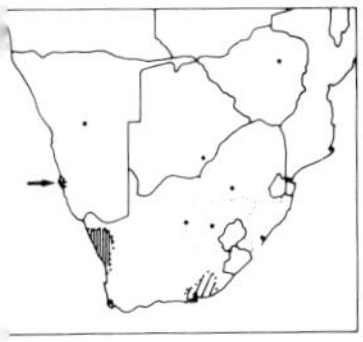

B.v. ventrale
B.v. occidentale

Southern and Namaqua Dwarf Chamaeleons
Bradypodion ventrale (Pl. 93)

(Endemic) 12-14 cm; max. 16 cm
Large dwarf chamaeleons. The flanks have small granules and scattered, flattened to convex tubercles, the largest ones forming 2-4 rows on the upper flanks. The head has a moderate, slightly raised casque and cranial crests. The gular region has 3-5 irregular grooves, separated by rows of enlarged tubercles. The gular crest has 12-17 scaly flaps, the first 4-7 being broader than they are long and overlapping. The dorsal crest is composed of 35-45 conical tubercles running in an interrupted series to near the end of the tail. The tail is shorter than the body in both sexes. Unaroused males, females and juveniles are predominantly mottled in grey, olive, cream and brown, and occasionally have three dark-edged, irregular bands on the back. An elongate, pale yellow to cream patch extends from the temporal region on to the forebody. The enlarged tubercles on the flanks are dark brown. The skin behind the casque is sometimes mustard-yellow. The gular region is pale yellow-green, with white grooves. Excited males are blue-black, with a bright rust stripe surrounding the flank tubercles.

Biology and breeding: When threatened, this chamaeleon rapidly loses its bright colours and retreats so as to position a branch between itself and danger. It may drop to the ground and hide among the leaf litter if escape or concealment in the foliage is impossible. One or two litters, each of 10-20 young, are born during summer. **Habitat:** Typical race in thick succulent valley bushveld and Fish River bush; western race in low shrubs and bushes. **Range:** E. Cape and Namaqualand. **Subspecies:** Two races are recognized. *B.v. ventrale* has its casque narrowed and pointed, and the dorsal crest prominent but interrupted; it occurs in E. Cape. In *B.v. occidentale*, the casque is not narrowed, and the dorsal crest is more or less continuous; it is found in W. Cape strandveld, with isolated populations on Kamiesberg, and at O'Kiep and Lüderitz Bay.

COMMON CHAMAELEONS
Chamaeleo

These are medium to large chamaeleons with prehensile tails, a narrow, compressed parietal bone, and diverticulate lungs. Males of some central African and Madagascan species have bizzare horns and crests that are used in display.

Found mainly in forest or well-wooded savannah, some species enter montane grassland or semi-desert. Most lay eggs, although the small, montane chamaeleons (*Chamaeleo bitaeniatus* and others) of E. Africa are viviparous.

There are about 50 species in the genus, restricted mainly to Africa and Madagascar, with isolated populations (sometimes treated as separate species) surrounding the Mediterranean and extending east to S. India and Ceylon. The Kenyan Miocene fossil *C. intermedius*, which is 14 million years old, is very similar to the Namaqua chamaeleon, *C. namaquensis* (page 186). Two species, neither of which is endemic, occur on the subcontinent.

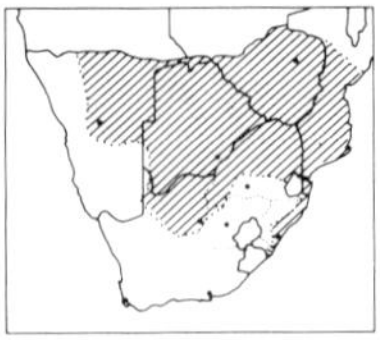

Flap-neck Chamaeleon *Chamaeleo dilepis* (Pl. 96)

20-24 cm; max. 35 cm

A large chamaeleon with a continuous crest on the throat and belly, a prehensile tail that is as long as the body, and occipital flaps (these are large in western populations, but small or rudimentary in the east). Coloration is varied, from pale yellow through green shades to brown. The belly crest is white, and there is usually a pale bar and several white spots on the sides. The interstitial skin of the male throat pouch is orange. **Biology and breeding:** This species is common in suitable habitat. The diet consists of insects, particularly grasshoppers and beetles. When threatened, it inflates its body, distends its throat, raises the occipital flaps and opens its mouth wide to expose the red-orange lining. It will bite readily. It is greatly feared by many tribes, and is the subject of much folklore, but is not poisonous. It is easier to find at night, when it turns blue-white in colour. Predators are mainly snakes (particularly the boomslang and twig snakes), but also include monkeys and birds (eg. the crowned hornbill).

Mating and egg-laying are energetic. In spring the male's gular skin turns pearl-white, and the female permits approach. Introductions are minimal, and the female is taken more or less by storm. Mating may last for up to an hour. Egg development takes 3-4 months. The female becomes bloated with 25-50 (max. 57) small eggs (13-15 x 8-9 mm); during this times, she is dull-coloured and very aggressive. She constructs a tunnel (15-30 cm long) in damp soil, in which to lay her eggs in late summer (March-May in Zululand). She may take up to 24 hours to complete laying. The eggs take approximately 150 days to hatch, but may take up to 300 days if a cold winter follows laying.
Habitat: Savannah woodland, entering coastal forest in Zululand.
Range: Tropical Africa, southwards to Natal, Transvaal, N. and E. Botswana to N. Cape, and N. Namibia. **Subspecies:** There is a great controversy over the races; 5-6 races have been described, but only *C.d. ruspolii* from Somalia is well defined. Southern African populations are sometimes referred to *C.d. quilensis*, but their status is uncertain.

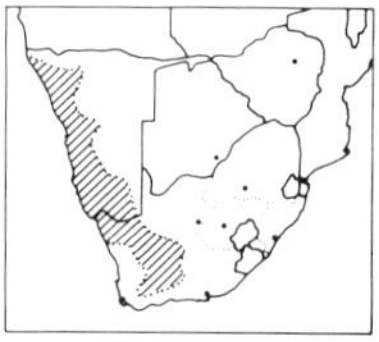

Namaqua Chamaeleon *Chamaeleo namaquensis* (Pl. 96)

18-21 cm; max. 25 cm

A large, ungainly chamaeleon with a robust head and a big mouth. The dorsal crest is composed of 12-14 very large, knob-like tubercles that decrease in size towards the tail. There is no gular or ventral crest. The tail is much shorter than the head and body, and is feebly prehensile. The body is dull green-grey to pinkish-maroon, with scattered dark spots and a row of 4-5 larger pale spots on the upper flanks. **Biology and breeding:** Mainly terrestrial, this chamaeleon lives in some of the hottest, most desolate regions. It eats anything that is small enough to swallow, including locusts, crickets, lizards, and even snakes (eg. Peringuey's adder). It has a big appetite, and may eat up to 200 beetles a day. A nasal salt gland rids the body of excess salt, allowing the chamaeleon to forage at the seaside.

Territories are defended, initially with display (side-to-side head bobs), but fight will break out if the transgressor does not retreat. Two or three clutches per year, each of 6-22 soft-shelled eggs (10-13 x 16-20 mm), are laid in a burrow 20-25 cm deep. A preferred site is at the foot of a windward slope of a large dune, and is defended by the female. Clutches laid in spring contain fewer but larger eggs than those laid in winter. The eggs take 90-100 days to hatch (and up to 115 days if they are laid in winter). Newly hatched males are smaller than females and reach maturity later (males at about 210 days, females at about 150). The young, which often hatch at night, have more dorsal tubercles than adults and climb bushes more readily. **Habitat:** Sandy regions (including coastal dunes) with scrub vegetation. **Range:** Western karroid areas, through Namaqualand and Namib Desert to S. Angola.

EAF
HAMAELEONS
hampholeon

These chamaeleons inhabit evergreen montane forests. The tail is short, and at most feebly prehensile. The parietal forms a broad plate over the temporal region. The lungs are simple, without diverticula. They are oviparous.

Eight species are recognized from E. and southern Africa, but only a single endemic species occurs on the subcontinent. They were previously placed in the genus *Brookesia*, which is now restricted to Madagascar.

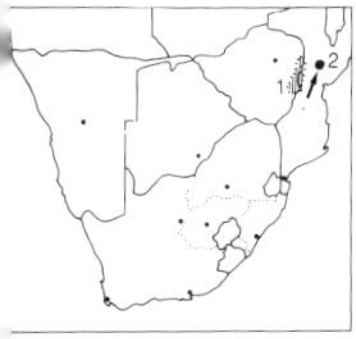

R.m. marshalli
R.m. gorongosae

Marshall's Leaf Chamaeleon *Rhampholeon marshalli* (Pl. 95)

(Endemic) 5-9 cm; max. 11 cm
A tiny chamaeleon with a short (50-75% of the body length), prehensile tail, a rostral appendage, a vertebral series of enlarged tubercles arranged in clumps, bicuspid claws, and feet with spinose soles. It does not have a gular or ventral crest, nor occipital flaps. In colour, the male is shades of brown, grey or olive-green, contrasting with the bright blue-green to black ventrum, and with a prominent row of pale tubercles along the chin, the sides of the lower body and most of the tail. Females are light green. **Biology and breeding:** This relatively sedentary chamaeleon lives on undergrowth in montane forests, feeding on small beetles, flies, etc. It resembles a small, dead leaf, and is therefore well camouflaged. It excavates a hole in leaf-litter, in which it lays 10-18 eggs (13 x 8 mm) in late summer. The embryos, which are well developed, hatch in 50-60 days. *R.m. gorongosae* males have been found riding on females' backs, probably to discourage other suitors. They mature in two years. **Habitat:** Wet montane evergreen forests. **Range:** Eastern border of Zimbabwe and adjacent Mozambique, with an isolated race on Gorongoza Mountain in Mozambique. **Subspecies:** Two races are recognized. *R.m. marshalli* males are only slightly smaller than the females; this race occurs in E. Zimbabwe and adjacent Mozambique. *R.m. gorongosae* males are half the size of the females and have an elongate proboscis; this race is restricted to Gorongoza Mountain in Mozambique.

YPICAL GECKOS
amily Gekkonidae

This family of unusual lizards has an amazing array of feet and eyes. The toe-tips of many species have groups of scales with minute hairs that catch in small cracks. These unique scales are arranged in rows or paired pads called scansors that allow them to 'stick' to seemingly smooth surfaces. Many species also have claws, which in some species are retractile between the scansors. The different species and genera are easily identified by their feet. Their eyes are usually large, with complicated pupils that dilate widely at night and close to pin-pricks during the day. Most lack movable eyelids; these have become fused and transparent so that each eye is covered by a spectacle. To keep this clean, they lick it with the long, fat tongue. Preanal and femoral pores are usually present in males.

Although a few geckos are diurnal, most are nocturnal. They can withstand much lower temperatures than most other lizards, and live in a large variety of different habitats, including cool mountain tops and temperate regions; they are most common in deserts, which may become very cold at night. Many have adapted well to urbanization and live in homes and factories, and have spread around the world by hiding in goods and ships. All are oviparous (the viviparous New Zealand geckos are now placed in a different family), and usually lay two relatively large, hard-shelled eggs; a single egg is laid by some geckos, particularly the American sphaerodactylines. Many species lay several clutches during a breeding season, and utilize communal egg-laying sites. They store calcium for the egg shells in special neck glands called endolymphatic sacs. Some genera reproduce parthenogenetically. Many live in colonies and have developed a range of different sounds to allow them to communicate in the dark. With the exception of a few colourful diurnal groups (eg. *Phelsuma*), most are drably coloured in buffs, greys and browns.

Typical geckos are among the most diverse lizards and are widely distributed throughout the world. The primitive gecko subfamilies

Eublepharinae and Diplodactylinae are now treated as separate families, and do not occur in southern Africa. Two gekkonid subfamilies are now recognized: the Teratoscincinae, which includes a single genus consisting of three species that are restricted to the near East; and the Gekkoninae, which distributed throughout the tropical regions and includes 607 species in 63 genera. Geckos are well represented in southern Africa, 14 genera (four of which are endemic) and 64 species (42 of which are endemic) being present.

Key to the southern African genera in the Gekkonidae

1.	Hind feet webbed	2
	Hind feet without webs	3
2.	Forefeet webbed	*Palmatogecko* (Web-footed gecko, page 209)
	Forefeet not webbed, but with adhesive scansors	*Kaokogecko* (Kaoko web-footed gecko, page 196)
3.	Toes not dilated, lacking enlarged terminal scansors	4
	Toes dilated, with enlarged terminal scansors	6
4.	Toes short, cylindrical, covered with minute scales, clawless in males, with minute claw in females	*Chondrodactylus* (Giant ground gecko, page 192)
	Toes elongate and strongly clawed	5
5.	Head and body flattened; toes not fringed; diurnal	*Narudasia* (Festive gecko, page 199)
	Head short, body cylindrical; toes fringed; nocturnal	*Ptenopus* (Barking geckos, page 211)
6.	Toes with large claw	7
	Toes clawless, or with minute, usually retractile claw	11
7.	Toe-tips dilated and with single series of scansors	*Homopholis* (Velvet geckos, page 195)
	Toe-tips with paired series of scansors	8
8.	Toe-tips dilated and with paired scansors separated by groove	9
	Toe-tips hardly dilated and with oblique, paired scansors	10
9.	Toes with single pair of scansors	*Phyllodactylus* (Leaf-toed geckos, page 210)
	Toes with 2-3 pairs of scansors	*Afroedura* (Flat geckos, page 189)

10. First toe well developed; pupils vertical; nocturnal .... *Hemidactylus* (Tropical house geckos, page 194)

First toe rudimentary; pupils round; diurnal .... *Lygodactylus* (Dwarf geckos, page 196)

11. Some toes much longer than others; diurnal 12

Toes of more or less equal size; nocturnal 13

12. First toe small; scansors at toe-tip divided; no femoral pores; nostrils tubular, each pierced between three nasals .... *Rhoptropus* (Namib day geckos, page 213)

First toe rudimentary; scansors at toe-tip undivided; long row of femoral pores; nostrils not tubular .... *Phelsuma* (Day geckos, page 209)

13. Three or more scansors at toe-tip .... *Pachydactylus* (Thick-toed geckos, page 199)

Two scansors at toe-tip .... *Colopus* (Kalahari ground gecko, page 193)

FLAT GECKOS *Afroedura*

These geckos are characterized by a large pair of adhesive scansors beneath the dilated toe-tip, separated by a small gap from 1-2 pairs of smaller scansors. Claws that are retractile between the scansors are present in both sexes. The head and body are flattened, and the eyes are large, with vertical pupils. The back is covered with small, flat, smooth, granular scales. Preanal pores are present in males, and the tail is usually segmented at the base and slightly longer than the body.

They are nocturnal, usually sheltering under exfoliating flakes on hard rock outcrops (eg. granite, gneiss and some sandstones) from sea level to the mountain tops. Some species are communal, and may be found in large aggregations (10-20 individuals) in a suitable crack. They lay two hard-shelled eggs, often in a communal egg-laying site. The eggs are slightly soft and sticky when first laid, but harden later and adhere together and to the rock.

The taxonomy is very confused; there are 8-10 species in the genus, and most are endemic to the subcontinent. One species occurs in Angola (*A. bogerti*), and another just crosses the Zambezi River (*A. transvaalica*).

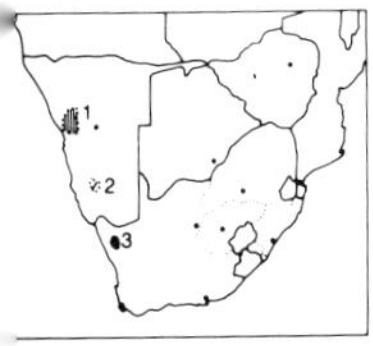

1. *A.a. africana*
2. *A.a. tirasensis*
3. *A.a. namaquensis*

African Flat Gecko *Afroedura africana* (Pl. 88)

(Endemic) 10-12 cm; max. 13 cm
A large flat gecko with an elongate, flattened body that is covered with small granules. The nasals are in contact behind the rostral. The toe-tips have three pairs of scansors. Males have 8-15 preanal pores in an angular series. The tail is segmented. The back is pale yellow to buff, with 5-6 wavy, dark brown bands. The tail is indistinctly banded (but regenerated tails are not banded). The belly is white-yellow. The dorsal bands are faint or replaced by blotches in the southern races. **Biology and breeding:** These lizards live under thin, exfoliating flakes on the shaded, overhanging surfaces of large granite boulders, where they are protected from the midday heat. They are also occasionally found under bark in large *Acacia* trees in dry riverbeds. Two

thin-shelled eggs (12-13 x 8-9,5 mm) are laid in rock cracks; these hatch in approximately 100 days. Predators include snakes (the spotted house snake and western keeled snake).

Habitat: Rocky desert and succulent karroid veld. **Range:** Isolated populations in central and S. Namibia and Little Namaqualand. **Subspecies:** Three races are recognized. *A.a. africana* does not have enlarged gular scales, and has 11-15 preanal pores; it occurs in SW. Damaraland. In *A.a. tirasensis*, the mental and adjacent lower labials are elongate, but the adjacent gular scales are not enlarged, and there are 9-10 preanal pores; it is found in Bethanie district of Namibia. *A.a. namaquensis* has enlarged gular scales and 8-10 preanal pores; it is found in Little Namaqualand.

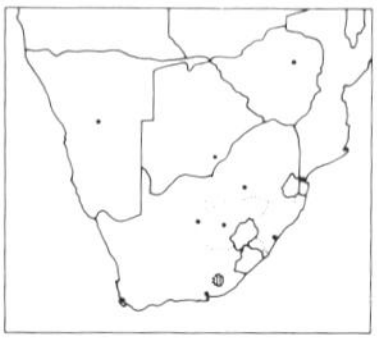

Amatola Flat Gecko *Afroedura amatolica* (Pl. 88)

(Endemic) 9-11 cm; max. 12 cm
This medium-sized species is very similar to the mountain flat gecko, *A. nivar* (page 191). The body is short and flattened. The nasals are not in contact behind the rostral, which enters the nostril. The scales on the back are flattened and overlapping. There are three pairs of scansors on the toe-tips. Males have 10-12 preanal pores in an angular series. The tail is segmented. The back is brownish-grey, with 7-8 wavy, zigzagging, dark brown bands tha may be broken up into scattered, irregular blotches. The belly is off-white. **Biology and breeding:** Up to 10 geckos may be found in a suitable crack on a granite outcrop. They hibernate in deep cracks to escape the snow that covers the mountains in winter. Two eggs (10-12 x 8-9,5 mm) are laid in a roc crack, and 10-30 eggs may be found in a communal site. Eggs laid in late summer may not hatch until the following summer. **Habitat:** Montane grassland. **Range:** Amatola and Katberg Mountains in E. Cape.

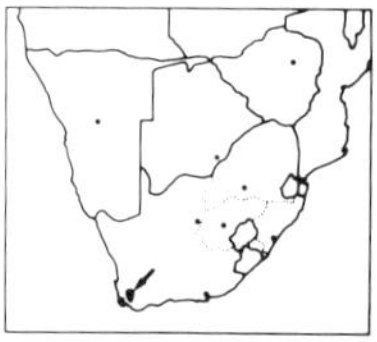

Hawequa Flat Gecko *Afroedura hawequensis* (Pl. 88)

(Endemic) 12-15 cm; max. 17 cm
A large flat gecko with a stout, flattened body that is covered with small granules. The nasals are not in contact behind the rostral. There are three pairs of scansors on all toes except the first. Males have 30-32 preanal pores in a curved series. The tail is segmented (leaf-like when regenerated). The back has 5-6 irregular, dark brown bands over a light grey-brown background, with scattered yellow spots at the edges of the dark bands and along the flanks. The belly is creamy white. **Biology and breeding:** These geckos live in narrow cracks in sandstone boulders, preferring shady positions. Living communally, up to five individuals may be found in the same crack, and they may share this with the smaller marbled leaf-toed gecko, *Phyllodactylus porphyreus* (page 211). Two large eggs (17 mm long) are laid in November-December. **Habitat:** Fynbos vegetation. **Range:** Cape fold mountains in SW. Cape.

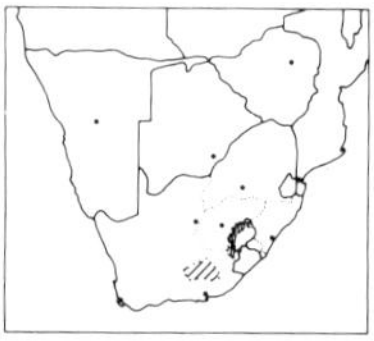

A.k. karroica
A.k. halli

Karoo Flat Gecko *Afroedura karroica* (Pl. 88)

(Endemic) 10-12 cm; max. 14 cm
A medium-sized flat gecko with a short, flattened body. The rostral is separated from the nostril, and there are no enlarged chin shields. The scales on the back are flattened and granular. There are three pairs of scansors beneath the toes. The tail is segmented. The back is light grey-brown to beige with numerous pale flecks and scattered, dark brown blotches that may rarely form irregular bands. The belly is off-white. **Biology and breeding:** Usually only a single individual or a pair is found under a suitable rock flake, and unlik other flat geckos they do not form large aggregations. Preferred flakes are those on the west side of large overhanging boulders of weathered sandstone

that catch the evening sun and are protected from seeping water. They emerge in the early evening to forage for ants and small beetles on rock faces. Clutches of two oval eggs (12-16 x 10-12 mm) are laid under a thin rock flake in a sunny position. These may hatch in 5-6 weeks, and hatchlings measure approximately 40 mm.

Habitat: Arid montane grassland. **Range:** E. Cape, from Aberdeen through isolated mountains of inland escarpment, to E. OFS and Lesotho. **Subspecies:** Two races are recognized. In *A.k. karroica*, the nasals are in contact behind the rostral, and there are 6-8 preanal pores in males; it occurs in inland mountains of E. Cape. *A.k. halli* does not have the nasals in contact behind the rostral, and there are only 4-6 preanal pores in males; it is found in W. Lesotho and adjacent regions.

Mountain Flat Gecko *Afroedura nivaria*

(Endemic) 9-10 cm; max. 11 cm
A medium-sized flat gecko that closely resembles the Amatola flat gecko, *A. amatolica* (page 190). It has a short, flattened body. The nasals are not in contact behind the rostral, which is separated from the nostril. The scales on the back are granular. There are three pairs of scansors below the toes. Males have 13-15 preanal pores in an angular series. The tail is segmented. The back is light brown, with darker mottlings that may form irregular, wavy, transverse bands. The belly is off-white. The tail has vague bars or blotches. **Biology and breeding:** These geckos live in rock cracks and under loose boulders lying on bedrock at altitudes well above the snow line (2 750 m). They eat beetles and grasshoppers. Two eggs (13 x 10 mm) are laid in January-March. Communal sites have been found with the weathered remains of 16 layers of old egg shells. **Habitat:** Montane grassland. **Range:** Drakensberg Mountains of Natal, OFS and Lesotho. **Subspecies:** None is recognized at present, although it is possible that the very similar Amatola flat gecko and Tembo flat gecko, *A. tembulica* (page 192), are isolated races of this species.

A.p. pondolia
A.p. langi
A.p. haackei
A.p. major
A.p. multiporis
A.p. marleyi

Pondo Flat Gecko *Afroedura pondolia* (Pl. 89)

(Endemic) 6-15 cm; max. 16 cm
This complex of small to large flat geckos is at present grouped in one species. They have elongate and flattened bodies. The nasals are usually not in contact behind the rostral (see Subspecies), which touches the nostril. The scales on the back range from small and granular (in *A.p. marleyi*) to large and bluntly keeled (in *A.p. haackei*). There are two pairs of scansors under the toes. Males may have 10-28 preanal pores (see Subspecies). The tail is smooth and not segmented. The dorsal coloration is varied (see Subspecies), but the belly is always off-white to translucent.

Biology and breeding: All the races are mainly nocturnal and rock-living, and usually occur singly or in pairs. There are, however, some important differences between the races. The typical race is common on the cliffs of the Transkei coast, and was once very common in houses along the Natal south coast, but is now being rapidly replaced by the more aggressive Moreau's tropical house gecko, *Hemidactylus mabouia* (page 194). The large race, *A.p. major*, is solitary and slow-moving. It clings upside down to the roof of a wide, horizontal crack in a shaded position on a granite rock face bordering a river. It is active during the day. *A p. multiporis* and *A.p. langi* inhabit rock outcrops, while the dwarf race, *A.p. marleyi*, lives among mountain summit rocks, and under the bark of trees. *A.p. haackei* also lives in rock outcrops and occasionally makes a rapid, clockwork-like 'churrrrr' sound. Two eggs (8 x 9 mm) are laid by the typical race. These hatch in 105-114 days; the hatchlings measure approximately 45 mm. The smaller eggs (5,6 x 8 mm) of *A.p. langi* are laid under stones and in rock cracks. **Habitat:** Varied; coastal bush and arid savannah. **Range:** S. Transkei, through Natal and Swaziland t Transvaal lowveld.

Subspecies: At least six races are recognized at present, some of which may be treated as full species (eg. *A.p. langi*). In addition, many recently discovered populations in E. and N. Transvaal are under investigation. *A.p. pondolia* is of medium size, has 13-16 preanal pores, and a blue-grey back with irregular, wavy, charcoal bands; it occurs in Transkei coastal regions to Durban, with scattered populations in Natal midlands. Lang's flat gecko, *A.p. langi*, is small, the nasals are in firm contact behind the rostral, there are 14-19 preanal pores, and the back is pale grey to dirty white, with s irregular, wavy, dark brown bands; it is found in E. Transvaal lowveld. Haacke's flat gecko, *A.p. haackei*, is large, has 28 preanal pores, and has mottled greyish markings on a lighter background; it is found in Nelspruit, an Khandizwe Hills in Kruger National Park (SA RDB, Restricted). *A.p. major* is very large, has 18-19 preanal pores, irregular blotches in dull shades of grey on the back, a metallic bluish sheen on the tail, and a thin, dark line from the nostrils through the eyes that forms a W-shape on the nape; it occurs in W. Swaziland. *A.p. multiporis* is medium-sized, has 17 preanal pores, and is greyish, with irregular coarse markings; it is found in Haenertsburg district of E. Transvaal (this area is now flooded by the Ebenezer Dam, and the race ma be extinct; SA RDB, Restricted). Marley's flat gecko, *A.p. marleyi*, is a dwarf form that has only 9-12 preanal pores, and a beige back with an extensive, fine, dark brown network with scattered, small, pale flecks; it is found in N. Zululand and Lebombo Mountains.

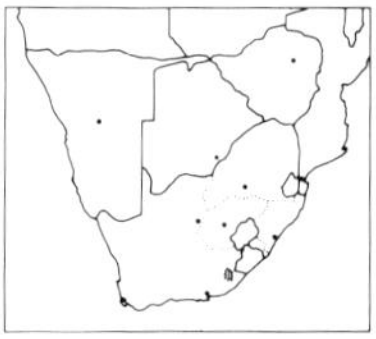

Tembo Flat Gecko *Afroedura tembulica*

(Endemic) 9-10 cm; max. 11 cm
This medium-sized species is very similar to the Amatola flat gecko, *A. amatolica* (page 190). The body is short and flattened, and covered with small granules. The nasals are not in contact behind the rostral, which enters the nostril. There are three pairs of scansors beneath the toes. Males have 6-9 preanal pores in an angular series. The tail is segmented. The back is greyish-brown with darker mottlings, and the belly is creamy white. **Biology and breeding:** This gecko is similar in habits and behaviour to the Amatola f gecko. **Habitat:** Rocky montane grassland. **Range:** Mountains around Queenstown in E. Cape.

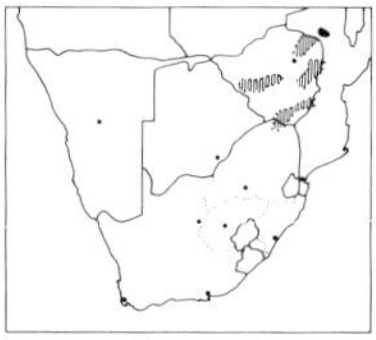

A.t. transvaalica
A.t. loveridgei

Transvaal Flat Gecko *Afroedura transvaalica* (Pl. 88)

10-12 cm; max. 13 cm
A medium-sized flat gecko with a slender, flattened body that is covered with small granules. The nasals are in contact behind the rostral. The chin shields are not enlarged. There are two pairs of scansors below the toes. Males have 5-11 preanal pores in an angular series. The tail is segmented. The back is pale grey-brown, with faint, irregular, dark crossbands, and the belly is white **Biology:** These geckos are often found in large aggregations beneath rock flakes in granite and sandstone outcrops. The diet includes bugs and grasshoppers. **Habitat:** Mesic savannah. **Range:** Isolated populations from N. Transvaal to Zambezi River. **Subspecies:** Two races are recognized. In *A.t. transvaalica*, the rostral borders the nostril; this race is found on Soutpansberg in N. Transvaal, through SE. Zimbabwe to Gorongoza Mounta and Mtoko district in Mozambique. *A.t. loveridgei* has the rostral excluded from the nostril; it is found on both banks of Zambezi River at Tete in Mozambique.

GIANT GROUND GECKO *Chondrodactylus*

This large, terrestrial, padless gecko is closely related to the thick-toed geckos, *Pachydactylus* (page 199). The genus consists of a single species, which is endemic to the subcontinent.

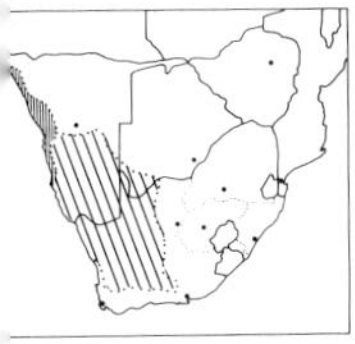

C.a. angulifer
C.a. namibensis

Giant Ground Gecko *Chondrodactylus angulifer* (Pl. 79)

(Endemic) 13-16 cm; max. 18 cm
This large gecko has a stout, cylindrical body, and a big head with a short snout and prominent, bulging eyes. The pupils are vertical. The back has scattered, enlarged, keeled tubercles. The feet are wide, with short toes that are not flared and lack scansors. Minute, fine claws are present in females, and males lack femoral and preanal pores. The tail is segmented (each segment bearing a transverse row of evenly spaced, enlarged tubercles), swollen and shorter than the body. Males grow larger than females and have a prominent hemipenial bulge. Coloration is varied. The back is pale orange to red-brown, with irregular, dark reticulations, overlaid with a variety of different patterns (see Subspecies). The belly is pink-white.

Biology and breeding: These terrestrial, nocturnal geckos spend the day in their retreat, usually a short burrow that they dig afresh each night, or an old scorpion burrow that they have widened. At night they emerge to hunt, and can often be found slowly stalking the sand flats. The diet is varied, and includes termites, moths, beetles and spiders, and also other small geckos. The female lays 1-2 large, almost round (18 x 16 mm), hard-shelled, fragile eggs in a chamber that it digs in sand. Gravid females are found in the wild from October to January, and hatchlings from December to April. In captivity, eggs can be laid at 14- to 40-day intervals, and it is probable that more than one clutch is laid per season in the wild. In captivity, the eggs hatch in 45-60 days, but probably take up to 90 days in the wild. Hatchlings are about 70 mm long. Predators include the horned adder, owls, bat-eared foxes and suricats. When alarmed, these geckos stand stiff-legged with the tail arched, scorpion-like, over their back. They hiss and lunge, and will bite if necessary. Although considered poisonous in some regions, they are quite harmless.

Habitat: Gravel plains, interdune spaces and sandy flats. **Range:** Through Namib Desert, S. Namibia and karroid regions of Cape. **Subspecies:** Two races are recognized. *C.a. angulifer* has stout limbs, ventral scales that increase in size on the sides, and a dorsal pattern consisting of 4-5 pale, dark-edged chevrons that extend as bars on to the tail and are more conspicuous in females (particularly on the shoulder region), but fainter in males (which usually have the pale bars reduced to a series of prominent, white dorsolateral spots); it occurs in S. Namibia into karroid areas of Cape. *C.a. namibensis* has slender limbs, ventral scales that decrease in size on the sides, and a pale dorsal pattern with enlarged, white-tipped tubercles that is the same in both sexes; it occurs in Namib Desert, from Lüderitz Bay to W. Kaokoveld.

ALAHARI GROUND ECKO
'olopus

This small, terrestrial, almost padless gecko is closely related to the thick-toed geckos, *Pachydactylus* (page 199), and is restricted to the Kalahari region.

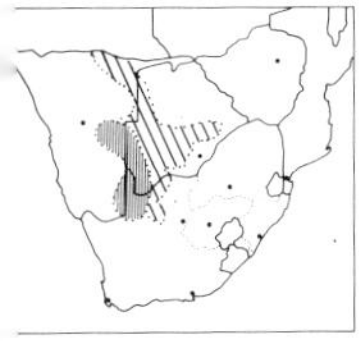

C.w. wahlbergii
C.w. furcifer

Kalahari Ground Gecko *Colopus wahlbergii* (Pl. 80)

(Endemic) 7-9 cm; max. 10 cm
A small, pretty gecko with an elongate, slender body and granular, overlapping scales on the back. The toes are small, with two undivided scansors, and minute claws on the hind feet of females. The prominent eyes have vertical pupils that each close to four pinholes. The nostrils are not tubular. Preanal and femoral pores are absent. The tail is stout and cylindrical, and almost equal in length to the body. Coloration is varied. The back is light orange to brown, usually with a complicated dorsal pattern (see Subspecies). The belly is chalky white. **Biology and breeding:** These geckos are found under cover or in short burrows. The diet is composed mainly of small insects (termites, grasshoppers, etc). They lay two hard-shelled eggs. Predators include snakes, owls and small carnivores. **Habitat:** Flat, sandy plains with

scattered vegetation. **Range:** Kalahari region of Botswana, extending into N. Cape and Namibia. It has not been recorded from Angola, although it reaches Caprivi Strip. **Subspecies:** Two races are recognized. *C.w. wahlbergii* has a rounded head and a dorsal pattern that consists of a vertebral series of large, pale, dark-edged spots that may fuse; it occurs in th Kalahari region, from N. Cape to Caprivi Strip. *C.w. furcifer* has a more pointe head, and a dark-edged, pale vertebral line on the back that divides on the neck; it is found in the dunes of W. and SW. Kalahari.

TROPICAL HOUSE GECKOS
Hemidactylus

These medium-sized geckos have large, flared toes, with large, retractile claws and prominent scansors that are usually arranged in pairs. The eyes a large, with vertical pupils and distinct upper eyelids. Scalation on the back is variable. Preanal and femoral pores are usually present.

These nocturnal geckos inhabit a wide range of habitats. The genus includes terrestrial, arboreal and rock-living species. All lay two hard-shelled eggs, but there are some all-female species that reproduce parthenogenetically. They have complicated behaviours; most vocalize. The vigorously defend their territories. Many species have adapted well to urban development, and have become commensal, living with man in his houses.

This genus is distributed widely throughout the Pacific region, S. Europe, Asia, central America and Africa. Some species have, and continue to, sprea through the tropical regions in the company of man and his goods. The genu contains over 70 species, only two of which reach southern Africa.

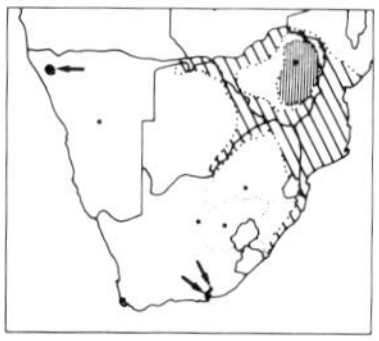

H.m. mabouia
H.m. tasmani

Moreau's Tropical House Gecko
Hemidactylus mabouia **(Pl. 89)**

12-16 cm; max. 17 cm
A medium-sized gecko with a flattened head and body, and large eyes with vertical pupils. The toe-tips are flared, with strong, retractile claws and 6-7 paired scansors. The back is covered with granular scales, with 10-18 irregular rows of weakly keeled tubercles. There are 22-50 preanal pores. The tail is stout and cylindrical but flattened below, with transverse rows of six enlarged tubercles. The back is pale grey to grey-brown, plain or with 4-5 faint, irregular, dark crossbands that fade if exposed to light for prolonged periods. The tail has about 10-12 dark bars. The belly is cream.

Biology and breeding: The typical race is mainly arboreal and is commor under tree bark and in the hollows of baobab trees. It frequently enters home *H.m. tasmani* prefers rock cracks and caves. They emerge in the evening to catch moths, beetles and cockroaches. Large specimens may eat small lizards. The males are very territorial, and fight vigorously; few have unbroken tails. They are normally silent, but can emit a quiet 'tik-tik-tik' (7-8 times). The eggs (8-10 x 10-13 mm) are adhesive and are often laid singly (but in pairs b *H.m. tasmani*), in nooks and crannies. Communal deposits of 50-60 eggs ma be found. Hatchlings measure 30-32 mm. **Habitat:** Varied; arid and mesic savannah, and coastal bush. **Range:** Natal coastal region (where its range is expanding southwards), through Transvaal lowveld and Mozambique to E. and W. Africa; also enters N. Namibia. Elsewhere, known from Madagasca and other islands in Indian Ocean, and east coast of S. and central America.

Subspecies: Two races are recognized, although *H.m. tasmani* may be a full species. *H.m. mabouia* is small (shorter than 6 cm snout-vent length) and mainly arboreal, the dorsal tubercles are small and weakly keeled, and the body is pale with faint crossbands; it occurs in coastal regions of Natal, through Transvaal lowveld, Mozambique, and border regions of Zimbabwe and E. Botswana. There is an isolated record in N. Namibia, and introduced individuals have been discovered in Port Elizabeth and East London. *H.m. tasmani* is large (longer than 6 cm snout-vent length) and mainly rock-living, the dorsal tubercles are large and strongly keeled, and the crossband on the body are conspicuous; it occurs in central and E. districts of Zimbabw

Flat-headed Tropical House Gecko
***Hemidactylus platycephalus* (Pl. 89)**

16-18 cm; max. 19 cm
A large house gecko with a stout, flat head and body. The eyes are large and have vertical pupils. The toe-tips are flared, with strong, retractile claws and 6-7 paired scansors. The back is covered with granular scales, with 8-12 irregular rows of small, conical tubercles. There are 45-57 preanal pores. The tail is stout and cylindrical but flattened below, with transverse rows of six enlarged tubercles. The back is pale grey to grey-brown, plain or with 4-5 faint, dark chevrons, and about 10 vague bars on the tail. The belly is uniform cream or pale yellow. **Biology and breeding:** These geckos are abundant on the trunks of baobab trees, but occasionally inhabit rock crevices and enter houses. The eggs (10-13 mm long) are cemented together in pairs in rock cracks or under bark. **Habitat:** Mopane and miombo woodland at low altitudes. **Range:** E. Africa, entering Mozambique and E. Zimbabwe.

VELVET GECKOS
Homopholis

These are medium to large, arboreal geckos with a soft skin that is covered with small, granular scales. The toe pads are dilated, with 8-12 undivided, chevron-shaped scansors, and small claws. The tail is relatively short and thick. Males have a pair of enlarged preanal scales with pores.

There are four species in the genus. *H. boivini* is restricted to Madagascar; *H. fasciata* occurs in E. Africa (with *H.f. erlangeri* restricted to Ethiopia and Somalia); and two species are endemic to the subcontinent.

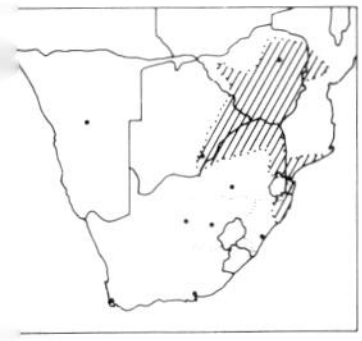

Wahlberg's Velvet Gecko *Homopholis wahlbergii* (Pl. 90)

(Endemic) 14-18 cm; max. 21,5 cm
A large velvet gecko with small dorsal scales that overlap and are in 80-87 scale rows at midbody. There are 26-30 scales between each eye and the anterior ear border. The preanal pore-bearing scales in males are enlarged and are usually separated by 2-3 small scales. The back is light to dark grey-brown, usually with irregular, pale and dark crossbars, and often with a series of pale vertebral blotches. Many specimens (particularly males) have a pair of broad, black dorsolateral stripes. The snout lacks a black band. The belly is cream, and is sometimes lightly spotted. **Biology and breeding:** These geckos are active during the day and night, but forage away from their retreat only at night. They shelter under bark, in holes in baobab trees, on hut roofs, and even in empty swallows' nests in caves and rock overhangs. Rock fissures, particularly on overgrown koppies and river banks, are favoured spots. The diet consists of large insects, particularly grasshoppers and cockroaches, but they also eat termites and millipedes. A pair of large (15-18 x 11-13 mm), hard-shelled eggs are laid in a rock crack or under bark. They will bite and squeal when first caught, and readily shed their tail. Despite this, they settle well in captivity and make fine pets.
Habitat: Varied; coastal bush, and mesic and arid savannah.
Range: Zimbabwe and S. Mozambique, into adjacent Zululand, N. Transvaal and E. Botswana.

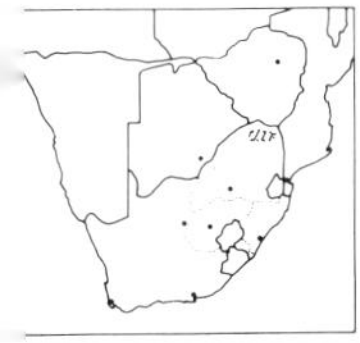

Muller's Velvet Gecko *Homopholis mulleri* (Pl. 90)

(Endemic) 9-11 cm; max. 12 cm
A small velvet gecko with small dorsal scales that do not overlap and are in 65-72 rows at midbody. There are 18-20 scales between each eye and the anterior ear border. The preanal pore-bearing scales in males are enlarged and usually in contact, but are sometimes separated by a small, wedge-shaped scale. The back is dark grey to light brown, with three large, silvery blotches on the forebody, followed by four white, black-margined chevrons. The flanks have 5-6 silvery blotches. The head is white, with minute

black spotting. The upper labials are white, bordered from eye to eye by a black band. The belly is white, with small dark spots and lines. **Biology and breeding:** This recently described species is poorly known; it is probably similar in habits and behaviour to Wahlberg's velvet gecko, *H. wahlbergii* (page 195). **Habitat:** Under bark and in holes in marula and knob-thorn trees in open, mixed mopane veld. **Range:** Restricted to N. Transvaal.

KAOKO WEB-FOOTED GECKO
Kaokogecko

This slender, medium-sized, terrestrial gecko is restricted to N. Namib Deser It is closely related to the web-footed geckos, *Palmatogecko* (page 209), but differs in having only its hind feet webbed.

The genus contains a single species.

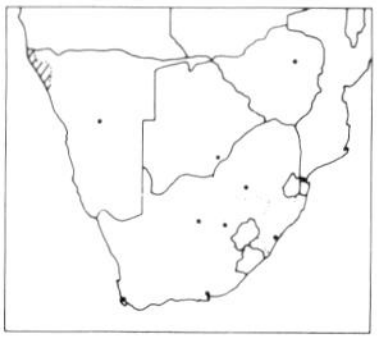

Kaoko Web-footed Gecko *Kaokogecko vanzyli* (Pl. 80)

8-10 cm; max. 11 cm

This beautiful and unusual gecko has a slender body with a soft, semi-transparent skin. The head is flattened, with a pointed snout. The eyes are large, with vertical pupils. The back is covered with minute granular scales. All the toes have two small scansors at the tip, and the toes of the hin feet are webbed and clawed in both sexes. Preanal and femoral pores are absent. The tail is cylindrical, unsegmented and shorter than the body. The back is light brown, with a series of 9-12 large, light purple to brown-pink blotches along the back and tail, and is scattered with dark brown spots that are concentrated around the borders of the dorsal blotches. The belly is whit **Biology:** This terrestrial, nocturnal species constructs a short burrow for its daytime shelter. The hole is excavated with the clawed forefeet, while the webbed hind feet push the loose sand to one side. During the day, the entrance to the burrow is closed with loose sand. In the early evening, they emerge to forage for food, often climbing nearby rocks. They eat termites an small beetles. **Habitat:** Gravel plains. **Range:** N. Namib Desert, just entering SW. Angola.

DWARF GECKOS
Lygodactylus

These geckos have a rudimentary inner toe, while the other toes have a large retractile claw and scansors that are paired and oblique. The head has a short, rounded snout, and the prominent eyes have round pupils and are encircled by distinct eyelids. The body is short and cylindrical, and is covere with small, granular scales. Males have preanal pores, but femoral pores are absent. The tail, which is equal in length to the body, is cylindrical and has modified adhesive scales at the tip which serve as a fifth foot.

These delightful dwarfs are very common in suitable habitat. They are diurnal and usually arboreal, although some species are rock-living. The adu male usually commands a shrub, low tree or boulder, sharing it with a numbe of females and subadults. They are well camouflaged; when threatened, the freeze or quickly run around the trunk, branch or rock, keeping it between themselves and danger. They eat ants and termites, and can be seen stationed next to their trails, picking off victims. Two hard-shelled eggs are la under bark or in a suitable cranny.

This ancient genus is distributed throughout subSaharan Africa and Madagascar, with two species also occurring in South America. There are about 50 species in the genus; nine species, three of which are endemic, occur in southern Africa.

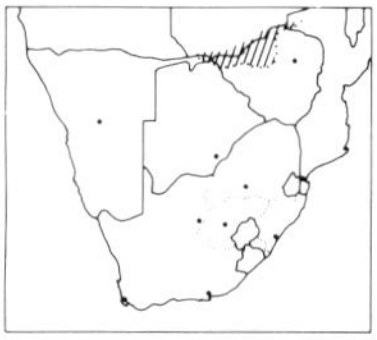

Angola Dwarf Gecko *Lygodactylus angolensis*

5-7 cm; max. 7,5 cm

A small dwarf gecko that is similar in appearance to the Cape dwarf gecko, *L. capensis* (page 197). It has a pair of lateral clefts in the mental. There are r soft spines above the eyes. Males have 7-10 preanal pores. The back is olive-brown, usually with a series of pale spots on the upper flanks. The belly and throat are cream. **Biology and breeding:** This species is similar in behaviour to the Cape dwarf gecko. **Habitat:** Mixed dry deciduous woodland and grassland. **Range:** Zambezi drainage basin, from Mozambique through W. Zimbabwe, N. Botswana, Caprivi Strip and NE. Namibia, to E. Angola.

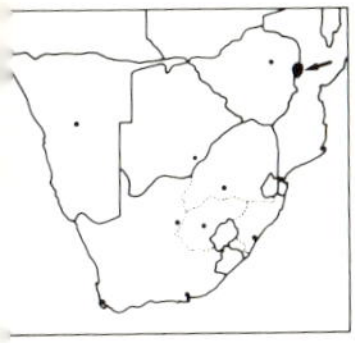

b. bernardi

Bernard's Dwarf Gecko *Lygodactylus bernardi* (Pl. 90)

6-7 cm; max. 8 cm
A dwarf gecko of robust build, with a pair of lateral clefts in the mental. There is a series of 4-7 soft spines above each eye. Males have 7-10 preanal pores. The back is olive-brown, with numerous pale spots. The throat is bluish-white, the belly is yellow, and the tail is orange to orange-brown below. **Biology and breeding:** They live in the cracks of sun-split boulders on hillsides. Two small eggs (7,5-8 x 6-6,5 mm) are laid in a rock crack. Large communal egg sites, which may contain the remains of hundreds of eggs, are common. **Habitat:** Montane grassland. **Range:** Inyanga district of E. Zimbabwe. **Subspecies:** Two races are recognized, but only the typical race occurs in the region. An isolated race, *L.b. bonsi*, occurs on Mulanje Mountain in Malawi.

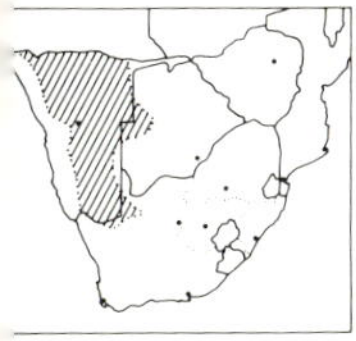

Bradfield's Dwarf Gecko *Lygodactylus bradfieldi* (Pl. 91)

5-6 cm; max. 6,1 cm
This small dwarf gecko is very similar to the Cape dwarf gecko, *L. capensis* (below), and is sometimes treated as a western race of that species. The subcaudal scales are irregular in size. There are 4-7 preanal pores in males. The back is grey-brown, with a pair of pale stripes bordered by narrow, black lines on the flanks. The throat often has dark spots. The belly is cream. **Biology and breeding:** This gecko is very similar in habits and behaviour to the Cape dwarf gecko. **Habitat:** Arid savannah. **Range:** Western part of the subcontinent, from N. Cape to S. Angola.

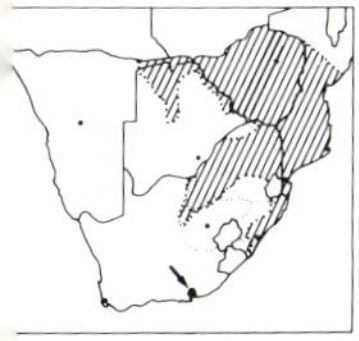

c. capensis

Cape Dwarf Gecko *Lygodactylus capensis* (Pl. 91)

6-7 cm; max. 8 cm
A small dwarf gecko with a pair of lateral clefts in the mental and no soft spines above the eyes. A postnasal is present. There are three subcaudal scales per tail segment. Males have 4-7 preanal pores that develop with sexual maturity. The back is grey-brown, with a dark streak from the snout to the shoulder that sometimes extends on to the flank, and a pale dorsolateral band that may break up into a series of light spots towards the tail. The throat is usually stippled with grey or dark brown. The belly is cream. **Biology and breeding:** These geckos prefer to forage in low scrub and on dead trees, and feed almost exclusively on termites. They reach sexual maturity in eight months, and usually live 15-18 months. Breeding is continuous, and communal egg sites are common. The hard-shelled eggs (5-6 x 6,5-7 mm) are usually laid in pairs (but sometimes singly) in rock cracks or under loose bark. They may take 2-5 months to hatch; hatchlings are approximately 25 mm long. **Habitat:** Well-wooded, dry savannah. **Range:** Eastern half of the subcontinent, from S. Transkei to E. Africa. A small colony thrives in the suburbs of Port Elizabeth, where they were introduced nearly 30 years ago. **Subspecies:** Three races are recognized, but only the typical form occurs on the subcontinent. It is replaced in Tanzania and N. Mozambique by *L.c. grotei*, and on Pemba Island by *L.c. pakenhami*.

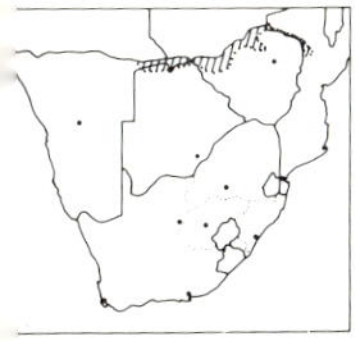

Chobe Dwarf Gecko *Lygodactylus chobiensis* (Pl. 91)

6-8 cm; max. 9 cm
A large dwarf gecko with a mental that lacks lateral clefts. The rostral is excluded from the nostril. There are no soft spines above the eyes. Males have

7-11 preanal pores. The back is blue-grey, with large, pale dorsal spots that extend on to the head and neck. The belly is yellow. The throats of males ma be all-black or pale yellow with two dark chevrons, which are faint or absent i females. **Biology and breeding:** These geckos prefer to forage high in trees and are common on baobab, acacia and mopane trees, and occasionally or houses. They feed on termites. They are relatively short-lived, growing to maturity, breeding and dying in 18 months. Two hard-shelled eggs (5 x 6,5 mm) are laid under bark, in disused termite mounds, etc. Females reach sexual maturity in 8-9 months, and breed continuously, laying clutches at seven- to eight-week intervals. Incubation lasts approximately 125 days; hatchlings measure 30-32 mm. Males reach sexual maturity in 9-10 months. **Habitat:** Mesic savannah. **Range:** Okavango Basin, extending along Zambe Valley to Tete, and on to Zimbabwe plateau and adjacent Zambia.

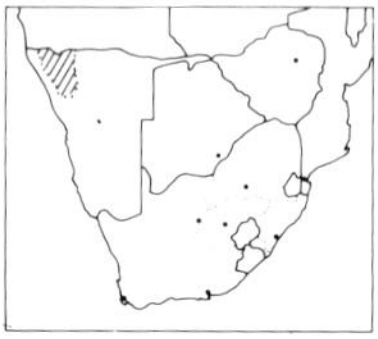

Lawrence's Dwarf Gecko *Lygodactylus lawrencei*

5-6 cm; max. 7 cm
A small dwarf gecko that resembles the spotted dwarf gecko, *L. ocellatus* (below). It has a large mental that lacks lateral clefts. The rostral is well separated from the nostril, and there are no soft spines above the eyes. Male have two well-developed preanal pores. There are no transversely enlarged subcaudals. The back is ashy-grey, with numerous thin, broken, darkish stripes, and often with a pale chestnut lateral stripe that extends from the nec on to the anterior half of the tail. The throat has two dark chevrons. The belly i white. **Habitat:** Rocky, arid savannah. **Range:** Kaokoveld in N. Namibia, extending into S. Angola.

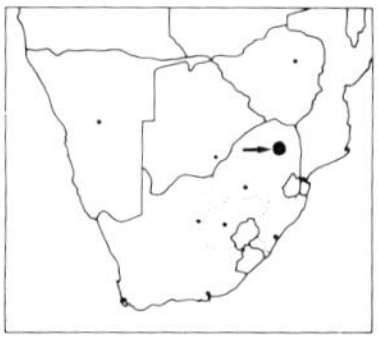

Methuen's Dwarf Gecko *Lygodactylus methueni* (Pl. 91)

(Endemic) 7-8 cm; max. 9,2 cm.
A large dwarf gecko with a large mental that lacks lateral clefts. There are eight enlarged chin shields, the rostral is in contact with the nostril, and there are no soft spines above the eyes. Males have 9-11 preanal pores. There are no transversely enlarged subcaudals. The back is olive to olive-grey, with ro of well-defined, pale-centred, reddish-brown spots that are occasionally absent or fused into an irregular band. The belly is yellow, and more intensel coloured towards the rear and under the tail. **Biology and breeding:** A threatened species (SA RDB, Vulnerable), much of its habitat is now under exotic pine plantation. They forage on the trunks of trees and adjacent rock outcrops. Two eggs are laid under stones or loose bark. **Habitat:** Remnant montane forest. **Range:** Endemic to Woodbush Forest in E. Transvaal.

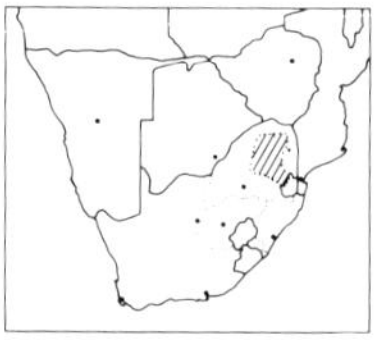

Spotted Dwarf Gecko *Lygodactylus ocellatus* (Pl. 91)

(Endemic) 5-7 cm; max. 7,5 cm
This medium-sized species is similar to Lawrence's dwarf gecko, *L. lawrenc* (above). The large mental lacks lateral clefts. There are 4-5 enlarged chin shields. The rostral is in contact with the nostril, and there are no soft spines above the eyes. Males have 7-10 preanal pores. There are no transversely enlarged subcaudals. The back is blue-olive to dark brown, with numerous reddish-brown to blackish, pale-centred spots that are arranged in rows or a irregularly scattered. The rump and tail are often tinged with chestnut-brown The chin is often yellowish, and it lacks dark chevrons. The belly is greenish-white to bluish-white. Males have dirty yellow limbs and tail, and ar pale brown below.

Biology and breeding: These geckos live among rocks and stones on exposed hillsides. Two oval, hard-shelled eggs (7,5 x 6,2 mm) are deposited under stones and in rock crevices. **Habitat:** Rocky hillsides. **Range:** Central and N. Transvaal and Swaziland.

Stevenson's Dwarf Gecko *Lygodactylus stevensoni* (Pl. 91)

(Endemic) 6-7 cm; max. 8 cm
A stout dwarf gecko with a pair of lateral clefts in the mental and no soft spines above the eyes. Males have 6-9 preanal pores. The back and flanks are blue-grey, with large, irregularly scattered, black spots. The throat bears dark chevrons. The belly is off-white.

Biology: Stevenson's dwarf geckos are mainly rock-living, but also shelter under the bark of dead trees. The bark snake is a major predator.
Habitat: Well-wooded granite hills. **Range:** SW. Zimbabwe and extreme N. Kruger National Park.

ESTIVE GECKO
arudasia

This very small gecko has long, slender, clawed toes that lack adhesive scansors. The genus contains a single species, which is endemic to central and S. Namibia.

Festive Gecko *Narudasia festiva* (Pl. 90)

(Endemic) 5-6 cm; max. 7 cm
A very small gecko with a flattened head and body. The eyes are large, with vertical pupils. The back is covered with small, granular scales. Preanal and femoral pores are absent. The tail is cylindrical and slightly longer than the body. The back is grey-purple to chestnut-brown, with scattered white and cream spots and a series of 5-6 narrow, zigzagging, black bands that are pale-edged behind. The tail is often more brightly coloured and is barred in black (but is uniform grey when regenerated). The belly is dirty white to yellow.
Biology and breeding: This very agile, diurnal gecko forages in the early morning and evening for ants and flies on the rock walls and boulders of mountain ravines and rocky slopes. It lays two hard-shelled eggs (7,6 x 6 mm) in rock cracks. **Habitat:** Arid mountain slopes. **Range:** Great Namaqualand to Damaraland.

HICK-TOED
ECKOS
achydactylus

This is a large group of characteristic southern African geckos. The toes have a series of undivided (rarely divided) scansors under the tips, and there are usually minute claws on the toes of the hind limbs. Many of the other geckos in the region (eg. the giant ground gecko, page 192; the Namib day geckos, page 213; the web-footed gecko, page 209; the Kaoko web-footed gecko, page 196; and the Kalahari ground gecko, page 193) are closely related to the thick-toed geckos. The body is covered with small, granular, non-overlapping scales, and many species also have scattered, enlarged, keeled tubercles on the back, tail and limbs. The prominent eyes have vertical pupils, and non-functional eyelids that are distinct and form a ridge around each eye. Femoral pores are absent, and preanal pores are present in only two species. They are usually camouflaged in buffs, browns and greys, with speckled, blotched or banded patterns. Juveniles are usually more brightly coloured, and in some species have very different colour patterns (eg. the velvety and Brandberg geckos, pages 200 and 205 respectively). Males have swollen hemipeneal pouches at the base of the tail that are often adorned on the sides with 2-6 enlarged spines.

These geckos occupy a wide variety of habitats, but are most common in the arid western parts of the region. All the species are nocturnal, although some of the bigger, rock-living species are often found at the entrance to their rock cracks during the day, and will eat any food that comes within range. At night they forage away from the safety of their retreat. All lay two hard-shelled eggs, and it is probable that they lay several clutches during a season. A few species have been reported to call in the wild, but most are silent, although they may give a distress call when handled.

There are at least 27 species in the genus, and possibly more. All but two species are found in southern Africa, and 17 of these are endemic.

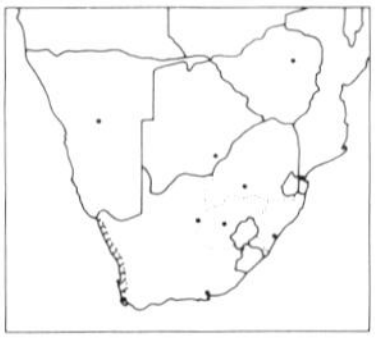

Austen's Gecko *Pachydactylus austeni* (Pl. 82)

(Endemic) 7-9 cm; max. 10 cm
A small gecko with a cylindrical body and a short snout. The rostral does not enter the nostril, which is pierced between two nasals. The scales on the bac are of uniform size, and lack enlarged tubercles. The middle row of scales below the toes and above the scansors is not enlarged. There are 3-4 scansors beneath the middle toes. The tail is cylindrical, unsegmented and slightly shorter than the body in adults. Coloration is very varied. The back m be pale grey, red-brown or dark brown, and patterned with either scattered white spots that may form irregular crossbands, or large, diffuse, dark blotches. A dark stripe may extend from the snout through the eye, and the upper eyelid may be bright yellow. The belly is uniform white. **Biology and breeding:** The habits of these common geckos are similar to those of the web-footed gecko, *Palmatogecko rangei* (page 209). They are terrestrial an nocturnal, sheltering during the day in a small burrow that they dig in loose sand. In spring, they lay two hard-shelled eggs in a hole dug in sand; these hatch in December. Schneider's adder is a major predator. **Habitat:** Sparse vegetated coastal dunes. **Range:** Coastal areas of W. Cape to Little Namaqualand.

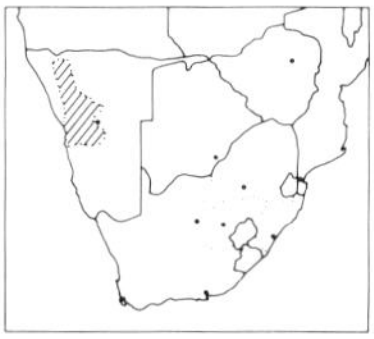

Velvety Gecko *Pachydactylus bicolor* (Pl. 85)

(Endemic) 9-11 cm; max. 12 cm
A beautiful, small gecko with a flattened head and body. The rostral, which is much broader than it is deep, does not enter the nostril. The nasorostrals are contact. The scales on the back are uniform, granular and abutting, and lack enlarged tubercles. The middle row of scales below the toes and above the scansors is enlarged. There are five scansors beneath the middle toe. The ta is hardly segmented and lacks bands of enlarged scales. Coloration is varie Juveniles have a jet-black body, bordered with white bands on the hips and neck, with those on the neck extending on to the lips; the head and tail are dusky brown. Adults have a light buff body, with irregular brown blotches an bars, and adjacent yellow-brown blotches; the tail is yellow-brown, with faint dark spots, and the belly is yellow-white. Subadults turn dusky chocolate, w scattered, small, white spots adjacent to darker blotches. **Biology and breeding:** These geckos are nocturnal and rock-living, favouring thin cracks in small, shattered rock outcrops. Two eggs (9,2 x 6 mm) are laid in a rock crack. **Habitat:** Rocky semi-desert. **Range:** NW. Damaraland and Kaokovel in Namibia.

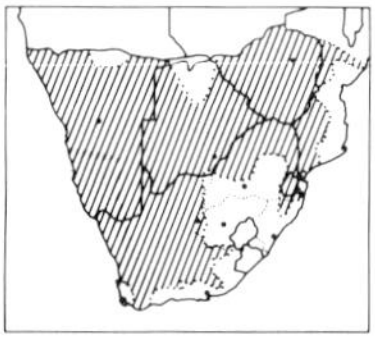

Bibron's Gecko *Pachydactylus bibronii* (Pl. 84)

15-19 cm; max. 21 cm
A large, stout gecko with very large, strongly keeled tubercles that are separated by granular scales on the back. The rostral is separated from the nostril, which is directed upwards and outwards. The middle row of scales below the toes and above the scansors is not or is only feebly enlarged. The are 10-12 scansors beneath the middle toes. The tail, which is about the san length as the body, has a fat base and is segmented, with regular, transvers rows of strongly keeled (almost spiny) scales. The back is dark buff to grey-brown or purplish-black, with 4-5 indistinct, dark, wavy crossbands an scattered, white tubercles. The belly is white, and the tail has 8-10 dark banc that are prominent in juveniles but fade in adults. **Biology and breeding:** Or of our most common geckos, they live on rock outcrops, under loose tree ba and on houses. They are gregarious and often live in colonies. As befits thei appearance, they are pugnacious and ever-willing to bite. Despite this, they make excellent, long-lived pets. A wide variety of prey is eaten, including grasshoppers, ants, termites, beetles and smaller lizards. Two eggs (16 x

14 mm) are laid in a rock crack, under bark, etc. **Habitat:** Varied; karroid veld, semi-desert, arid and mesic savannah. **Range:** Widely distributed; absent from parts of E. Cape and W. Cape Peninsula, Transkei, most of Natal, and highveld of OFS and Transvaal. Elsewhere, to Angola and Tanzania.

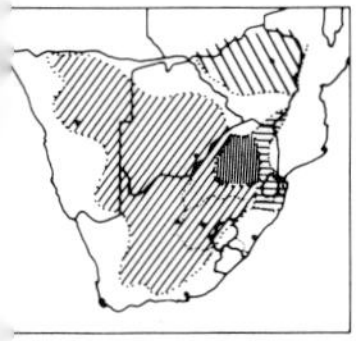

P.c. capensis
P.c. vansoni
P.c. oshaughnessyi
P.c. affinis

Cape Gecko *Pachydactylus capensis* (Pls. 82 and 83)

8-12 cm; max. 14 cm
A small to medium-sized gecko with a robust body and 16-22 rows of enlarged, weakly keeled tubercles that are separated by granular scales on the back. The nostril is well separated from the rostral. The middle row of scales below the toes and above the scansors is not enlarged. There are five scansors beneath each middle toe. The tail is slightly flattened, with transverse rows of 6-8 enlarged, keeled tubercles, and is slightly longer than the body; regenerated tails lack tubercles, are carrot-shaped and are rarely longer than the body. The back is typically light brown to grey-brown above, flecked with small black and white spots. The belly is white. Northern races often have conspicuous white bands. **Biology and breeding:** Terrestrial, they are commonly found under calcrete blocks (in the Kalahari), rotting logs and in disused termitaria, but occasionally inhabit low rock cracks and houses. At night, they emerge to forage for small insects. Clutches of two eggs (11-14 x 9-10 mm in *P.c. capensis*; 9,5 x 7 mm in *P.c. affinis*; and 7 x 5,5 mm in *P.c. vansoni*) are laid in old termitaria or under stones in September-December. Incubation takes 90-110 days, and hatchlings (35-50 mm) are found in November-January. They have many predators, including snakes and small carnivores. **Habitat:** Varied; karroid veld, grassland, and mesic savannah. **Range:** Cape Province, through most of the subcontinent, to S. Angola, Malawi and Zambia. An isolated race (*P.c. katanganus*) occurs in Shaba Province in Zaire. **Subspecies:** There are five races, four of which occur in the region. The Cape gecko (*P.c. capensis*) has a tail with transverse rows of strongly keeled tubercles, and irregular black blotches and pale spots on the back; it is found in Karoo and most of the central parts of the subcontinent. Van Son's gecko (*P.c. vansoni*) has a tail with transverse rows of weakly keeled tubercles, a series of 5-8 white crossbands (which are occasionally absent) on the back, and 1-2 semicircular white lines behind the head; it is found in N. Natal, NE. and E. Transvaal and SE. Zimbabwe.
The Transvaal gecko (*P.c. affinis*) has a slightly flattened head and body, and lacks tubercles on top of the head. Juvenile Transvaal geckos are purple-brown above, with 5-6 narrow, dark-edged, yellow-white crossbands; adults are grey-brown to reddish-brown above, with scattered, dark brown to blackish spots, and usually also with scattered, smallish, white spots. It occurs in central and N. Transvaal. O'Shaughnessy's gecko (*P.c. oshaughnessyi*) has a tail without keeled tubercles, and 3-5 broad yellow crossbands on the back; it occurs in the northern half of Zimbabwe, to Malawi and Zambia.

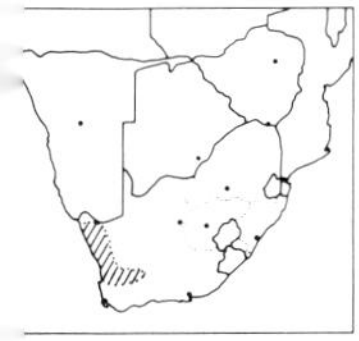

Western Cape Gecko *Pachydactylus labialis* (Pl. 82)

(Endemic) 6-8 cm; max. 9 cm
This gecko has 16-18 rows of enlarged, keeled tubercles, separated by granular scales on the back. The rostral does not enter the nostril, and the nasorostrals are separated by a granule. The middle row of scales below the toes and above the scansors is not enlarged. There are five scansors beneath the middle toes. The tail, which is slightly longer than the body, is cylindrical and segmented, with transverse rows of six enlarged, keeled tubercles. Adults are orange-brown to dark greyish-brown, with large, diffuse paler blotches. The tubercles are often dark-tipped. The belly is creamy white, and the tail is usually barred with dark brown. Juveniles are often dark brown, with paler tubercles. **Biology:** They live under stones on sandy soil. **Habitat:** Succulent karroid veld. **Range:** W. Cape, from Little Namaqualand to Calvinia.

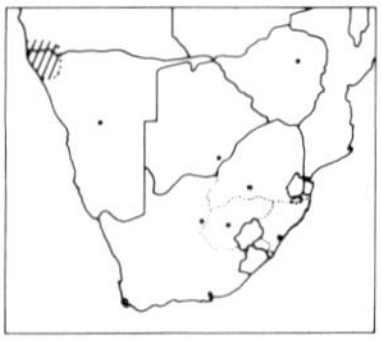

Angolan Banded Gecko *Pachydactylus caraculicus*

6-8 cm; max. 9 cm
A small gecko with a flattened body and short, stout limbs. Each nostril is pierced between three nasals, and the nasorostrals are in contact. The scale on the back are uniform, with a broad band of small, feebly keeled scales down the back, and smooth and larger scales on the sides. The toes are feebly dilated and have three scansors. The tail is cylindrical, distinctly segmented and tapers to a fine point; it is slightly longer than the body. Juveniles are reddish-brown above, with five ivory-white, dark-edged bands and a pale yellow tail with narrow brown crossbars; the belly is plain white. In adults, the white bands become less distinct and may become dusted with fine dark speckles; the tail develops white bars above and becomes dark below. In southern populations only the neck band remains distinct, the banc on the body breaking up into white spots. **Biology and breeding:** These geckos live in rocky areas, sheltering in rock cracks and under boulders. Hatchlings (28-30 mm) emerge in May. **Habitat:** Bushy Karoo-Namib shrubland. **Range:** SW. Angola, extending just into N. Kaokoveld in Namibia

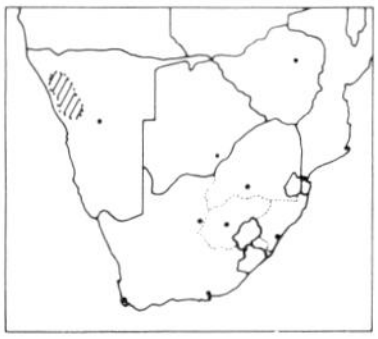

Banded Gecko *Pachydactylus fasciatus* (Pl. 86)

(Endemic) 9-11 cm; max. 12 cm
A small, slender, flattened gecko with 18 regular rows of enlarged, rounded and keeled tubercles that are separated by granular scales on the back. The nasorostrals are in contact. The middle row of scales below the toes and above the scansors is enlarged. There are 6-9 scansors beneath the middle toes. The tail, which is longer than the body, is thin, segmented and has transverse rows of six enlarged, keeled tubercles. The back is light brown to yellowish-brown, with three regular, dark brown crossbands on the body anc brown stripe that encircles the head. The belly is dirty white, and the tail has 8-10 dark bands. Juveniles are more distinctly marked. **Biology and breeding:** This species is poorly known. It is probably very similar to Weber's gecko, *P. weberi* (page 208). **Habitat:** Semi-desert. **Range:** Great Namaqualand to Damaraland.

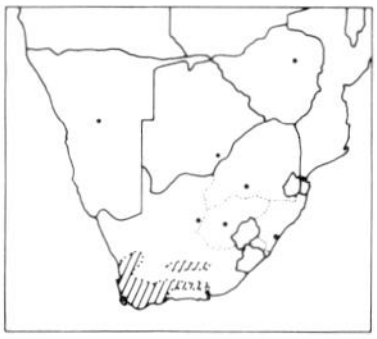

Ocellated Gecko *Pachydactylus geitje* (Pl. 81)

(Endemic) 6-7,5 cm; max. 8 cm
A small gecko with a short, cylindrical body, and uniform, granular scales on the back. The rostral does not enter the nostril, and the nasorostrals are separated. The middle row of scales below the toes and above the scansors enlarged. There are 4-5 scansors beneath the middle toes. The tail is cylindrical, unsegmented and covered with smooth, slightly overlapping scales; regenerated tails are fat and covered with irregular scales. Coloration is varied. The back may be greyish-brown to dark brown, with small, scattered, dark-edged, white or yellow spots that are sometimes arranged in rows. A dark brown stripe runs from the snout through the eye. The belly is white, and sometimes speckled with brown, particularly on the sides. Inland populations often lack the typical ocelli, having instead diffuse, pale blotche **Biology and breeding:** These secretive and gentle geckos hide among debris and under stones, and feed at night on small insects. They appear to require moist conditions. When at rest, they curl the fat tail around them like a contented cat; when disturbed, they stand stiff-legged with the head and tail held high. Two eggs (8-9,5 x 7-8 mm) are laid among debris, and hatch in 60 days, but may take much longer (up to 122 days) if incubated at low temperatures. Hatchlings measure 30 mm. **Habitat:** Varied; coastal strandveld, fynbos and rocky grassland associated with inland escarpment. **Range:** SW. Cape, extending along Cape fold mountains to Port Elizabeth, and inland along escarpment mountains to Cradock.

Koch's Gecko *Pachydactylus kochii* (Pl. 80)

(Endemic) 8-10 cm; max. 11 cm
A long, slender gecko with long limbs and uniform, granular scales on the back. The nasorostrals are in contact. There are three scansors beneath the middle toes. The tail is unsegmented and slender, and tapers to a fine point. In colour, they are greyish-white above, with a slight lavender tinge, and with five bright reddish-brown crossbands. The belly is chalky white, and the tail has 12 red-brown bars. **Biology and breeding:** Few details are known about this species. It is probably very similar to its close relative, the Marico gecko, *P. mariquensis* (page 204). The first specimens were collected at night under quartz lumps, or while they were moving slowly over open sand.
Habitat: Open sandy plains with scattered boulders. **Range:** Vicinity of Cape Cross in Namibia.

Button-scaled Gecko *Pachydactylus laevigatus* (Pl. 84)

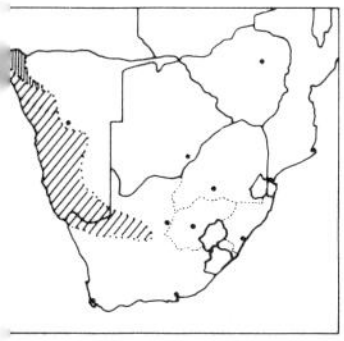

P.l. laevigatus
P.l. fitzsimonsi

13-16 cm; max. 17 cm
A large, stout gecko with very large, smooth or faintly keeled tubercles that are separated by granular scales on the back. The rostral is separated from the nostril, which is directed almost vertically upwards. The middle row of scales below the toes and above the scansors is not or is only feebly enlarged. There are 10-12 scansors beneath the middle toes. The tail, which is shorter than the body, has a fat base, is segmented, and has regular, transverse rows of strongly keeled (almost spiny) scales. The back is pale creamy olive to olive grey-brown, with 3-4 dark, wavy crossbands, usually edged behind with bright white spots. The belly is white. **Biology and breeding:** This species is very similar in habits and behaviour to Bibron's gecko, *P. bibronii* (page 200), but is not as common or as social, and is restricted to rock outcrops (although occasionally found on houses). The female lays 2-3 clutches, each of two large eggs (18-20 x 14-16 mm), in a small hole dug in the sand (possibly also in rock cracks) in August-December. Incubation takes 60-80 days, and the hatchlings (60-65 mm) appear in December-March. **Habitat:** Semi-desert and arid savannah. **Range:** N. Cape to S. Angola. **Subspecies:** Two races are recognized, and both occur in the region. In *P.l. laevigatus*, the dorsal tubercles are round and evenly spaced; it is found in Little Namaqualand to Kaokoveld. *P.l. fitzsimonsi* has irregular, polygonal and abutting dorsal tubercles; it occurs in Damaraland in Namibia, into Angola.

Spotted Gecko *Pachydactylus maculatus* (Pl. 81)

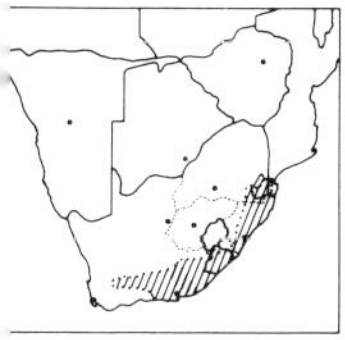

(Endemic) 8-10 cm; max. 11 cm
A small, gentle gecko with a fat body and a rounded snout. The enlarged tubercles on the back are small and conical, and only slightly larger than the few surrounding granular scales. The nasorostrals are usually separated by 2-4 granules. There are 3-4 scansors beneath the middle toes. The tail is cylindrical, unsegmented and grows fat (particularly when regenerated); original tails are slightly longer than the body, while regenerated tails are shorter. The back is grey to greyish-brown, with four rows of elongate, blackish spots that are not white-edged and occasionally fuse into irregular crossbands. The belly is off-white, and usually extensively flecked with brown.
Biology and breeding: They are terrestrial, occasionally entering rocky areas, but normally found under debris, in rotting logs, old termitaria, or under loose bark on dead trees. A favoured retreat in Cape coastal regions is an empty giant land snail's shell. Up to 10 geckos may pack into a single shell during winter. They feed at night on small insects and spiders. Predators are numerous, and include the large Natal hunting spider, which often takes only the shed tail. These geckos live for 3-4 years, and most adults lose their tail at least once, but sometimes up to three times, which emphasizes its usefulness

in defence. Sexual maturity is reached within a year. Females are gravid from September to February, and lay 2-3 clutches, each of two hard-shelled eggs (8 x 10 mm), in dry sand at the base of grass clumps in well-drained positions or in crannies in debris or dead logs. Communal egg sites containing up to 1: eggs have been found in old land snail shells. Incubation takes 100-120 days in the wild (80-90 days in captivity); hatchlings are 30-35 mm long. **Habitat:** Varied; fynbos and coastal bush to arid karroid veld. **Range:** Inland escarpment of Cape from Sutherland, and coastal region from Knysna, through E. Cape, Transkei and Natal to Swaziland and extreme SE. Transvaal

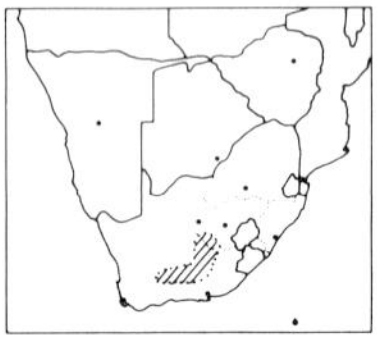

Golden Spotted Gecko *Pachydactylus oculatus* (Pl. 81)

(Endemic) 8-9 cm; max. 10 cm

A small gecko with a fat body and a slightly pointed snout. The enlarged tubercles on the back are of varied sizes, conical, and are larger than the many surrounding granular scales. The nasorostrals are usually separated b 1-2 granules. There are 3-4 scansors beneath the middle toes. The tail is cylindrical, unsegmented and grows fat (particularly when regenerated); original tails are slightly longer than the body, while regenerated tails are shorter. The back is pale brown to beige, with four rows of rounded, brown to reddish-brown spots that are distinctly white-edged and largest along the backbone. The belly is white. Juveniles are more intensely coloured, and the dark dorsal spots are black-edged, not white-edged. **Biology and breeding:** This species is very similar to the spotted gecko, *P. maculatus* (page 203), but is found only on rock outcrops, favouring horizontal cracks in small boulders or flat rocks lying on exposed bedrock. Predators include snakes and larger lizards. Two hard-shelled eggs (10-11 x 8-9 mm) are laid i a rock crack. **Habitat:** Karroid veld, occasionally entering montane grasslan and fynbos. **Range:** Cape inland escarpment to S. OFS.

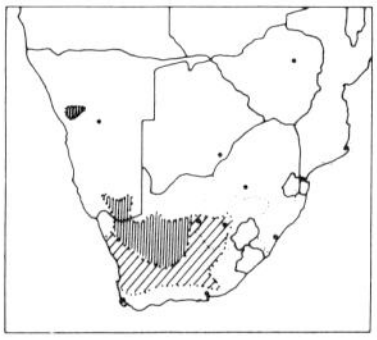

P.m. mariquensis
P.m. latirostris

Marico Gecko *Pachydactylus mariquensis* (Pl. 80)

(Endemic) 8-10 cm; max. 11 cm

A small, slender, thin-legged gecko with a short snout. It has uniform, granula scales on the back, and lacks enlarged tubercles. The rostral does not enter the nostril, which is pierced between three nasals. The middle row of scales below the toes and above the scansors is not enlarged. There are 3-4 scansors beneath the middle toes. The tail is cylindrical, unsegmented and thin, and usually slightly shorter than the body. The back is grey to pinkish-buff, with 5-6 wavy, reddish-brown, dark-edged crossbands. These may break up into irregular, vague, pale and dark blotches, particularly on the flanks. The belly is creamy white, and sometimes pink-tinged. The original ta has 5-7 dark-edged, brown crossbands (regenerated tails are grey-brown, with scattered black flecks). **Biology and breeding:** They are terrestrial, livir on sandy, flat plains, and sheltering during the day under stones or in old scorpion holes. They emerge at night to forage for small insects. Most activit occurs on warm spring nights, when the males look for mates. At this time th vocalize, giving a short, slow pulse of clicks ('wek . . . wek . . . wek . . '). The female lays 1-2 clutches, each of two hard-shelled eggs (10 x 7,5 mm), in sandy soil under a rock slab in October-March. **Habitat:** Flat sandy plains w sparse vegetation. **Range:** Western arid region of South Africa, with scattere populations in S. and central Namibia. **Subspecies:** There are two races, bo of which occur in the region. In *P.m. mariquensis*, the nasorostrals are in contact, and the scales on the back are granular and abutting; it occurs in E. Cape (Addo), north to S. OFS and west to Ceres and Little Namaqualand. *P.m. latirostris*, the nasorostrals are separated by granular scales, and the scales on the back are flattened and slightly overlapping; it occurs in centra and N. Karoo to Great Namaqualand, with an isolated population in Namibia around Spitzkoppe and south-west of Brandberg.

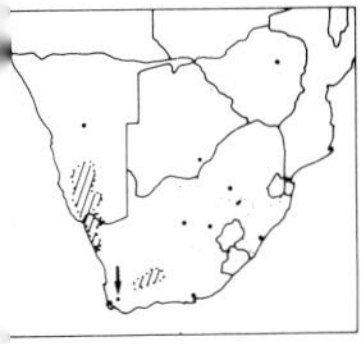

Namaqua Gecko *Pachydactylus namaquensis* (Pl. 84)

(Endemic) 12-15 cm; max. 16 cm
A large, stout gecko with rough but very delicate skin. The enlarged tubercles on the back are separated by granular scales. The rostral enters the nostril. The middle row of scales beneath the toes is slightly enlarged. There are 10-11 scansors beneath the middle toes. The tail, which is slightly shorter than the body, has a fat base, and is distinctly segmented, with transverse rows of 6-8 pointed tubercles. The back may be pale pinkish, light greyish-pink, chocolate-brown or light olive-grey, usually with indistinct, paler and darker markings. The belly is creamy white to yellowish-white. **Biology and breeding:** These geckos live in rock cracks, usually on large outcrops or summit rock bands, and rarely on isolated boulders. They may bask at the crack entrance during the day, but are active mainly at night. They eat grasshoppers, beetles, and moths and their caterpillars. They have a startling defence: when gripped, they twist violently, tearing off large areas of the skin, and often slipping away. The flayed gecko is a frightening sight but the skin is quickly regenerated. Two large, round, hard-shelled eggs (16 mm dia.) are laid in a rock crack. Communal egg sites may contain 40 eggs. **Habitat:** Karroid succulent veld. **Range:** Great and Little Namaqualand, with a population in Karoo around Beaufort West and an isolated record from Langeberg.

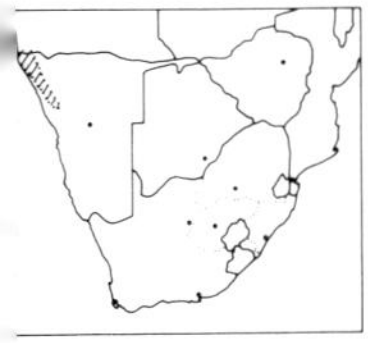

Kaokoveld Rock Gecko *Pachydactylus oreophilus* (Pl. 85)

9-11 cm; max. 12 cm
A medium-sized, flattened gecko with a large head and 16-18 rows of feebly keeled, enlarged tubercles that are separated by granular scales on the back. There are 11-12 upper labials. The lower forelimbs lack keeled tubercles. There are 6-7 scansors beneath the middle toes. The tail is longer than the body and is segmented, with transverse rows of strongly keeled tubercles above. The head is pinkish-grey, with dark brown lips and two dark, curved lines running from the eye around the back of the neck. The back is grey-brown, with irregular, dark blotches that are concentrated on either side of the backbone. The belly is mainly white. **Biology:** They emerge from rock cracks and under exfoliating flakes at night, to forage on vertical rock faces. **Habitat:** Semi-desert. **Range:** From Brandberg in Namibia to S. Angola.

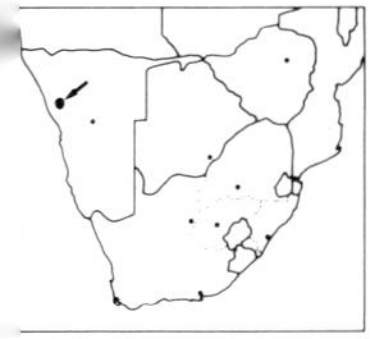

Brandberg Gecko *Pachydactylus gaiasensis*

(Endemic) 9-11 cm; max. 12,5 cm
A medium-sized, flattened gecko with a large head and 14 rows of strongly keeled, enlarged tubercles that are separated by granular scales on the back. The lower forelimbs have bluntly keeled tubercles, and females have minute claws on the toes. There are 6-8 scansors beneath the middle toes. The tail is shorter than the body and is segmented, with transverse rows of strongly keeled tubercles above, and enlarged, smooth scales below. Adults are light grey above, with mauvish-brown, irregular blotches that may form vague crossbars, and a thin, grey vertebral stripe. The belly is greyish-white. Juveniles have a brown head and jet-black body and forelimbs, bordered at the front and back by well-defined, white bands. The tail and hind limbs are greyish-brown, and the belly is brownish-white. **Biology:** These terrestrial geckos emerge at night from underneath sandstone boulders to forage slowly on the sand. **Habitat:** Arid, sandy veld. **Range:** Vicinity of Brandberg in Namibia.

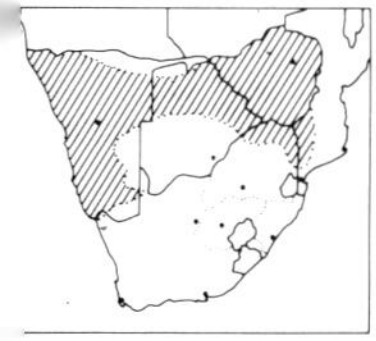

Speckled Gecko *Pachydactylus punctatus* (Pl. 81)

6-8 cm; max. 9 cm
A beautiful, gentle, small gecko. It has small, granular, flattened and overlapping scales on the back, and lacks enlarged tubercles. The rostral

does not enter the nostril. The middle row of scales below the toes and above the scansors is enlarged. There are 3-4 scansors beneath the middle toes. The tail is cylindrical, unsegmented and covered with large, overlapping scales. Coloration is very varied. The back may be pale grey to purple-brown, with numerous dark spots that are sometimes pale-centred and may fuse to form an irregular reticulation, with or without scattered, light spots. There is usually a dark streak from the nostril through the eye, which is edged with white-yellow. The belly is creamy white, and the tail is sometimes barred. **Biology and breeding:** These terrestrial geckos are very common in suitable habitat. They live under rotting logs, stones, or other cover, and emerge at night to hunt for small invertebrates. Two eggs (8-9 x 6-7 mm) are laid in midsummer on sand under logs, stones or among debris. Hatchlings measure 40 mm. **Habitat:** Varied; arid desert, but preferring dry savannah. **Range:** Through much of the northern parts of the region, from E. Transvaal to N. Botswana and Zimbabwe, throughout Namibia, to Angola and Katanga in the north and Richtersveld in the south.

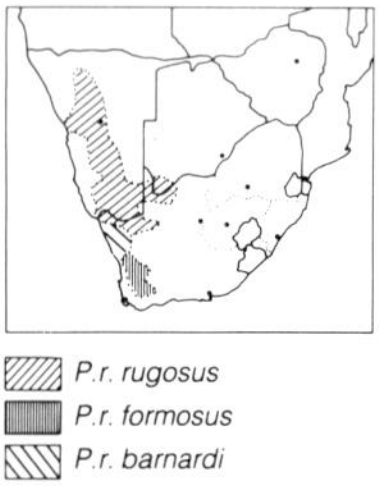

Rough-scaled Gecko *Pachydactylus rugosus* (Pl. 83)

(Endemic) 9-10 cm; max. 11 cm
An unusual, small, round-bodied gecko, usually with irregular rows of very enlarged tubercles that are separated by granular scales on the back. The middle row of scales below the toes and above the scansors is not enlarged. There are five scansors beneath the middle toes. The tail is shorter than the body and is segmented, with a transverse series of 8-10 enlarged tubercles and spiny subcaudals (southern races lack spiny subcaudals). Coloration is varied (see Subspecies). Juveniles of all races are more brightly coloured. **Biology:** The typical race is usually found under loose bark on dead trees in dry riverbeds. The other races often occur on rock outcrops, sometimes in horizontal rock cracks, but often in dead twigs and debris accumulated around stones on granite bedrock. When threatened, they arch the tail over the back. **Habitat:** Semi-desert and succulent karroid veld. **Range:** W. Cape to Kaokoveld in Namibia. **Subspecies:** Three races are recognized, and all occur in the region. The back of *P.r. rugosus* is covered with very large tubercles, and the scales on the belly are rough. It has an olive green-brown back, with four wavy, cream to yellow-brown crossbands, and sometimes with a white stripe on the side of the head. The tail has narrow white bars and scattered white tubercles. The belly is greyish-white, with small brown flecks. It is found in the inland regions of Namibia, through N. Cape to Calvinia. *P.r. formosus* also has large tubercles on the back, but the belly scales are smooth. In colour, it is brown to reddish-brown above, with five wavy, pale yellowish, dark-edged crossbands, the first completely encircling the head. The tail is broadly banded, and the belly is off-white and finely speckled with brown. It is found in W. Cape, from Ceres to Little Namaqualand. *P.r. barnardi* is similar to *P.r. formosus*, but the dorsal crossbands are disrupted into paired blotches. It is intermediate in colour and pattern between the other two races, as befits its geography; it occurs in Little Namaqualand.

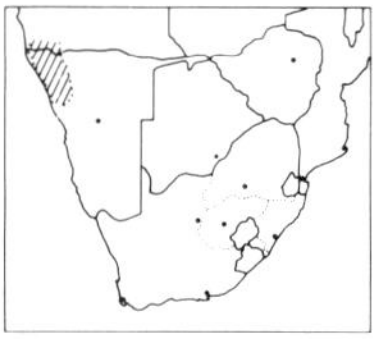

Large-scaled Gecko *Pachydactylus scutatus* (Pl. 85)

6-8 cm; max. 9 cm
A small, round-bodied gecko, with large and strongly keeled scales on the back, except for a strip of smaller, smoother scales along the backbone. The rostral enters the nostril. The middle row of scales below the toes and above the scansors is enlarged. There are 5-7 scansors beneath the middle toes. The tail is cylindrical and segmented, with transverse rows of 4-8 keeled tubercles, and is slightly longer than the body. Adults are light greyish-brown above, with a few, scattered, dark blotches. A dark, pale-centred stripe runs from each nostril through the eye and around the back of the head. The belly

dirty white. Juveniles are paler, have an olive-yellow tail and lack the dark spots on the back. **Biology and breeding:** These geckos are nocturnal and terrestrial, living under rocks and occasionally under exfoliating flakes on low boulders. The white, oval eggs (7-8 x 5-6 mm) are laid in pairs under rock flakes. The species was once reported to lay four eggs, but this is improbable. **Habitat:** Arid veld. **Range:** Restricted to N. Namibia, extending into S. Angola.

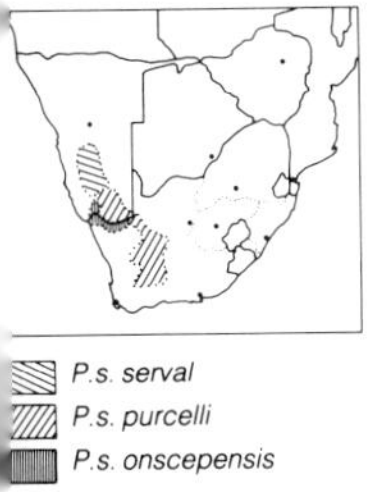

Western Spotted Gecko *Pachydactylus serval* (Pl. 86)

(Endemic) 7-9 cm; max. 10 cm
A small gecko with a flattened head and body. The scalation on the back is variable (see Subspecies). The rostral, which is only slightly broader than it is deep, enters the nostril. The nasorostrals are in contact. The middle row of scales below the toes and above the scansors is enlarged. There are 5-6 scansors beneath the middle toes. There are no large tubercles on the hind limbs. The tail is cylindrical and segmented, with transverse rows of 6-8 spinose tubercles, and is slightly longer than the body. Coloration is varied. The back is pale olive to greyish-brown, with dark brown blotches (small and scattered in the southern race, larger and fusing in the typical race). The tubercles on the tail are golden, and the belly is white. Populations along the lower Orange River are light orange-brown, with diffuse darker crossbands and golden tubercles on the body and tail. **Biology and breeding:** These geckos are found in isolated populations, where they are usually very common. They live in small, hard-rock cracks (granite, sandstone and dolerite) and under exfoliating flakes. Two small, hard-shelled eggs (9-11 x 6-7 mm) are laid in rock cracks. Hatchlings measure 35-40 mm. **Habitat:** Semi-desert and succulent karroid veld. **Range:** S. Karoo (Rietbron) to lower Orange River, and to Great Namaqualand; absent from W. Cape and Little Namaqualand. **Subspecies:** Three races are recognized, and all occur in the region. *P.s. serval* has scattered, enlarged tubercles, that are not keeled or raised, on the back; it is found in S. Namibia. The back of *P.s. purcelli* is covered with granular scales and lacks enlarged tubercles; it is found in the karroid areas of South Africa. *P.s. onscepensis* has scattered, raised, keeled tubercles on the back; it occurs along lower Orange River valley.

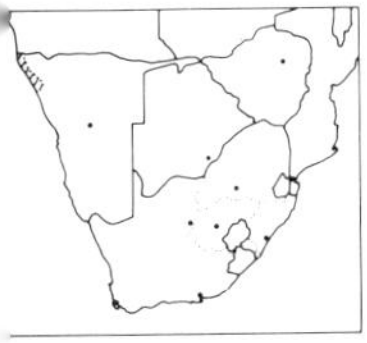

San Steyn's Gecko *Pachydactylus sansteyni*

(Endemic) 7-8 cm; max. 8,5 cm
A small, slender gecko with 6-7 irregular rows of small tubercles that are separated by granular scales on the back. The rostral enters the nostril, and there are 10-12 upper labials. The middle row of scales below the toes and above the scansors is enlarged. There are 4-5 scansors beneath the middle toes. The hind limbs have enlarged, conical, keeled tubercles. The tail is segmented, with transverse rows of 4-6 spinose tubercles, and is shorter than the body. Adults are mauvish-brown above, with irregular, dark brown spots and blotches. A dark brown stripe runs through the eye, from the snout to the ear. The belly is off-white. Juveniles have finer spotting on the body and the tubercles are inconspicuous. **Biology:** They are terrestrial, living under sandstone boulders on sandy soil, and emerging at night to feed on small insects. Juveniles may enter rock cracks. **Habitat:** Semi-desert. **Range:** Vicinity of Kuidas, S. Kaokoveld in Namibia, north to Cunene River mouth.

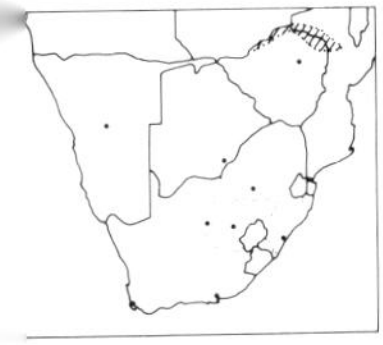

Tete Gecko *Pachydactylus tetensis*

16-18 cm; max. 21 cm
A very large gecko that, unlike all other southern African *Pachydactylus* species, has 8-14 preanal pores in males. The back has enlarged, strongly keeled tubercles that may form four irregular rows on either side of the backbone; they are separated by granular scales. There are 11-14 scansors

beneath the middle toes. The tail has a pair of slightly enlarged scales. The back is uniform pale grey above, and the belly is white. **Biology:** Usually found in wide, shaded rock cracks, occasionally inhabiting hollow baobab trunks. They are gregarious, and several specimens are often found side by side in a crack. **Habitat:** Mopane woodland. **Range:** Lower Zambezi valley.

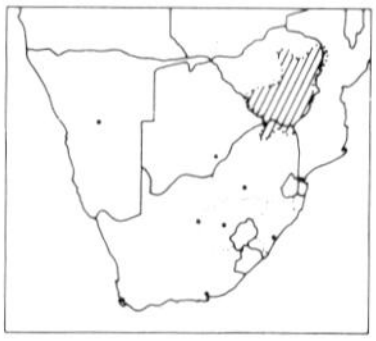

Tiger Gecko *Pachydactylus tigrinus* (Pl. 82)

(Endemic) 10-12 cm; max. 13 cm
A small, strongly flattened gecko with scattered, small, flattened, feebly keeled tubercles on the back, that are separated by granular scales. The rostral does not enter the nostril. The middle row of scales below the toes and above the scansors is not enlarged. There are 4-5 scansors beneath the middle toes. The tail is strongly segmented, and lacks enlarged tubercles; regenerated tails are carrot-shaped. Coloration is varied. Juveniles are purple-brown to black above, with 5-6 narrow white or yellow crossbands (these may be disrupted into spots in northern populations). Adults from N. Transvaal retain the juvenile 'tiger' pattern; in SW. Zimbabwe, the ground colour is lighter and the white crossbands are dark-edged, occasionally breaking up into lines of white spots with dark spots between them; and in N. Zimbabwe and the Matopos, the pale markings are obscured by numerous large, black blotches. The belly is brownish in southern populations, and usually white in the north. **Biology:** Rock-living, these geckos inhabit narrow crevices in granite and sandstone outcrops, often with the Transvaal flat gecko, *Afroedura transvaalica* (page 192). **Habitat:** Mesic savannah. **Range:** Zimbabwe and adjacent N. Transvaal, E. Botswana and W. Mozambique.

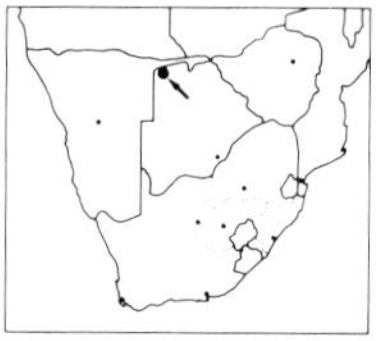

Tsodilo Gecko *Pachydactylus tsodiloensis*

(Endemic) 9-11 cm; max. 12 cm
A medium-sized, flattened gecko with a similar colour pattern to Weber's gecko, *P. weberi* (below). The rostral does not enter the nostril, and the nasorostrals are in contact behind the rostral. There are 14 upper labials, the last forming a large oval scale at the angle of the jaw. The back is covered with granular scales, with 18 rows of enlarged tubercles. The middle row of scales below the toes and above the scansors is enlarged. There are seven scansors beneath the middle toes. The tail is swollen at the base and segmented, with transverse rows of eight enlarged, keeled tubercles; it is slightly longer than the body. Adults are light greyish-brown above, with 5-6 off-white, brown-edged crossbars. The head is marked with dark brown spots and vermiculations, with a pale, dark-edged stripe passing from behind each nostril, running through the eyes, and fusing on the top of the head. The belly is white. In some specimens the crossbars are indistinct. Juveniles are similar to adults but the head is unmarked. **Biology:** These nocturnal geckos inhabit quartzite and dolomitic limestone outcrops, rarely leaving the rock to cross the sandy flats. They eat ants and small insects. **Habitat:** Mesic savannah. **Range:** Endemic to Tsodilo Hills in N. Botswana.

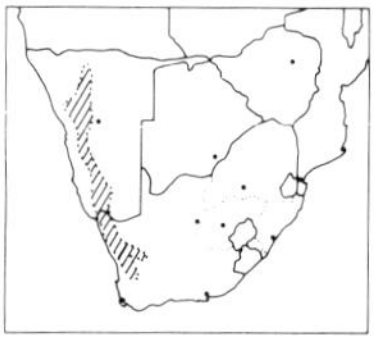

Weber's Gecko *Pachydactylus weberi* (Pl. 86)

(Endemic) 9-10 cm; max. 11 cm
A small, slender, flattened gecko with 22 irregular rows of small tubercles that are separated by granular scales on the back. The rostral does not enter the nostril. The middle row of scales below the toes and above the scansors is enlarged. There are 5-7 scansors beneath the middle toes. There are enlarged tubercles on the hind limbs. The tail, which is slightly longer than the body, is segmented, with regular, transverse rows of 6-8 enlarged, keeled tubercles. In southern populations the juveniles are golden-brown, with four light brown to cream, dark-edged crossbands; the tail merges to white at the tip and has

10-12 black bands. Southern adults are dull brown, with irregular dark brown blotches, and golden tubercles that give a speckled appearance, particularly on the tail. In northern populations, the dark edges to the juvenile crossbands are greatly expanded to form a rich reddish-brown body with three golden crossbands (the first on the back of the neck); the tail is banded in light and dark brown, with rings of golden tubercles. Northern adults have wavy, dark brown blotches above that may form irregular crossbars, and the tail retains faint juvenile bands; the body and tail are speckled with golden tubercles. Populations from Great Namaqualand and central Namibia are intermediate between these phases. **Biology and breeding:** These very agile geckos live in cracks in hard rock (granite, sandstone, quartzite, etc.), and are active at night. They eat mainly moths and spiders. Two hard-shelled eggs (9-10 x 6-7 mm) are laid in a rock crack. They take 80-90 days to hatch; hatchlings are 35-40 mm long. Communal egg sites may contain up to 60 eggs. **Habitat:** Succulent karroid veld. **Range:** Central Namibia, through Namaqualand to W. Cape. **Subspecies:** Although some races have been described (eg. *P.w. acuminatus*, *P.w. werneri*, etc.), none is recognized at present.

EB-FOOTED
ECKO
almatogecko

This medium-sized, slender gecko has a large head with immense, jewel-like eyes. The body is covered with minute, granular scales. Despite its appearance, it is closely related to the thick-toed geckos, *Pachydactylus* (page 199).

The genus contains a single species, which is endemic to the Namib Desert.

Web-footed Gecko *Palmatogecko rangei* (Pl. 79)

10-12 cm; max. 14 cm

A beautiful and bizarre gecko with an elongate, cylindrical body and thin legs. The large eyes, which appear bloodshot, have vertical pupils that each close to two pinholes in bright light. The nostrils are tubular. The toes of the forelimbs and hind limbs are joined with webbing, except for the toe-tips, which have a large, claw-like scale. There are no adhesive scansors. The tail is thin and unsegmented. The body is almost semi-transparent. Males are slightly smaller than females, and have a prominent hemipenial bulge at the tail base, which bears a row of 6-8 spines on the sides. The back is fleshy-pink to pink-brown, with irregular, reticulated dark blotches that extend on to the tail. There is a dark brown band across the snout and a blue-black patch above the eyes. The belly and sides are chalky white, and the other ventral regions are transparent pink. **Biology and breeding:** Terrestrial and nocturnal, these geckos spend the day in a tunnel (up to 50 cm long) that they dig in fine sand or silt. They emerge at sunset. The diet comprises small insects, spiders, etc. They obtain all their moisture from condensing fog. When threatened, they adopt a raised, stiff-legged posture, but do not squeak. Males may fight and bite one another in territorial disputes. The tail is rarely lost, and then always from near the base; it is readily regenerated. The skin is shed in pieces and then eaten. Predators include owls, snakes and the Namib golden mole. (SA RDB, Peripheral.) Clutches of two large, fragile, hard-shelled eggs (21 x 10 mm) are laid from November to March in a small tunnel dug in sand. They hatch in about 90 days; hatchlings measure 30-35 mm. **Habitat:** Wind-blown sands of coast and desert. **Range:** Namib – Richtersveld to S. Angola.

AY GECKOS
elsuma

This group of geckos has reduced inner toes. All but the distal scansors on the toe-tips are undivided. The eyes are large, and usually have round pupils. Males have femoral and preanal pores in a continuous series. These beautifully coloured geckos are diurnal, and mainly arboreal.

There are approximately 25 species in the genus, distributed throughout the islands of the Indian Ocean, with a few species reaching the east coast of Africa. One isolated species, previously placed in a separate genus (*Rhoptropella*), occurs in Namaqualand.

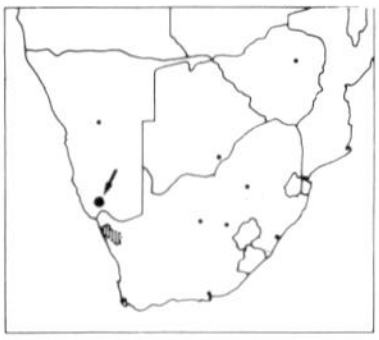

Namaqua Day Gecko *Phelsuma ocellata* (Pl. 90)

(Endemic) 6-7 cm; max. 8 cm
A small day gecko with a small inner toe. The flared toe-tips have 7-8 undivided scansors and no claw. The nostrils are not tubular, and each pierces the first upper labial. The eyes are large, and have vertical pupils and prominent eyelids that cannot close and that form a continuous ring around the eye. The scales on the back are smooth and granular, but the belly scales are overlapping. Males have a long series of 24-31 femoral pores that are continuous across the preanal region. The back is light brown to greyish-brown, with scattered pale and dark spots. The belly is cream to blue-white, often tinged with pink. **Biology and breeding:** These very active geckos are found running and jumping between boulders on rocky hillsides and outcrops. They occasionally forage on succulent bushes or fig trees. The shelter under exfoliating rock flakes, where they are very well camouflaged. (SA RDB, Restricted.) Small beetles, aphids and other insects are eaten. Two small, hard-shelled eggs (8,5 x 6,5 mm) are laid under a rock flake in September. They may lay additional clutches later in the season. **Habitat:** Succulent karroid veld. **Range:** N. Little Namaqualand, through Richtersveld to SW. Namibia.

LEAF-TOED GECKOS *Phyllodactylus*

These small to medium-sized geckos are distinguished by a pair of leaf-shaped scansors under each toe-tip, and a small, retractile claw. The pupils are vertical. The body is covered with small, usually smooth, granular scales. Femoral pores are absent.

They are nocturnal and usually rupicolous, although some species live under tree bark or have become terrestrial. They are common in suitable habitats. They lay two hard-shelled eggs. Females often have very large endolymphatic sacs on the sides of the neck, where they store calcium for us in egg shell production.

This large genus contains 50-60 species, patchily distributed through Africa, tropical America, Europe, Madagascar and Australia. Four endemic species, one of which has a problematic status, occur on the subcontinent.

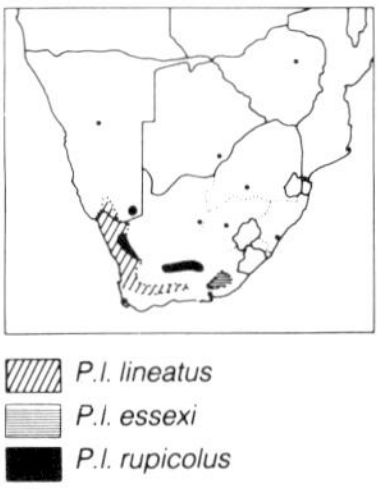

Striped Leaf-toed Gecko *Phyllodactylus lineatus* (Pl. 87)

(Endemic) 4-6 cm; max. 6,5 cm
A very small, variable species, the taxonomy of which is confused (see Subspecies). The back is covered with small, flattened, partly overlapping scales. The first upper labial enters the nostril, and the rostral has a cleft in its centre. The anterior chin shields are enlarged (but not in *P.l. essexi*). Males have 4-6 preanal pores. The tail is cylindrical, not segmented, and is the sam length as the body. Coloration is varied (see Subspecies). **Biology and breeding:** They are nocturnal, sheltering during the day in a retreat, but emerging to forage at sunset. The typical race is usually found under debris, tree bark or in small rock piles. The other races are restricted to large rock outcrops, where they select thin, often vertical cracks, and in winter retreat in deeper cracks to avoid frost. The diet is composed of small insects, particularly termites. Two hard-shelled, round eggs (7-8 mm long) are laid under bark or in a rock crack. They may lay several clutches during a season **Habitat:** Coastal fynbos, succulent and transitional karroid veld, montane grassland, and entering Fish River bush. **Range:** E. Cape, through Cape fold mountains and inland escarpment, to Great Namaqualand in Namibia. **Subspecies:** Three races are recognized, but the differences in habits, habitats and morphology indicate that several species may be confused. *P.l. lineatus* is small, has a rounded head and body, enlarged central chin shields, and is light grey, usually with a dark, striped pattern; it occurs on the coastal plain and valleys from SW. Cape to Great Namaqualand. *P.l. essexi* also small, but has a flattened head and body, central chin shields that are n

enlarged, and a pink-brown back with irregular, dark, transverse stripes; it is found on rock outcrops of Suurberg and summits of E. Cape fold mountains. *P.l. rupicolus* is large, has a flattened head and body, slightly enlarged central chin shields, and a pink-brown back with a mottled pattern; it occurs on rocky summits of Cedarberg and inland Cape escarpment.

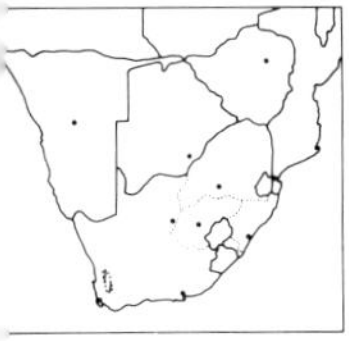

Small-scaled Leaf-toed Gecko
Phyllodactylus microlepidotus (Pl. 87)

(Endemic) 9-13 cm; max. 16 cm
A large, very attractive leaf-toed gecko with a body that is covered with minute, flattened scales. The central chin shields are not enlarged, and the first upper labial enters the nostril. The rostral has a cleft in its centre. The tail is round, not segmented and is longer than the body. The back is grey to slate, with a blackish, reticulate pattern. The belly is off-white. **Biology:** Restricted to mountain summits, these geckos live in large rock cracks in sandstone outcrops or under the bark of dead Clanwilliam cedar trees. **Habitat:** Montane fynbos. **Range:** W. Cape fold mountains, from Cedarberg to Ceres.

Péringuey's Leaf-toed Gecko *Phyllodactylus peringueyi*

(Endemic) 4-5 cm; max. 6 cm
A small species with a long neck and scattered, large, keeled tubercles on the back. The nostril is pierced between the rostral and four nasal scales. The rostral lacks a median cleft. It is pale brown, with two dark brown stripes, and the belly is whitish. **Habitat:** Unknown. **Range:** Known from two specimens, reputed to have come from near Port Elizabeth and Little Namaqualand in 1910. No other specimens have been found, despite many searches. The species is probably not African, but it has not been successfully matched with any overseas species. It is one of the herpetological mysteries of the region.

Marbled Leaf-toed Gecko *Phyllodactylus porphyreus* (Pl. 87)

(Endemic) 8-9 cm; max. 10 cm
A medium-sized species with a flattened head and body. The back is covered with small granules. There are no enlarged central chin shields, and the first upper labial does not enter the nostril, which is pierced between the rostral and three nasal scales. The rostral has no cleft in its centre. The tail is round, unsegmented and longer than the body. The back is greyish to light brown, rarely plain, sometimes with a prominent, pale dorsal stripe, but usually mottled in reddish-brown to dark brown. There is often a dark stripe through the eye. The belly is off-white to cream, and sometimes speckled on the sides. **Biology and breeding:** Nocturnal, they live in varied habitats, including cracks in rock outcrops, under bark on dead trees and in rotting logs. They are common on houses in Cape Town and along the S. Cape coastal region, where they are often seen around outdoor lights, feeding on insects. They are not territorial, and as many as 24 individuals may live in the same retreat. Two hard-shelled eggs (9-11 x 7-8 mm) are laid beneath tree bark or in a rock crack, or sometimes even among rubbish. Communal sites may contain up to 30 eggs. These hatch in 50-60 days; hatchlings are 30-40 mm long. Growth is rapid, and they mature in about two years. **Habitat:** Coastal and montane fynbos; also entering cold evergreen forest and urban situations. **Range:** W. and S. Cape, from Nieuwoudtville to Cape St. Francis. Also on Dassen Island.

ARKING GECKOS
enopus

These are small, unusual burrowing geckos that lack scansors on the toes, but have a fringe of scales to aid digging. They are the only local geckos with movable upper eyelids.

Terrestrial and nocturnal, they dig elaborate burrow systems with concealed escape holes. They live alone, but form diffuse colonies. At night they forage for insects, walking slowly across the sand and pausing often. Males call on summer nights at their burrow entrance, attracting mates and proclaiming their territory. The call is a chain of clicks ('ceek-ceek-ceek . . .') that vary in number and pitch between the species. One or two hard-shelled, white eggs are laid in a shallow pit in spring. Predators include owls, snakes and meercats. When caught, these geckos squeak and may bite.

The genus contains three species, all of which are endemic. Two are restricted to the Namib Desert, while the other is widely distributed in the western arid regions.

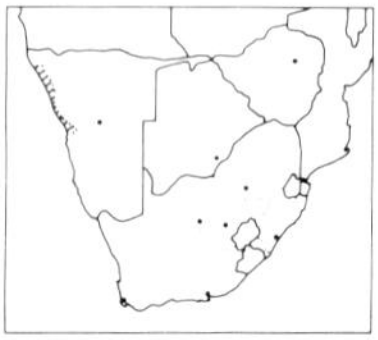

Carp's Barking Gecko *Ptenopus carpi* (Pl. 79)

(Endemic) 7-9 cm; max. 10,4 cm
A slender barking gecko with long legs and weakly fringed toes. There are 105-140 scale rows at midbody. The nostrils are not swollen, and lack valves The back is creamy white, with fine, orange-brown speckles and reticulations There are 3-5 dark brown crossbars on the back and 5-9 on the tail. The belly is white, with a yellow heart shape on the throat that is absent in juveniles and seasonal in females. Specimens from N. Namib Desert are darker. **Biology and breeding:** These geckos live in shallow burrows with a few side tunnels that they dig in very compact soil. They are very well camouflaged. If disturbed, they raise the body on stiff legs, puff out their yellow throat, hiss ar may even bite. The call is a monotonous series of slow, low-pitched clicks (3-20, but usually 12). **Habitat:** Flat, barren gravel plains with an annual rainf of less than 125 mm. **Range:** Central and N. Namib Desert, from Kuiseb Rive north to Rocky Point.

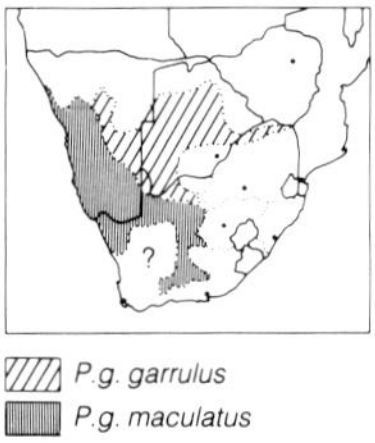

Common Barking Gecko *Ptenopus garrulus* (Pl. 79)

(Endemic) 6-8 cm; max. 10 cm
A small barking gecko with swollen nostrils and strongly fringed toes. There are 110-190 scale rows at midbody. The back colour varies (see Subspecies The belly is white. Males have an orange or yellow heart shape on the throat. **Biology and breeding:** These geckos become active for a short period around sunset, and their calls signal sunset in the desert. The call varies regionally from one to 13 clicks, but usually consists of five. They do not inha mobile dunes. In rocky areas, they often live in a sand hummock trapped at the base of a bush. After dark, they emerge and walk slowly in search of prey which includes termites, ants and small beetles. If disturbed, they freeze and are effectively camouflaged. **Habitat:** Desert and semi-desert on various soi substrates, preferring flat, stable, sandy soil with sparse vegetation cover. **Range:** Western arid regions of southern Africa, from Little Karoo to N. Transvaal. **Subspecies:** Two races are recognized. *P.g. garrulus* has mo than 160 scale rows at midbody, and is finely speckled in reddish-brown, sometimes with diffuse, darker spots; it occurs in Kalahari sands of Namibia, Botswana and adjacent regions. *P.g. maculatus* has less than 160 scale row at midbody, and the colour varies with the substrate, but is usually greyish-yellow, with black crossbars, five pairs of pale spots on the flanks ar 5-10 dark bars on the tail; it is found in Namibia (including some areas in Namib Desert), extending through Namaqualand to Great and Little Karoo.

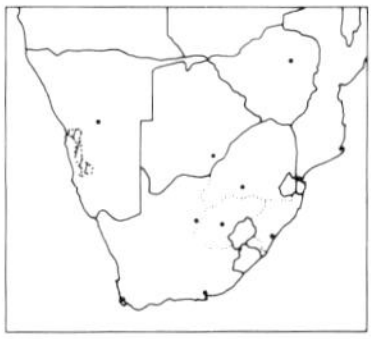

Koch's Barking Gecko *Ptenopus kochi* (Pl. 79)

(Endemic) 8-10 cm; max. 12 cm
A stout burrowing gecko with large bulging eyes, swollen nostrils, and minut body scales in 187-222 rows at midbody. The flattened toes are fringed with elongate scales, and the body is slightly longer than the tail. The body is reddish-brown with dark speckles, particularly on the sides. An irregular row light spots (which may be yellow in adult males) occurs on the flanks. The be

and inner surfaces of the limbs are white. The throat is white in females and juveniles, and sulphur-yellow in males. **Biology:** These geckos live in extensive, elaborate burrows that they dig in the fine sand of interdune spaces and silt of dry riverbeds. The burrows may be up to 90 cm long and 40 cm deep. **Habitat:** Desert. **Range:** South-central Namib Desert.

AMIB DAY ECKOS
hoptropus

These are medium-sized geckos with long legs. The feet have four long, thin toes (the fifth toe is small and rudimentary) with flaired tips that have a series of 5-13 undivided scansors, and sometimes a rudimentary claw. The body is squat and covered with small, granular scales. The head is distinct. The nostrils are swollen, and the eyes are large, with vertical pupils and immovable eyelids that form a complete ring around the eye. The lower jaw has elongate chin shields. Males lack femoral pores.

These diurnal geckos run around on vertical rock faces, or on broken, rocky ground. They are sit-and-wait hunters, hiding in a shaded vantage point and attacking any insects that come within range. Two hard-shelled eggs are laid in a rock crack or in sand under a slab. Communal egg-laying sites are frequently used.

At present, the genus contains six species, five of which are found on the subcontinent. Two are endemic.

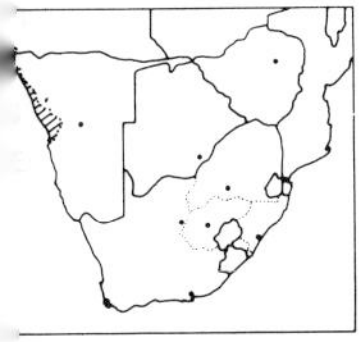

Namib Day Gecko *Rhoptropus afer* (Pl. 92)

8-10 cm; max. 11 cm
A small species with relatively short, stout toes. There are 5-6 scansors under the middle toes. There are no preanal pores. The middle row of scales under the tail is not enlarged. The scales on the back are small and rounded. The tail is slightly shorter than the head and body. The back is light olive to grey-brown, with scattered red-brown and pale spots. The tail is banded in dark brown and dirty yellow, and the chest and belly are bluish-white. The throat, anal region and lower surfaces of the limbs and tail are bright yellow. **Biology and breeding:** These small day geckos prefer dry gravel plains with sheet rock and exfoliating flakes. They are very active, darting rapidly between rock slabs, where they are perfectly camouflaged. They live on the hottest Namib plains, sheltering in cracks from the midday sun. If there is a cool breeze, they climb up to a vantage point and lift the body high to cool in the wind. They feed on ants and small beetles. Two white, hard-shelled, oval eggs (10-12 x 6-8 mm) are laid under rock slabs. The eggs often stick to the stone, but may be found loose in the sand. **Habitat:** Rocky desert. **Range:** Namib Desert, from Kuiseb River to S. Angola.

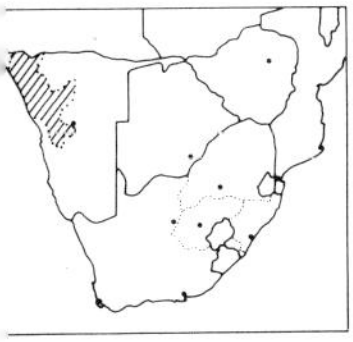

Barnard's Namib Day Gecko *Rhoptropus barnardi* (Pl. 92)

6-8 cm; max. 8,5 cm
A small species with long, slender toes, the fourth one having eight undivided scansors beneath it. Males have 4-7 preanal pores arranged in a row. The middle row of scales under the tail is not enlarged. The scales on the back are slightly keeled and tubercular. The tail is flattened and segmented at its base (but smooth when regenerated), and is slightly longer than the head and body. The back is pale grey to pinkish-brown or dark brown, with scattered pale and dark spots that may form irregular bands. The belly is pale bluish to white.

Biology and breeding: This day gecko lives in regions of higher rainfall than other Namib day geckos, preferring small rock outcrops and ridges of varied rock types. It is rarely found together with the larger, rupicolous species, possibly because of competition. Clutches of two eggs (11-12 x 9-10 mm) are laid in May-June in rock cracks. Communal nesting sites, containing up to 200 eggs, may be found. **Habitat:** Semi-desert. **Range:** Extends inland in the western half of Namibia, from Damaraland, north to S. Angola.

Kaokoveld Namib Day Gecko *Rhoptropus biporosus* (Pl. 92)

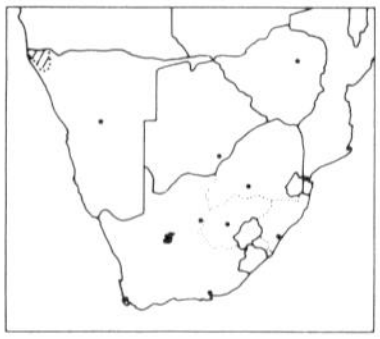

(Endemic) 6-7,5 cm; max. 8 cm
This species is similar to Barnard's Namib day gecko, *R. barnardi* (page 213 but has a slimmer build, a rounder snout, a thinner tail and only two preanal pores. The toes are long and slender, with 11 undivided scansors beneath th fourth one. The middle row of scales under the tail is not enlarged. The scales on the back are slightly tubercular. The tail is flattened and segmented at its base (but smooth when regenerated), and is slightly shorter than the head ar body. The back is light grey to fawn, with irregular spots and markings. The limbs and tail have indistinct, dark bars, and the belly is white to cream. **Biology:** These geckos are active on low rock outcrops and boulders, preferring flat surfaces. **Habitat:** Semi-desert. **Range:** Recorded only from th vicinity of Orupembe in Kaokoveld.

Boulton's Namib Day Gecko *Rhoptropus boultoni* (Pl. 92)

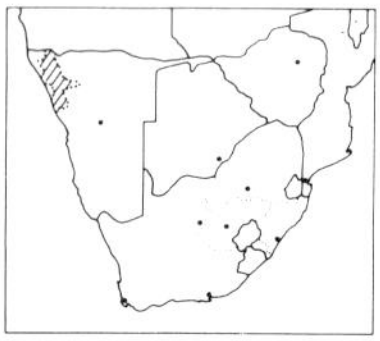

R.b. boultoni

11-14 cm; max. 16 cm
This largish, stocky species has long, slender toes, with 13 undivided scansors beneath the fourth one. Males have 5-8 preanal pores arranged in a row. The middle row of scales under the tail is enlarged, at least on the distal half, and the scales on the back are smooth and only feebly tubercular. The t is flattened, thick and segmented at its base, and slightly shorter than the head and body. The back is dark sooty-grey to olive-brown, with large, scattered, irregular maroon to dull brick-red blotches. The belly is dull bluish-grey to slate. **Biology and breeding:** These geckos forage on vertica granite boulders, feeding on ants, spiders and beetles. The males are pugnacious and defend their territories against intruders. Females in June contained two large eggs (11,5 x 8,5 mm). **Habitat:** Rocky desert. **Range:** Damaraland in Namibia, north to S. Angola. **Subspecies:** Three race are recognized, with only the typical race occurring on the subcontinent. It is replaced in Angola by *R.b. benguellensis* (in Benguela Province) and *R.b. montanus* (in Huila).

Bradfield's Namib Day Gecko *Rhoptropus bradfieldi* (Pl. 92)

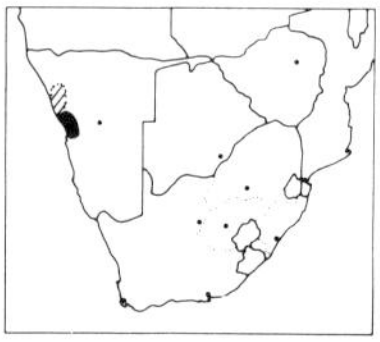

R.b. bradfieldi
R.b. diporus

(Endemic) 9-12 cm; max. 13 cm
This species is similar to Boulton's Namib day gecko, *R. boultoni* (above), bu has 11 undivided scansors beneath the fourth toe, and the tail is slightly long than the head and body. The back is sooty-brown to blackish, with faint trace of crossbands or narrow zigzags. The tail segments are edged with black. Th belly is slate-grey to bluish. **Biology and breeding:** During the heat of the da these geckos hang on the shaded vertical surfaces of large, dark rocks, dashing out to seize any food that comes within range. The diet comprises ants, moths and beetles. Adults give high pitched squeaks when approachir one another; it is possible that this is a territorial signal. Two eggs (9 x 15 mm) are laid in a rock crack, where they stick firmly to the surface. Favoured cracks may be used by many individuals over many years, and accumulate 'beds' of old egg shells. **Habitat:** Semi-desert. **Range:** Kuiseb River to Twyfelfontein in W. Damaraland, but absent from W. Namib-Naukluft Desert Park. **Subspecies:** Two races are recognized. *R.b. diporus* males ha two enlarged preanal scales with pores (absent in females); it is found on Brandberg, to S. Kaokoveld. *R.b. bradfieldi* males lack preanal pores; it occurs in the region south of Brandberg.

CROCODILIANS

Order Crocodylia

Crocodiles, alligators and the gharial are the last vestiges of the ruling reptiles, the archosaurs, which dominated the earth's history for over 150 million years. Crocodilians are, in fact, more closely related to birds than they are to other living reptiles. Their many unusual features include a heart that is four-chambered, with complete division of the ventricles, thus permitting more efficient blood oxygenation; an extra eyelid, the nictitating membrane, which sweeps dirt from the eyeball; an improved limb articulation that permits a better gait (the 'high walk'), so that some crocodilians can even gallop, albeit for short distances; a hard palate in the roof of the mouth; a longitudinal cloacal aperture; and a single penis. The archosaurs reached their heyday in the Jurassic era (190-130 million years ago), with the emergence of aquatic, terrestrial and even arboreal forms (the last-named known from Lesotho fossils). They have changed very little in the last 65 million years. All living forms are aquatic, and are distributed throughout the world's tropical regions.

Only 22 species remain, and all are endangered. All are now protected. Three living families are recognized: the Gavialidae, the Alligatoridae, and the Crocodylidae, which is the only family present in Africa.

:ROCODILES
amily Crocodylidae

Crocodiles are recognizable by having the fourth mandibular tooth visible when the jaw is closed (it is hidden in a socket in alligators). The first fossils occur in the late Cretaceous period of Europe and N. America. Many living species grow to a large size, but claims of nine- to 10-metre giants are exaggerated.

Of the 14 living species (in three genera), only three occur in Africa.

ROCODILES
'rocodylus

All true crocodiles are very similar in appearance, the species differing mainly in the shape of the snout.

The larger species (the saltwater crocodile and the Nile crocodile) both take large game, including man, while the smaller species (Johnston's crocodile and Morelet's crocodile) feed mainly on fish and small mammals.

Fossils are known from the Upper Cretaceous period. The 12 living species are distributed throughout the tropical regions of the world; they are poorly represented in Amazonia, where they are replaced by the caimans. Two species occur in Africa. The African slender-snouted crocodile, *C. cataphractus,* is found in W. and central Africa, inhabiting the larger rivers. The Nile crocodile is the only species that enters the subcontinent.

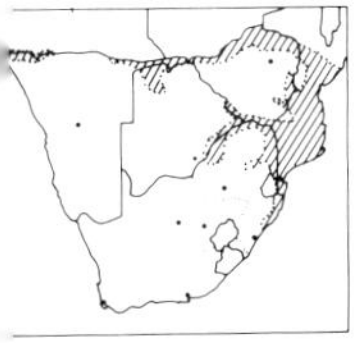

Nile Crocodile *Crocodylus niloticus* (Pl. 96)

250-350 cm; max. 590 cm

Nile crocodiles may exceptionally exceed 1 000 kg in weight. The jaws are long, and have prominent teeth. The eyes and valved nostrils are situated on top of the head. The skin is covered with geometrically arranged, horny plates, many of which are keeled and bony. The plates on top of the head are fused to the skull. The hind feet are webbed. The tail is 40% of the total body length, rectangular in cross-section, and has two raised dorsal keels. The young are greenish, with irregular black markings over the back and sides, and the thro and belly are uniform straw-yellow. Adults are darker, being uniform olive to grey, with a yellow or cream belly.

Biology and breeding: Young crocodiles dig a burrow (sometimes communally) up to 3 m long in which they shelter for the first 4-5 years of life. They spend a lot of time out of water and eat small prey. Subadults take up residence in swamps and backwaters, eating fish, terrapins, birds and small mammals. Adults grasp prey with a fast, sideways swipe of the head. The tai may be used to knock over vegetation to dislodge nestling birds, or to direct fish to within striking distance of the jaws. They feed regularly on fish, particularly catfish, but also ambush game coming to drink. This is seized an pulled into the water to drown. Antelope are usually taken, but even zebra ar buffalo may be overcome. Man is considered fair game; attacks (and fatalitie are still relatively common. Large food items are softened by biting. If too larg to be swallowed whole, prey is torn to bits by the crocodile seizing a mouthfu and spinning on its long axis. Carrion is readily taken. Cooperative behaviou in feeding and breaking up prey is known. On hot days, they come ashore o sand bars to bask. At high temperatures they lie with mouth agape, losing excess heat by evaporation. They swim effortlessly, using the broad, flattene tail. The webbed hind feet allow careful manoeuvring during mating and whe preparing to ambush food. The valved nostrils and gular flap at the back of t mouth enable them to feed underwater. They have lived for up to 60 years in captivity, and very large wild specimens may live to 100 years. Hunting has been the cause of the decline in the numbers of crocodiles (SA RDB, Vulnerable).

Crocodilians are surprisingly attentive parents, and nest construction and the care of the young is very advanced. Sexual maturity is reached in 12-15 years at about 2-3 m (70-100 kg). At the start of the breeding season (May), males develop a dominance hierarchy. Courtship is elaborate. Mating takes place in the water in July-August. The female selects a suitable sunny sand bank that is above floodwater level and which has good drainage and cover nearby. She will use it, unless disturbed, for the rest of her life. At night, usua in November, she digs a hole (30-45 cm deep) with her hind legs, and lays 16-80 white, hard-shelled eggs (70-78 x 50-56 mm, 85-125 g). The nest site defended against predators and other crocodiles, and during this period the female does not eat, but may go to the water to drink. The male remains in th vicinity, but is not allowed near the nest mound by the female. 84- 90 days later the hatchlings, while still in the egg, give a high-pitched cheeping noise that is audible 20 m away. The female carefully opens the nest and takes all the young into her mouth. The hatchlings (280-320 mm) are taken to the wat washed and released. They remain close together in a 'crèche area' for 6-8 weeks. The sex of the hatchlings is dependent on the egg incubation temperature; unlike chelonians (page 21), females are produced at lower temperatures (26-30 °C) and males at higher temperatures (31-34 °C).

Habitat: Larger rivers, lakes and swamps, but also into river mouths, estuaries and mangrove swamps. **Range:** Okavango Basin, Cunene River and major rivers draining to the east coast, south to Tugela River (and historically to East London). Viable populations now restricted mainly to gam reserves. Elsewhere, on Madagascar (indicating limited marine excursions) and throughout Africa (except at high altitudes and in deserts). Now extinct other Indian Ocean islands and in coastal Mediterranean regions.
Subspecies: A number have been proposed, based on minor differences in scalation, but all are of doubtful validity.

denotes species which are on record as having caused human fatalities

denotes species which can inflict serious bites, but which have not been recorded as having caused human fatalities

Great Escarpment (Valley of Desolation, Graaff-Reinet)

ahari hardveld with mesic acacia savannah (Lobatse, Botswana)

Succulent karroid veld (near Springbok, Little Namaqualand)

mib Desert (Swakopmund, Namibia)

Lowveld arid savannah (E. Transvaal)

Granite inselbergs with euphorbia trees (near Pietersburg, N. Transvaal)

Cape fold mountains with fynbos (Cedarberg, SW. Cape)

Maputaland with coastal dune thicket and swamp (Kosi Bay, NE. Natal)

Rocky desert (near the Brandberg, Namibia)

ndescribed species of dwarf chamaeleon

Undescribed species of leaf-toed gecko

Puff adder (unusual striped phase)

Plated lizard (blue-throated phase)

own house snake (xanthic phase)

Puff adder/Gaboon adder (hybrid)

Angulate tortoise (page 27)

Angulate tortoise (light phase) (page 27)

Leopard tortoise (page 26)

Leopard tortoise (light phase) (page 26)

Leopard tortoise (egg and hatchling; note egg tooth) (page 26)

ell's hinged tortoise (subsp. *belliana*) (page 30)

Speke's hinged tortoise (subsp. *spekii*) (page 30)

arrot-beaked tortoise (male) (page 24)

Natal hinged tortoise (female) (page 31)

arrot-beaked tortoise (female) (page 24)

Greater padloper (page 23)

Greater padloper (hatchling) (page 23)

Nama or Berger's padloper (page 25)

Karoo or Boulenger's padloper (page 24)

Speckled padloper (subsp. *signatus*) (page 25)

Speckled padloper (subsp. *cafer*) (page 25)

eometric tortoise (page 28)

Geometric tortoise (juvenile) (page 28)

ent tortoise (subsp. *tentorius*) (page 29)

Serrated or Kalahari tent tortoise (page 28)

nt tortoise (subsp. *verroxii*) (page 29)

Tent tortoise (subsp. *trimeni*) (page 29)

Green turtle (page 33)

Hawksbill turtle (page 34)

Olive ridley turtle (page 35)

Loggerhead turtle (adult) (page 35)

Loggerhead turtle (hatchling) (page 35)

eatherback turtle (adult) (page 32)

eatherback turtle (hatchling) (page 32)

Zambezi soft-shelled terrapin (page 37)

Jile soft-shelled terrapin (page 36)

Nile soft-shelled terrapin (page 36)

American red-eared terrapin (juvenile) (page 38)

American red-eared terrapin (adult) (page 38)

Marsh or Helmeted terrapin (page 39)

Marsh or Helmeted terrapin (juvenile) (page 39)

Marsh or Helmeted terrapin withdrawing head sideways (page 39)

stern hinged terrapin (subsp. *castanoides*) (page 41)

rrated hinged terrapin (page 40)

Okavango hinged terrapin (page 41)

ashona hinged terrapin (page 41)

Pan hinged terrapin (subsp. *subniger*) (page 40)

Gaboon adder (subsp. *gabonica*) (page 99)

Puff adder (subsp. *arietans*) (page 98)

Puff adder (subsp. *arietans*) (page 98)

Puff adder (subsp. *arietans*, dark phase) (page 98)

Berg adder (E. Transvaal) (page 99)

Berg adder (S.W. Cape) (page 99)

Berg adder (E. Cape) (page 99)

Horned adder (N. Cape) (page 100)

Horned adder (Etosha, S.W.A./Namibia) (page 100)

Horned adder (N. Transvaal) (page 100)

Many-horned adder (subsp. *cornuta*) (page 100)

Desert mountain adder (page 101)

Lowland swamp viper (page 102)

Many-horned adder (subsp. *albanica*) (page 100)

Plain mountain adder (Compassberg) (page 101)

Péringuey's adder (page 102)

Namaqua dwarf adder (page 101)

Plain mountain adder (Cedarberg) (page 101)

Common or Rhombic night adder (Zimbabwe) (page 97)

Common or Rhombic night adder (E. Cape) (page 97)

Snouted night adder (page 98)

Common or Rhombic egg eater (page 84)

ast African egg eater (subsp. *medici*) (page 84)

Dwarf beaked snake (page 68)

Western keeled snake (page 65)

Bark or Mopane snake (subsp. *viperinus*) (page 68)

Variegated or Spotted slug eater (page 63)

Mopane snake (subsp. *nototaenia*) (page 68)

Mozambique shovel-snout (page 66)

Sundevall's shovel-snout (subsp. *sundevallii*) (page 66)

otted house snake (Karoo, Cape) (page 59)

Spotted house snake (E. Transvaal) (page 59)

African rock python (subsp. *natalensis*) (page 51)

chieta's dwarf python (page 50)

Spotted or Rhombic skaapsteker (subsp. *rhombeatus*) (page 69)

Dwarf beaked snake (page 68)

Many-spotted snake (page 65)

Mole snake (juvenile) (page 63)

Bird or Twig snake (subsp. *capensis*) (page 87)

Elongate quill-snouted snake (subsp. *inornatus*) (page 81)

Bicoloured quill-snouted snake (subsp. *bicolor*) (page 81)

Shield-nose snake (subsp. *intermedius*) (page 90)

wo-striped shovel-snout (page 66)

Fisk's house snake (page 59)

Spotted harlequin snake (page 80)

Eastern tiger snake (subsp. *semiannulatus*) (page 86)

mib tiger snake (Karoo, Cape) (page 86)

Namib tiger snake (Namib Desert) (page 86)

Cape cobra (juvenile) (page 93)

Cape cobra (adult, speckled phase) (page 93)

Coral snake (subsp. *lubricus*) (page 89)

Rinkhals (banded phase) (page 94)

Egyptian cobra (subsp. *annulifera*, banded phase) (page 92)

Black-necked spitting cobra (subsp. *nigricincta*) (page 93)

Günther's garter snake (page 90)

Angolan garter snake (subsp. *semiannulata*) (page 91)

Boulenger's garter snake (subsp. *boulengeri*, juvenile) (page 91)

Sundevall's garter snake (subsp. *media*, juvenile) (page 91)

Sundevall's garter snake (subsp. *decosteri*, juvenile) (page 91)

Sundevall's garter snake (subsp. *decosteri*, adult) (page 91)

Grey-bellied grass snake (subsp. *variabilis*) (page 70)

Rhombic skaapsteker (subsp. *rhombeatus*, striped phase) (page 69)

Striped skaapsteker (page 70)

Dwarf sand snake (page 72)

Stripe-bellied sand snake (subsp. *subtaeniatus*) (page 72)

Jalla's sand snake (page 72)

Cape sand snake (subsp. *leightoni*) (page 71)

ork-marked sand snake (subsp. *trinasalis*) (page 71)

Namib sand snake (subsp. *namibensis*) (page 71)

aroo sand snake or Whip snake (page 71)

Cross-marked or Montane grass snake (striped phase) (page 73)

Olive grass snake (page 73)

Gerard's black and yellow burrowing snake (subsp. *gerardi*) (page 79)

Striped harlequin snake (page 80)

Kalahari purple-glossed snake (page 78)

Eastern purple-glossed snake (subsp. *microphthalma*) (page 78)

Bicoloured quill-snouted snake (subsp. *lineatus*) (page 81)

Bicoloured quill-snouted snake (subsp. *lineatus*) (page 81)

ape file snake (subsp. *capensis*) (page 62)

rnate green snake (page 83)

Striped swamp snake (subsp. *bangweolicus*) (page 65)

Yellow-bellied sea snake (page 96)

Aurora house snake (hatchling) (page 59)

ellmich's wolf snake (page 61)

Semiornate snake (page 82)

Cape centipede eater (page 76)

Reticulated centipede eater (page 76)

Black centipede eater (page 76)

South-western shovel-snout (page 67)

Shield-nose snake (subsp. *scutatus*) (page 90)

Coral snake (subsp. *infuscatus*) (page 89)

Forest cobra (page 92)

Egyptian cobra (subsp. *annulifera*, plain phase) (page 92)

Eygptian cobra (subsp. *anchietae*) (page 92)

Cape cobra (plain phase) (page 93)

Mozambique spitting cobra or M'fezi (page 93)

Common slug eater (subsp. *lutrix*, with new-born young) (page 63)

Mole snake (plain phase) (page 63)

Southern brown egg eater (page 84)

Brown house snake (page 58)

Swazi rock snake (page 60)

Transvaal quill-snouted snake (page 81)

Variegated wolf snake (page 60)

Sundevall's shovel-snout (subsp. *lineata*, speckled phase) (page 66)

East African shovel-snout (subsp. *stuhlmanni*) (page 67)

Sundevall's shovel-snout (subsp. *lineata*, striped phase) (page 66)

Natal green snake (subsp. *occidentalis*) (page 83)

Green water snake (page 83)

Spotted bush snake (page 82)

Spotted bush snake (page 82)

Western green snake (page 83)

Green mamba (page 95)

Boomslang (male, speckled phase) (page 87)

Boomslang (green phase) (page 87)

Boomslang (female, olive phase) (page 87)

Boomslang (juvenile) (page 87)

Bird or Twig snake (subsp. *mossambicanus*) (page 87)

Cross-barred or Marbled tree snake (subsp. *aulica*) (page 86)

Sundevall's garter snake (subsp. *media*, adult) (page 91)

Aurora house snake (page 59)

Yellow-bellied house snake (page 60)

Olive house snake (page 59)

Common brown water snake (page 58)

Mulanje water snake (subsp. *mlanjensis*) (page 57)

usky-bellied water snake (subsp. *laevissimus*) (page 57)

Whyte's water snake (subsp. *obscuriventris*) (page 58)

erald or Red-lipped snake (page 85)

rotse water snake (page 85)

Olive marsh snake (page 64)

Forest marsh snake (subsp. *sylvatica*) (page 64)

Short-snouted grass snake (subsp. *brevirostris*) (page 72)

Cross-marked or Montane grass snake (plain phase) (page 73)

Leopard grass snake (subsp. *leopardinus*) (page 72)

Western sand snake (page 71)

Rufous beaked snake (subsp. *rostratus*) (page 68)

Rufous beaked snake (subsp. *rostratus*) (page 68)

Black mamba (page 95)

Black mamba (page 95)

Black spitting cobra (subsp. *woodi*, adult) (page 93)

Black spitting cobra (subsp. *woodi*, hatchling) (page 93)

Mole snake (black phase) (page 63)

Rinkhals (plain phase, shamming death) (page 94)

Angola file snake (page 62)

Black file snake (page 62)

Cape wolf snake (subsp. *capense*) (page 60)

Eastern wolf snake (page 61)

Dwarf wolf snake (page 61)

Lined olympic snake (page 69)

Natal black snake (page 77)

Lined olympic snake (page 69)

Boulenger's garter snake (subsp. *boulengeri*, adult) (page 91)

Sundevall's garter snake (subsp. *longicauda*) (page 91)

Southern or Bibron's burrowing asp (dark belly phase) (page 75)

Southern or Bibron's burrowing asp (light belly phase) (page 75)

Duerden's burrowing asp (page 75)

Duerden's burrowing asp (subsp. *duerdeni*) (page 75)

Common purple-glossed snake (page 78)

Natal purple-glossed snake (page 78)

Delalande's blind snake (page 47)

Bibron's blind snake (page 47)

Slender blind snake (page 46)

Beaked blind snake (page 47)

Schlegel's blind snake (subsp. *petersii*, spotted phase) (page 47)

Schlegel's blind snake (subsp. *schlegelii*, striped phase) (page 47)

Fornasini's blind snake (page 47)

Flower-pot snake (page 46)

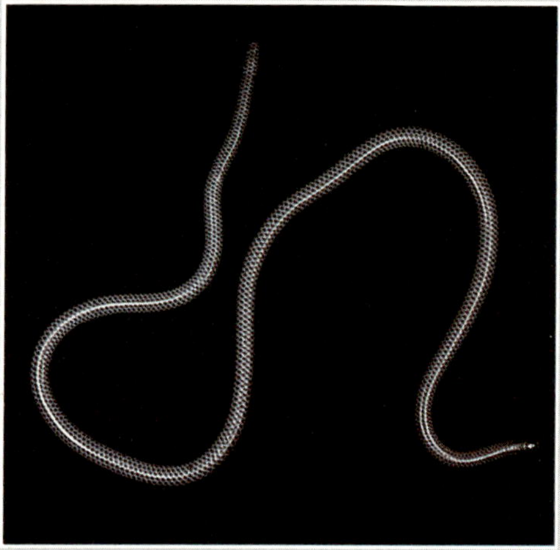

Slender thread snake (page 49)

Peters's thread snake (subsp. *scutifrons*) (page 49)

Long-tailed thread snake (page 48)

Distant's thread snake (page 49)

Western thread snake (page 50)

Cape spade-snouted worm lizard (subsp. *capensis*) (page 107)

Angolan spade-snouted worm lizard (page 106)

Slender spade-snouted worm lizard (subsp. *sphenorhynchus*) (page 107)

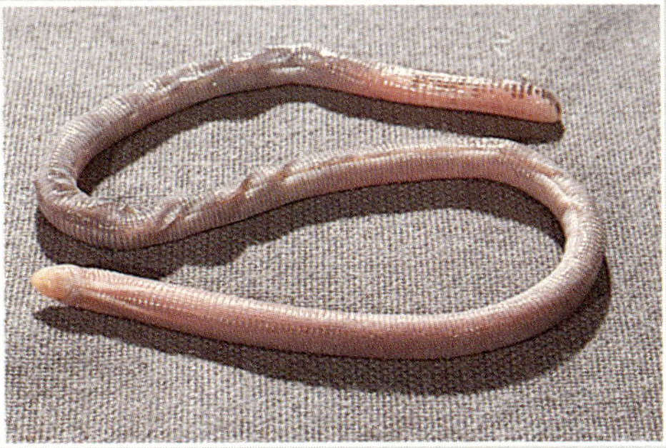

Blunt-tailed worm lizard (page 108)

Leonhard's spade-snouted worm lizard (page 106)

Cape spade-snouted worm lizard (subsp. *rhodesiana*) (page 107)

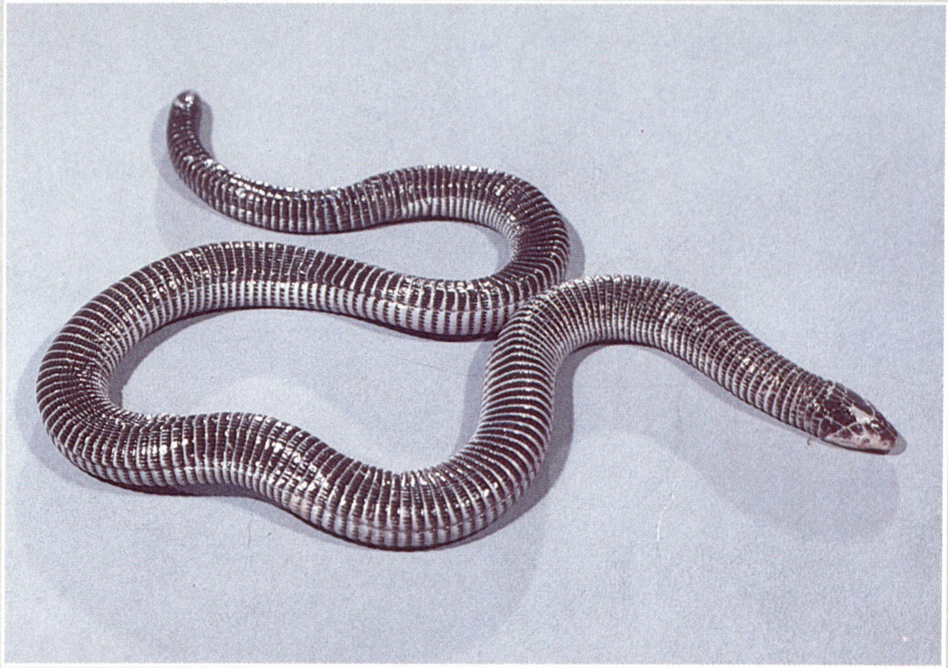

Black round-headed worm lizard (page 106)

Kalahari round-headed worm lizard (page 105)

Violet round-headed worm lizard (page 105)

Lang's round-headed worm lizard (subsp. *langi*) (page 104)

Swynnerton's round-headed worm lizard (page 105)

PLATE 42

?rain's blind legless skink (page 116)

Boulenger's blind legless skink (page 118)

omi's blind legless skink (page 118)

Meyer's blind legless skink (page 118)

?oastal legless skink (orange phase) (page 114)

Giant legless skink (page 115)

Thin-tailed legless skink (subsp. *gracilicauda*) (page 114)

Percival's legless skink (subsp. *occidentalis*) (page 115)

Coastal legless skink (dark phase) (page 114)

Percival's legless skink (subsp. *tasmani*) (page 115)

Cuvier's blind legless skink (page 117)

Cape legless skink (subsp. *orientalis*) (page 114)

Cape legless skink (subsp. *meleagris*) (page 114)

Striped legless skink (subsp. *tristis*) (page 114)

Kalahari burrowing skink (page 125)

FitzSimons's burrowing skink (page 124)

Woodbush legless skink (page 115)

Cregoi's blind legless skink (subsp. *cregoi*) (page 117)

Striped blind legless skink (subsp. *subtaeniatus*) (page 117)

Striped blind legless skink (subsp. *lineatus*) (page 117)

Golden blind legless skink (subsp. *fitzsimonsi*) (page 116)

lewitt's dwarf burrowing skink (page 121)

Gronovi's dwarf burrowing skink (page 122)

impopo dwarf burrowing skink (subsp. *limpopoensis*) (page 123)

Zululand dwarf burrowing skink with new-born babies (page 120)

Algoa dwarf burrowing skink (page 120)

mith's dwarf burrowing skink (subsp. *inornatus*) (page 122)

Smith's dwarf burrowing skink (subsp. *mossambicus*) (page 122)

Lowveld dwarf burrowing skink (page 120)

Silvery dwarf burrowing skink (page 120)

Striped dwarf burrowing skink (page 121)

Kasner's dwarf burrowing skink (page 122)

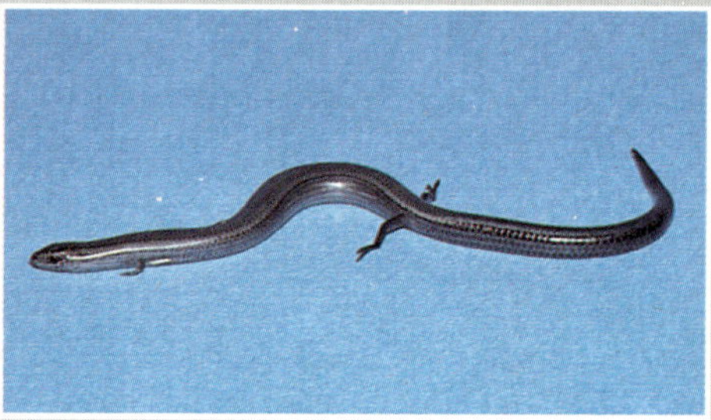
Cape dwarf burrowing skink (page 121)

PLATE 48

lontane dwarf burrowing skink (page 123)

Western dwarf burrowing skink (page 121)

lbert's burrowing skink (page 124)

Mozambique writhing skink (page 126)

undevall's writhing skink (subsp. *sundevallii*) (page 127)

Cape grass lizard (subsp. *anguina*) (page 156)

Large-scaled grass lizard (subsp. *macrolepis*) (page 156)

Transvaal grass lizard (page 156)

African long-tailed seps (subsp. *fitzsimonsi*) (page 153)

African long-tailed seps (subsp. *africanus*) (page 153)

PLATE 50

;ommon long-tailed seps (subsp. *tetradactylus*) (page 155)

;eyer's long-tailed seps (page 154)

Arnold's skink (subsp. *mlanjensis*) (page 119)

hort-legged seps (subsp. *laevicauda*) (page 154)

Wahlberg's snake-eyed skink (page 134)

uton's skink (page 126)

Red-sided skink (subsp. *smithii*, male) (page 129)

Red-sided skink (subsp. *depressa*) (page 129)

Cape skink (page 128)

Hoesch's skink (page 129)

Western three-striped skink (page 132)

Boulenger's skink (page 128)

Variable skink (page 133)

Variegated skink (subsp. *variegata*) (page 134)

Variegated skink (subsp. *punctulata*) (page 134)

Wedge-snouted skink (page 127)

Striped skink (subsp. *striata*, Natal) (page 132)

Kalahari tree skink (page 132)

Striped skink (subsp. *punctatissima*, Johannesburg) (page 132)

Striped skink (subsp. *wahlbergii*, N.E. Botswana) (page 132)

Western rock skink (subsp. *sulcata*, male) (page 133)

Western rock skink (subsp. *sulcata*, male and female, striped phase) (page 133)

Ovambo tree skink (page 128)

Five-lined or Rainbow skink (subsp. *margaritifer*, male) (page 130)

Five-lined or Rainbow skink (subsp. *margaritifer*, female) (page 130)

Angolan blue-tailed skink (page 131)

Angolan blue-tailed skink (page 131)

Shovel-snouted lizard (page 136)

Wedge-snouted desert lizard (page 140)

Smith's desert lizard (page 140)

Reticulated desert lizard (plain phase) (page 141)

Reticulated desert lizard (reticulated phase) (page 141)

Small-scaled desert lizard (page 141)

Spotted desert lizard (page 141)

Bushveld lizard (adult) (page 137)

Knox's desert lizard (page 141)

Bushveld lizard (hatchling) (page 137)

Namaqua sand lizard (Karoo) (page 146)

Namaqua sand lizard (Namib Desert) (page 146)

Short-headed sand lizard (male and female, colour phases) (page 144)

Short-headed sand lizard (male) (page 144)

Husab sand lizard (page 147)

Western sand lizard (subsp. *undata*) (page 146)

Western sand lizard (subsp. *rubens*) (page 146)

Western sand lizard (subsp. *inornata*) (page 146)

Cape sand lizard (page 145)

Spotted sand lizard (subsp. *pulchella*) (page 145)

Spotted sand lizard (subsp. *lineoocellata*) (page 145)

Burchell's sand lizard (female) (page 144)

Burchell's sand lizard (hatchling) (page 144)

Burchell's sand lizard (male) (page 144)

Spotted sandveld lizard (page 142)

Ornate sandveld lizard (subsp. *taeniolata*) (page 143)

Ornate sandveld lizard (subsp. *ornata*) (page 143)

Striped sandveld lizard (subsp. *livida*) (page 143)

Striped sandveld lizard (subsp. *tessellata*) (page 143)

PLATE 60

Blue-tailed sandveld lizard (page 142)

Delalande's sandveld lizard (page 143)

Common rough-scaled lizard (page 139)

Cape rough-scaled lizard (page 138)

Blue-tailed tree lizard (subsp. *laevis*) (page 137)

Common mountain lizard (subsp. *montana*) (page 148)

Cape mountain lizard (male) (page 148)

Cape mountain lizard (female) (page 148)

Cottrell's mountain lizard (page 147)

Southern rock lizard (page 139)

Soutpansberg rock lizard (page 139)

Nile monitor (subsp. *niloticus*) (page 173)

Nile monitor (subsp. *niloticus*, hatchling) (page 173)

Nile monitor (subsp. *niloticus*, adults mating) (page 173)

Rock monitor (subsp. *albigularis*) (page 172)

Rock monitor (subsp. *albigularis*, hatchling) (page 172)

Rough-scaled plated lizard (subsp. *major*) (page 151)

Giant plated lizard (subsp. *validus*) (page 153)

Desert plated lizard (page 149)

Namaqua plated lizard (page 152)

Kalahari plated lizard (subsp. *auritus*) (page 151)

Black-lined plated lizard (page 152)

Yellow-throated plated lizard (page 151)

Short-legged seps (subsp. *seps*) (page 154)

Dwarf plated lizard (page 150)

Giant girdled lizard or Sungazer (page 159)

Large-scaled girdled lizard (page 159)

Tropical girdled lizard (subsp. *tropidosternum*) (page 163)

Cape girdled lizard (subsp. *cordylus*, coastal form) (page 158)

Tropical girdled lizard (subsp. *jonesi*) (page 163)

Tasman's girdled lizard (page 162)

Cape girdled lizard (subsp. *cordylus*, inland form) (page 158)

Cape girdled lizard (subsp. *niger*) (page 158)

Transvaal girdled lizard (subsp. *vittifer*) (page 163)

Zimbabwe girdled lizard (page 162)

Herero girdled lizard (page 162)

Dwarf girdled lizard (page 160)

McLachlan's girdled lizard (page 160)

Peers's girdled lizard (page 161)

Lawrence's girdled lizard (page 159)

Namaqua girdled lizard (page 160)

Armadillo girdled lizard (page 157)

Armadillo girdled lizard (defence) (page 157)

Jordan's girdled lizard (juvenile) (page 162)

Karoo girdled lizard (Central Karoo) (page 161)

Karoo girdled lizard (N.W. Cape) (page 161)

Karoo girdled lizard (N.E. Karoo) (page 161)

Karoo girdled lizard (S. Namibia) (page 161)

Warren's girdled lizard (subsp. *perkoensis*) (page 164)

Warren's girdled lizard (subsp. *warreni*) (page 164)

Warren's girdled lizard (subsp. *barbertonensis*) (page 164)

Warren's girdled lizard (subsp. *vandami*) (page 164)

Warren's girdled lizard (subsp. *depressus*) (page 164)

Warren's girdled lizard (subsp. *regius*) (page 164)

Blue-spotted girdled lizard (page 158)

Spiny crag lizard (page 171)

Graceful crag lizard (subsp. *capensis*) (page 169)

Graceful crag lizard (subsp. *robertsi*) (page 169)

Cape crag lizard (subsp. *microlepidotus*) see **C** at right (page 171)

Belly colours of Cape crag lizard subspecies (page 171)

Cape crag lizard (subsp. *namaquensis*) see **D** top right (page 171)

Cape crag lizard (subsp. nov.) see **A** top right (page 171)

Cape crag lizard (subsp. *fasciatus*) see **B** top right (page 171)

)rakensberg crag lizard (subsp. *melanotus*) (page 170)

›rakensberg crag lizard (subsp. *subviridis*, male) (page 170)

Drakensberg crag lizard (subsp. *subviridis*, female) (page 170)

Drakensberg crag lizard (subsp. *subviridis*, male) (page 170)

)rakensberg crag lizard (subsp. *transvaalensis*) (page 170)

Dwarf flat lizard (subsp. *minor*) (page 166)

Cape flat lizard (page 165)

Dwarf flat lizard (subsp. *guttatus*) (page 166)

FitzSimons's flat lizard (page 165)

Relict flat lizard (page 168)

Common flat lizard (subsp. *intermedius*) (page 166)

Common flat lizard (subsp. *wilhelmi*) (page 166)

Common flat lizard (subsp. *rhodesianus*) (page 166)

Common flat lizard (subsp. *natalensis*) (page 166)

Common flat lizard (subsp. *natalensis*, showing belly) (page 166)

Tree agama (page 179)

Southern rock agama (subsp. *atra*, breeding male) (page 176)

Southern rock agama (subsp. *atra*, male) (page 176)

Southern rock agama (subsp. *atra*, female) (page 176)

Southern rock agama (subsp. *atra*, with juvenile eagle) (page 176)

Southern rock agama (subsp. *knobeli*) (page 176)

Namibian rock agama (subsp. *planiceps*, male) (page 178)

Namibian rock agama (subsp. *planiceps*, female) (page 178)

Kirk's rock agama (page 178)

Spiny agama (subsp. *hispida*, male) (page 177)

Spiny agama (subsp. *hispida*, female) (page 177)

Spiny agama (subsp. *makarikarica*) (page 177)

Etosha agama (page 177)

Ground agama (subsp. *aculeata*) (page 175)

Ground agama (subsp. *armata*) (page 175)

Ground agama (subsp. *distanti*) (page 175)

Anchieta's agama (page 175)

Giant ground gecko (subsp. *angulifer*) (page 193)

Koch's barking gecko (page 212)

Carp's barking gecko (page 212)

Common barking gecko (subsp. *maculatus*) (page 212)

Web-footed gecko (page 209)

Web-footed gecko (page 209)

Kaoko web-footed gecko (page 196)

Kalahari ground gecko (subsp. *wahlbergii*) (page 193)

Koch's gecko (page 203)

Marico gecko (subsp. *mariquensis*) (page 204)

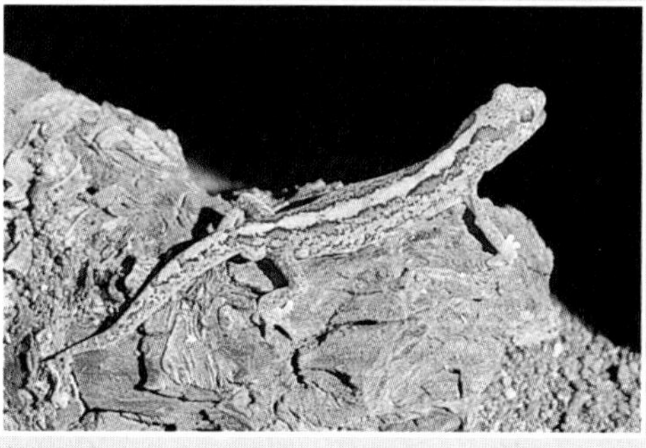

Marico gecko (subsp. *mariquensis*, striped phase) (page 204)

Marico gecko (subsp. *latirostris*) (page 204)

Speckled gecko (spotted phase) (page 205)

Speckled gecko (grey phase) (page 205)

Ocellated gecko (spotted phase; S.W. Cape) (page 202)

Spotted gecko (page 203)

Ocellated gecko (mottled phase; Beaufort West) (page 202)

Golden spotted gecko (page 204)

Austen's gecko (page 200)

Western Cape gecko (page 201)

Tiger gecko (hatchling) (page 208)

Tiger gecko (adult) (page 208)

Cape gecko (subsp. *affinis*) (page 201)

ape gecko (subsp. *capensis*) (page 201)

Cape gecko (subsp. *oshaughnessyi*) (page 201)

ape gecko (subsp. *vansoni*) (page 201)

Rough-scaled gecko (subsp. *formosus*) (page 206)

ough-scaled gecko (subsp. *rugosus*) (page 206)

Bibron's gecko (page 200)

Bibron's gecko (dark phase) (page 200)

Button-scaled gecko (subsp. *laevigatus*, sloughing) (page 203)

Namaqua gecko showing skin torn in defence (Central Karoo) (page 205)

Button-scaled gecko (subsp. *fitzsimonsi*) (page 203)

Namaqua gecko (green phase; Little Namaqualand) (page 205)

Large-scaled gecko (page 206)

Kaokoveld rock gecko (page 205)

Velvety gecko (hatchling) (page 200)

Velvety gecko (juvenile) (page 200)

Velvety gecko (adult) (page 200)

Western spotted gecko (subsp. *purcelli*) (page 207)

Western spotted gecko (subsp. *serval*) (page 207)

Banded gecko (page 202)

Weber's gecko (Namaqualand) (page 208)

Banded gecko (juvenile) (page 202)

Weber's gecko (Central Namibia) (page 208)

Striped leaf-toed gecko (subsp. *rupicolus*) (page 210)

Striped leaf-toed gecko (subsp. *lineatus*) (page 210)

Marbled leaf-toed gecko (page 211)

Marbled leaf-toed gecko (striped phase) (page 211)

Small-scaled leaf-toed gecko (page 211)

Karoo flat gecko (subsp. *karroica*) (page 190)

Amatola flat gecko (page 190)

Transvaal flat gecko (subsp. *transvaalica*) (page 192)

African flat gecko (subsp. *africana*) (page 189)

Hawequa flat gecko (page 190)

Pondo flat gecko (subsp. *pondolia*) (page 191)

Pondo flat gecko (subsp. *major*) (page 191)

Pondo flat gecko (subsp. *langi*) (page 191)

Moreau's tropical house gecko (subsp. *mabouia*) (page 194)

Moreau's tropical house gecko (subsp. *tasmani*) (page 194)

Flat-headed tropical house gecko (page 195)

Muller's velvet gecko (page 195)

Wahlberg's velvet gecko (spotted phase) (page 195)

Bernard's dwarf gecko (subsp. *bonsi*) (page 197)

Namaqua day gecko (page 210)

Festive gecko (page 199)

Wahlberg's velvet gecko (page 195)

Cape dwarf gecko (subsp. *capensis*) (page 197)

Bradfield's dwarf gecko (page 197)

Spotted dwarf gecko (page 198)

Chobe dwarf gecko (page 197)

Methuen's dwarf gecko (page 198)

Stevenson's dwarf gecko (page 199)

Namib day gecko (page 213)

Bradfield's Namib day gecko (subsp. *bradfieldi*) (page 214)

Bradfield's Namib day gecko (subsp. *diporus*) (page 214)

Kaokoveld Namib day gecko (page 214)

Boulton's Namib day gecko (subsp. *boultoni*) (page 214)

Barnard's Namib day gecko (page 213)

Natal Midlands dwarf chamaeleon (page 184)

Cape dwarf chamaeleon (page 183)

Drakensberg dwarf chamaeleon (page 182)

Southern dwarf chamaeleon (subsp. *ventrale*) (page 185)

Namaqua dwarf chamaeleon (subsp. *occidentale*) (page 185)

Karoo dwarf chamaeleon (page 182)

Robertson dwarf chamaeleon (page 182)

Transkei dwarf chamaeleon (page 181)

Knysna dwarf chamaeleon (in threat display) (page 181)

Transvaal dwarf chamaeleon (page 184)

Blackheaded dwarf chamaeleon (page 182)

Setaro's dwarf chamaeleon (page 183)

Smith's dwarf chamaeleon (page 184)

Zululand dwarf chamaeleon (page 183)

Smith's dwarf chamaeleon (grey phase, female) (page 184)

Marshall's leaf chamaeleon, adult male (left), female (right) (subsp. *marshalli*) (page 187)

Flap-neck chamaeleon (subsp. *dilepis*) (page 186)

Namaqua chamaeleon (page 186)

Flap-neck chamaeleon (subsp. *dilepis*, juvenile) (page 186)

Nile crocodile (page 215)

Nile crocodile (in water) (page 215)

Glossary

Abdominal: Pertaining to the region of the abdomen; a scute on the plastron of a chelonian shell (see illustration, page 23).
Adaptation: A morphological, physiological or behavioural feature, evolved over a period of time, that particularly suits an organism (or group of related organisms) to its way of life.
Anal: Pertaining to the region of the anus; a scute on the plastron of a chelonian shell (see illustration, page 23); a scale in front of the cloacal opening in squamates (see illustration, page 44).
Anapsid: Descriptive of a skull in which there are no openings in the temporal region.
Annulus (pl. annuli): A ring on the scute of a chelonian shell, representing a period of growth; a body segment of an amphisbaenian (worm lizard).
Anterior: The front.
Apical pit: A sense organ near the tip of the body scales of some squamates.
Aposematic: Warning; usually bright body coloration, advertising that a species is dangerous or poisonous.
Aquatic: Living in water.
Arboreal: Living in or among trees.
Arribada: A simultaneous, mass emergence of sea turtles on to a small beach to lay their eggs.
Areola (pl. areolae): The central region of a chelonian scute, which may be raised or hollow.
Autotomy: The voluntary shedding of a part of the body. (Caudal autotomy: the voluntary shedding of the tail.)
Axillary: Pertaining to the region of the armpit; a scute on the plastron of a chelonian shell (see illustrations, pages 23 and 39).
Bicarinate: See Carinate.
Bicuspid: See Cuspid.
Biped: An animal with two feet.
Bridge: The part of a chelonian shell where the carapace joins the plastron.
Buttock tubercle: The enlarged, conical scale found on the rear upper part of the hind leg of some tortoises.
Carapace: The hard structure covering all or part of an animal's body; the upper surface of a chelonian shell.
Carinate: Having ridges, as in some lizard scales. (Bicarinate: having two ridges. Tricarinate: having three ridges. Quinquecarinate: having five ridges.)
Casque: The helmet-like structure found on the back of the head of some chamaeleons.
Caudal: Pertaining to the tail.
Chelonian: A shield reptile (tortoises, turtles and terrapins).
Chin shields: The enlarged scales found beneath the head of some lizards and snakes.
Circumtropical: Encircling the earth, in the area between 23°30'N and 23°30'S.
Class: The taxonomic category ranking below 'phylum' and above 'order'.
Cloaca: The common chamber into which the urinary, digestive and reproductive systems discharge their contents, and which opens to the exterior.
Clutch: Collective noun for all the eggs laid by a single female at one time.
Commensal species: Species living together, all usually benefiting from the association.
Conical: Descriptive of a raised scale that narrows to a pointed centre.
Coronoid: A bone in the lower jaw of some reptiles.
Costal: A scute on the carapace of a chelonian shell (see illustrations, pages 23 and 31).
Cranial: Pertaining to the skull. (Cranial crests are found on top of the head.)
Crepuscular: Active at dawn and dusk.
Cryptic: Hidden or camouflaged.
Cuspid: Having tooth-like projections. (Biscuspid: having two cusps. Tricuspid: having three cusps.)
Cycloid: Descriptive of a scale with an evenly curved, free border, like the scales of many fish.
Diverticulum (pl. diverticula): An elongate sac with only one opening. (Diverticulate lungs have numerous blind branches rather than a single, large sac.)
Distal: Furthest from the body.
Dorsal: Pertaining to the upper surface of the body.
Dorsolateral: The upper surface of the body, bordering the backbone.
Family: The taxonomic category ranking below 'order' and above 'genus'.
Fauna: The animal life of a locality or region.
Femoral: Pertaining to the upper part of the hind limb (the thigh).
Frontal: A scale on the head of a reptile (see illustrations, pages 48, 52 and 111).
Frontonasal: A scale on the head of a reptile (see illustration, page 111).
Frontoparietal: A scale on the head of a reptile (see illustration, page 111).
Genus (pl. genera): The taxonomic category ranking below 'family' and above 'species'.

Glandular: Descriptive of the modified secretory scale found on the thigh or in front of the cloaca of some lizards.
Girdles: The supporting structure of the limbs, e.g. the hips and shoulders.
Granular: Descriptive of small, usually non-overlapping, scales.
Gular: Pertaining to the throat region; a plate on the plastron of a chelonian shell (see illustration, page 23).
Hemipenis (pl. hemipenes; adj. hemipenial): One of the grooved copulatory structures present in male squamate reptiles.
Herbivorous: Eating plant matter.
Hinge: A flexible joint in the shell of some chelonians that allows the front or rear of the shell to close.
Imbricate: Descriptive of an overlapping scale.
Incubation: Keeping eggs warm to ensure continuous development.
Inframarginal: A scute on the plastron of a sea turtle shell (see illustration, page 31).
Intergular: A scute on the plastron of a terrapin shell (see illustration, page 39).
Internasal: A scale on the head of a reptile (see illustration, page 52).
Interparietal: A scale on the head of a reptile (see illustrations, pages 48 and 111).
Interspaces: The spaces between blotches.
Interstitial skin: The (usually thin) skin lying between the scales.
Introduced: A species brought from areas where it occurs naturally to areas where it has not previously occurred; alien or 'exotic' (foreign) species are always introduced.
Invertebrate: An animal that lacks a backbone.
Juxtaposed: Description of scales that touch but do not overlap.
Keel: A prominent ridge, occurring on the back of some turtles and on the dorsal scales of some snakes.
Keratin (adj. keratinized): A hard, tough, non-soluble skin protein; horns, nails and claws are made of keratin.
Labial: Pertaining to the lip region; a scale on the head of a reptile (see illustrations, pages 46, 48, 52 and 111).
Lamella (pl. lamellae): Any thin, plate-like or scale-like structure.
Lateral: Pertaining to the sides of the body.
LD_{50}: The lethal dose (LD) that will kill half (50%) of the animals into which it is injected; used as an indication of venom toxicity.
Loreal: Pertaining to the region on the side of the head, between the nostril and the eye; a scale on the head of a reptile (see illustrations, pages 52 and 111).
Marine: Living in the sea.
Mandibular: A bone in the lower jaw.
Marginal: A plate on the edge of the carapace of a chelonian shell (see illustrations, pages 23 and 31).
Maxillary: A bone in the upper jaw.
Medial: Pertaining to the region near the middle of the body.
Melanistic: Darker or blacker than 'normal'.
Mental: A scale on the head of a reptile (see illustration, page 111).
Mimic: A species that resembles a different species, usually a distasteful or inedible one.
Mucronate: Descriptive of body scales that are strongly overlapping and drawn into a spine.
Nasal: Pertaining to the region of the nose; a scale on the head of a reptile (see illustrations, pages 46, 48 and 111).
Nasorostral: A scale, sometimes resulting from the fusion of the nasal and the rostral, on the head of a reptile.
Nocturnal: Active at night.
Nuchal: A scute at the front of the carapace of a chelonian shell (see illustration, page 23).
Occipital: Pertaining to the region at the back of the skull; a scale on the head of a reptile (see illustration, page 111).
Oceanic: Living in the open seas.
Ocellus (pl. ocelli): An eye-like, ring-shaped spot.
Ocular: Pertaining to the region of the eye. (The subocular is the scale below the eye on the head of a reptile; see illustration, page 111).
Orbit: The eye socket.
Order: The taxonomic category ranking below 'class' and above 'family'.
Osteoderm: A very small bone in the skin of some reptiles.
Osteological: Pertaining to the bones.
Oviduct: The tube which carries eggs from the ovary.
Oviparous: Reproduction by eggs which hatch outside the female's body.
Palatine: A bone in the roof of the mouth.
Papilla (pl. papillae): A small skin projection that may be sensitive to touch.
Parietal: Pertaining to the region on the crown of the head; paired bone forming part of the roof and sides of the skull; scale on the head of a reptile (see illustrations, pages 52 and 111).
Parthenogenetic: Pertaining to the ability of females to develop fertile eggs without mating with a male.
Parturition: The act or process of birth.
Pectoral: Pertaining to the region of the body where the forelimbs originate; a scute on the plastron of a chelonian shell (see illustration, page 23).
Pheromone: A substance produced and discharged by an organism which induces a response in another individual of the same species.

Pineal eye: A primitive, light-sensitive area on the top of the head of some lizards and extinct reptiles. It is not used for vision, but controls seasonal breeding.
Plastron (adj. plastral): The lower surface of the chelonian shell.
Polymorphism: The occurrence of more than one type of individual in a species.
Pore: A minute opening or passage.
Posterior: The rear or back part.
Pre-: In front of.
Prefrontal: A scale on the head of a reptile (see illustrations, pages 46, 48, 52 and 111).
Ptosis: Drooping of the eyelids.
Race: A population of a species which is distinguishable from the rest of that species; a subspecies.
Recurved: Descriptive of a tooth that bends backwards.
Reticulate: Resembling a network.
Retractile: A part that may be drawn inwards.
Riverine: Living in or near rivers.
Rupicolous: Living on and among rocks.
Rostral: Pertaining to the rostrum (nose); a scale at the front of the nose of a reptile (see illustrations, pages 46, 48, 52 and 111).
SA RDB: South African Red Data Book.
Scale: A thin, flattened, plate-like structure that forms part of the surface covering of various vertebrates, especially fishes and reptiles.
Scansors: Specialized pads composed of thousands of minute hairs that catch in cracks and allow the animal to climb vertical surfaces; found on the toe-tips of many geckos.
Scat: Faecal pellet.
Scute: Any enlarged scale on a reptile; the horny plates of a chelonian shell.
Sedentary: Not free-living.
Septum (pl. septa): A partition separating two cavities or masses of tissue.
Sub-: Beneath.
Subcontinent: The African continent south of a line joining the Cunene and Zambezi rivers.
Sublingual: Pertaining to the area beneath the tongue.
Submarginal: Pertaining to the area near the margin.
Subspecies: See Race.
Sulcus: A surface groove; a groove in the hemipenis of squamates along which sperm flows.
Supra-: Above.
Supraciliary: Pertaining to the region above the eyelid; a scale on the head of a lizard (see illustration, page 111).
Sutures: The junction of two parts which are immovably connected.
Sympatric: Living in the same region.
Tarsal: Pertaining to the lower part of the hind limb, above the foot.
Taxonomy: The science of classification; the arrangement of animals and plants into groups based on their natural relationships.
Temporal: Pertaining to the region to the side of the forehead; a scale on the head of a reptile (see illustrations, pages 52 and 111).
Terrestrial: Living on the ground.
Tricuspid: See Cuspid.
Tubercle: A small, rounded protuberance.
Tympanic: Pertaining to the region of the ear; a scale on the head of a lizard (see illustration, page 111).
Tympanum: The ear drum.
μg/kg: Abbreviation for the amount of venom which will kill an animal, related to its mass. Sea snake venom has a toxicity of 130 μg/kg; 0,013 g ('μg' is the symbol for a microgram, one-millionth of a gram) will kill a 100-kg animal.
Unicuspid: Having one cusp. See Cuspid.
Ventral: Pertaining to the under surface of the body.
Vertebral: Pertaining to the region of the backbone; a scute on the carapace of a chelonian shell (see illustrations, pages 23 and 31).
Vestigial: Being smaller and of a more simple structure (a remnant) than in an evolutionary ancestor.
Viviparous: Reproduction by giving birth to live young which develop in the mother's body.
Xanthic: Having only red pigmentation in the skin.
Zonary: Descriptive of the concentric zones of pigment found on a scute of a chelonian shell.

Suggested further reading

Auerbach, R. (1988). *Reptiles and Amphibians of Botswana.*

Boycott, R.C. and Bourquin, O. (1988). *The South African Tortoise Book – a guide to southern African tortoises, terrapins and turtles.* Southern Book Publishers, Johannesburg.

Branch, W.R. (Ed.) (1988). *South African Red Data Book: Reptiles and Amphibians.* South African Scientific Programmes Report, CSIR, Pretoria.

Broadley, D.G. (1983). *FitzSimons' Snakes of Southern Africa.* Delta Books (Pty) Ltd, Johannesburg and Cape Town.

Halliday, T., and Adler, K. (Eds.) (1986). *The Encyclopaedia of Reptiles and Amphibians.* George Allen and Unwin, London.

Jacobsen, N. (1985). *Ons Reptiele.* CUM-Boeke, Roodepoort.

Patterson, R. and Bannister, A. (1987). *South African Reptile Life.* C. Struik Publishers, Cape Town.

Pienaar, U. de V., Haacke, W.D., and Jacobsen, N. (1983). *The Reptiles of the Kruger National Park.* National Parks Board, Pretoria.

Pooley, A. (1982). *Discoveries of a Crocodile Man.* William Collins Sons and Co. Ltd, London and Johannesburg.

Societies and their publications

The Herpetological Association of Africa has an international membership and is based in South Africa. It produces a journal (for scientific notes and major articles) and a newsletter (for announcements, husbandry hints and news items). Details of membership can be obtained from the Secretary, Herpetological Association of Africa, c/o Department of Herpetology, National Museum, P.O. Box 266, Bloemfontein, 9300 South Africa.

The Society for the Study of Amphibians and Reptiles is an international society based in North America. It produces the *Journal of Herpetology* for scientific publications and the *Herpetological Review* for more general articles, announcements and book reviews. An irregular facsimile series reprints classic works that are out of print. Details of membership can be obtained from the Society for the Study of Amphibians and Reptiles, Department of Zoology, Ohio University, Athens, Ohio 45701, United States of America.

The Herpetologists' League is an international society based in North America. It produces a journal, *Herpetologica*, for small scientific publications and *Herpetological Monographs* for longer, detailed studies. Details of membership can be obtained from the Herpetologists' League, Department of Biology, Elmhurst College, Elmhurst, Illinois 60126, United States of America.

Museums and institutions involved in herpetological research

The following centres have herpetologists on their staff who can identify specimens and answer questions.

Museums

J.R. Ellerman Museum, Department of Zoology, University of Stellenbosch, Stellenbosch, 7600 South Africa.
National Museum, P.O. Box 266, Bloemfontein, 9300 South Africa.
Natural History Museum of Zimbabwe, P.O. Box 240, Bulawayo, Zimbabwe.
Port Elizabeth Museum, P.O. Box 13147, Humewood, 6013 South Africa.
South African Museum, P.O. Box 61, Cape Town, 8000 South Africa.
State Museum, P.O. Box 1203, Windhoek, South West Africa/Namibia.
Transvaal Museum, P.O. Box 413, Pretoria, 0001 South Africa.

Nature Conservation Departments

Cape Department of Nature and Environmental Conservation, Jonkershoek Nature Conservation Station, Private Bag 5014, Stellenbosch, 7600 South Africa.
Natal Parks Board, P.O. Box 662, Pietermaritzburg, 3200 South Africa.
Department of Agriculture and Nature Conservation, Private Bag 13306, Windhoek, South West Africa/Namibia.
Transvaal Division of Nature Conservation, Private Bag X209, Pretoria, 0001 South Africa.

Snake and Crocodile Parks

FitzSimons Snake Park, P.O. Box 10457, Marine Parade, Durban, 4056 South Africa.
Kwena Gardens Crocodile Farm, P.O. Box 234, Sun City, Bophuthatswana.
Port Elizabeth Museum and Snake Park, P.O. Box 13147, Humewood, 6013 South Africa.
Transvaal Snake Park, P.O. Box 97, Halfway House, 1685 South Africa.

Photographic credits

All photographs in this book are by Bill Branch except for those listed below.

Numbers in brackets refer to photographs counted from left to right and top to bottom of each plate.

J. Akester: Pl. 16 (3); Pl. 67 (3); Pl. 82 (4); Pl. 83 (2).

G. Alexander: Pl. 47 (6).

A. Bannister: Pl. 16 (5); Pl. 19 (6); Pl. 21 (6); Pl. 23 (6); Pl. 24 (5); Pl. 74 (1).

H. Berger-Dell'mour: Pl. 40 (3); Pl. 41 (2); Pl. 49 (2, 3); Pl. 51 (2); Pl. 56 (4, 5, 6); Pl. 58 (1, 2, 3, 4, 5); Pl. 61 (3, 4); Pl. 67 (4); Pl. 80 (3); Pl. 90 (5); Pl. 92 (4, 5).

R. Boycott: Pl. 2 (1, 4); Pl. 5 (1); Pl. 9 (1); Pl. 14 (1); Pl. 18 (1); Pl. 35 (4, 6).

D. Broadley: Pl. 25 (2).

J. Coates-Palgrave: Pl. 21 (2); Pl. 24 (1); Pl. 33 (5); Pl. 36 (5); Pl. 39 (3); Pl. 41 (4); Pl. 42 (1, 5); Pl. 53 (1); Pl. 61 (5); Pl. 78 (1); Pl. 89 (6); Pl. 95 (6).

A. de Villiers: Pl. 6 (5, 6); Pl. 7 (4, 6); Pl. 10 (1); Pl. 48 (3, 4, 5, 6).

M. Griffin: Pl. 16 (2); Pl. 20 (6); Pl. 24 (3); Pl. 38 (3); Pl. 43 (1, 4); Pl. 80 (1).

W. Haacke: Pl. 11 (3, 4); Pl. 15 (5); Pl. 16 (4); Pl. 17 (6); Pl. 25 (3); Pl. 29 (1, 2); Pl. 36 (1); Pl. 37 (1, 3); Pl. 41 (3); Pl. 42 (3); Pl. 45 (4, 5); Pl. 46 (2); Pl. 50 (3); Pl. 78 (2); Pl. 86 (3, 5).

L. Hoffmann: Pl. 40 (7).

G.R. Hughes: Pl. 8 (3); Pl. 9 (2).

N. Jacobsen: Pl. 23 (1); Pl. 46 (1); Pl. 62 (6); Pl. 74 (4); Pl. 89 (3).

P. le Fras Mouton: Pl. 57 (3).

Johan Marais: Pl. 12 (5); Pl. 18 (6); Pl. 20 (4); Pl. 22 (2); Pl. 24 (6); Pl. 37 (5); Pl. 39 (4); Pl. 45 (3); Pl. 49 (4); Pl. 52 (2); Pl. 70 (1); Pl. 74 (3, 5); Pl. 82 (3); Pl. 90 (4).

C. Mattison: Pl. 40 (2).

C. McCartney: Pl. 60 (3).

G. McLachlan: Pl. 85 (1, 2).

National Parks Board: Pl. 11 (5); Pl. 23 (7); Pl. 24 (4); Pl. 26 (3); Pl. 34 (5, 6); Pl. 41 (5); Pl. 46 (5); Pl. 48 (2); Pl. 61 (1).

C. Schlettwein: Pl. 52 (4); Pl. 55 (4, 5).

A. Schoeman: Pl. 9 (4, 5); Pl. 56 (1); Pl. 64 (3).

C. Stuart: Pl. 26 (2); Pl. 42 (4); Pl. 81 (1); Pl. 83 (3).

S. Spawls: Pl. 18 (5); Pl. 29 (3); Pl. 33 (4); Pl. 38 (4); Pl. 54 (4); Pl. 57 (4); Pl. 60 (1, 2); Pl. 64 (5); Pl. 78 (4); Pl. 80 (2); Pl. 90 (2); Pl. 91 (4).

C. Tilbury: Pl. 11 (1); Pl. 22 (1); Pl. 25 (5); Pl. 28 (2); Pl. 32 (6); Pl. 37 (2); Pl. 38 (2); Pl. 44 (1); Pl. 51 (3); Pl. 53 (4, 5); Pl. 54 (1); Pl. 55 (1); Pl. 61 (2); Pl. 64 (1); Pl. 65 (1); Pl. 75 (2); Pl. 76 (3, 4); Pl. 77 (1, 3); Pl. 78 (3); Pl. 81 (2); Pl. 85 (3, 5); Pl. 86 (2); Pl. 89 (4); Pl. 90 (3, 6); Pl. 92 (6); Pl. 93 (2, 3, 5); Pl. 94 (2, 3); Pl. 95 (1, 2, 4); Pl. 96 (4).

Index

Numbers marked in **bold type** against the English names denote pages on which the main species accounts appear.

NOTES

NOTES